SCIENCE AND TECHNOLOGY STRATEGIES FOR DEVELOPMENT IN INDIA AND CHINA—A COMPARATIVE STUDY

SCIENCE AND TECHNOLOGY STRATEGIES FOR DEVELOPMENT IN INDIA AND CHINA— A COMPARATIVE STUDY

Edited by

V.P. KHARBANDA
ASHOK JAIN

HAR-ANAND
PUBLICATIONS PVT LTD

Har-Anand Publications Pvt Ltd
364-A, Chirag Delhi, New Delhi-110017

PRINTED IN INDIA

Published by Ashok Gosain & Ashish Gosain for Har-Anand Publications Pvt Ltd
Typeset at S.S. Computers, Delhi and printed at Printline, New Delhi.

Preface

India and China, after adoption of policies for liberalization and globalization during the late 1980's, are in the process of moulding their science and technology strategies in order to develop the capability to face the new situation of intense international competition. The S&T infrastructure built up during the past 50 years for a planned economy is being slowly transformed to suit the conditions of a market economy with the adoption of concepts like market competition, less state-intervention, increased public and private R&D, Intellectual Property Rights, etc. in their science, technology and development policies and programmes. Technology Innovation has become a key word for industrial growth and economic development. Over the last five decades, these countries have had more or less similar conditions of development and have rich experience in linking science and technology to development. As such, these countries are distinctly in a different class as compared to the other developing countries and carry vast opportunities for bilateral cooperation and world peace. With the ongoing formation of regional groupings of the countries and formation of new international regimes after the cold war, it is important that these two Asian giants come together and share their experiences in reaping the benefits of science and technology, not to dominate the world but to prevent domination. It is, therefore, important for the two countries to understand and assess their experiences of managing Science and Technology Strategies.

The present book is the direct outcome of a joint bilateral project entitled "A Comparative Study of Science and Technology Development Strategies in India and China", undertaken by the National Institute of Science Technology & Development Studies (NISTADS), New Delhi, India and Central South University of

Technology (CSUT) , Changsha, China and a Joint Workshop on the above subject organised in 1995. Apart from invited articles from some selected experts from India and China, the book primarily contains the articles presented during this Workshop. This has required assistance of a number of institutions and personalities in India and China. We sincerely thank all of them. We are indebted to the Council of Scientific and Industrial Research (CSIR), India and the National Science Foundation of China (NSFC), China for the cooperation extended by them in making the project feasible, which was approved under the CSIR-NSFC Agreement -1992. We are thankful to the ministries of External Affairs of both the countries, Department of Science and Technology, India and the Central South University of Technology, China, for their assistance in facilitating exchange of materials and in organizing Workshop, discussions and visits to various institutions in India and China.

Thanks are also due to Ms Krishna Verma for assistance in typing the manuscript and Mr. Pradeep Banerjee of National Institute of Science Communication (NISCOM), New Delhi, for designing the cover page.

Editors

Abbreviations and Acronyms

ABB	—	Asea Brown Boveri Ltd
AICTE	—	All India Council of Technical Education
AMMS	—	Academy of Military Medical Sciences
ASEAN	—	Association of South East Asian Nations
ATIRA	—	Ahmedabad Textile Industry Research Association
BARC	—	Bhabha Atomic Research Center
BITS	—	Birla Institute of Technology and Science
BMU	—	Beijing Medical University
BTIS	—	Biotechnology Information System
CAAS	—	Chinese Academy of Agricultural Sciences
CAMS	—	Chinese Academy of Medical Sciences
CAPM	—	Chinese Academy of Preventive Medicine
CAS	—	Chinese Academy of Sciences
CASs	—	Centers of Advanced Studies
CCMB	—	Center for Cellular and Molecular Biology
CCST	—	Cabinet Committee on Science and Technology
COSIST	—	Committee on Strengthening of Infrastructure in Science and Technology
COST	—	Committee on Science and Technology
CPO	—	Chinese Patent Office
CR	—	Cultural Revolution
CSIR	—	Council of Scientific and Industrial Research
CSUT	—	Central South University of Technology
DAA	—	Development, Adaptation and Absorption
DAE	—	Department of Atomic Energy
DBT	—	Department of Biotechnology
DIC	—	District Industrial Center
DNES	—	Department of Non-Conventional Energy Sources
DOC	—	Department of Coal
DOD	—	Department of Ocean Development
DOE	—	Department of Electronics
DOEn	—	Department of Environment
DOM	—	Department of Mines

DOS	—	Department of Space
DRDO	—	Defense Research Development Organization
DSIR	—	Department of Scientific and Industrial Research
DST	—	Department of Science and Technology
ECIL	—	Electronics Corporation of India Ltd
EEC	—	European Economic Commission
ELISA	—	Enzyme Linked Immuno Sorbent Assay
EP	—	Education Policy
FARM	—	Farmer centered Agriculture Resource Management
FCA	—	Foreign Collaboration Agreements
FITT	—	Foundation for Innovation and Technology Transfer
FYP	—	Five Year Plan
GATS	—	General Agreement on Trade in Services
GATT	—	General Agreement on Tariffs and Trade
GLF	—	Great Leap Forward
GNFC	—	Gujarat Narmada Fertilizer Corporation
GNP	—	Gross National Product
HCG	—	Human Chorionic Gonadotropin
HGT	—	Home Grown Technologies
HRD	—	Human Resource Development
HTDZ	—	High Technology Development Zones
ICAR	—	Indian Council of Agricultural Research
ICBM	—	Inter Continental Ballistic Missiles
ICGEB	—	International Center for Genetic Engineering and Biotechnology
ICICI	—	Industrial Credit and Investment Corporation of India
ICMR	—	Indian Council of Medical Research
IDBI	—	Industrial Development Bank of India
IFCI	—	Industrial Finance Corporation of India
IGCAR	—	Indira Gandhi Center for Atomic Research
IIC	—	Indian Industrial Commission
IIRB	—	Industrial Intelligence and Research Bureau
IISc	—	Indian Institute of Science
IIT	—	Indian Institute of Technology
IMTECH	—	Institute of Microbial Technology
IPD	—	India Patent Design Act
IPM	—	Institute of Policy and Management
IPR	—	Intellectual Property Rights
IPT	—	In Plant Training
ITT	—	Interface of Technology Transfer
ITU	—	International Telecommunication Union

KVIB	—	Khadi and Village Industries Board
MAD	—	Mutual Assured Destruction
MECON	—	Metallurgical and Engineering Consultants
MFERT	—	Ministry of Foreign Economic Relations and Trade
MFM	—	May Fourth Movement
MOE	—	Ministry of Education
MOU	—	Memorandum of Understanding
MPRS	—	Mobile Post Doctoral Research Stations
MPT	—	Ministry of Posts and Telecommunications
MTCC	—	Microbial Type Culture Collection
NCBI	—	National Center for Biotechnology Information
NCL	—	National Chemical Laboratory
NCST	—	National Committee on Science and Technology
NEP	—	New Education Policy
NFATCC	—	National Facility for Animal Tissue & Cell Culture
NFPTCR	—	National Facility for Plant Tissue Culture Repository
NFSC	—	National Federation of Scientific Societies
NIC	—	National Informatics Center
NICs	—	Newly Industrialized Countries
NII	—	National Institute of Immunology
NIS	—	National Innovation System
NISSAT	—	National Information System for Science and Technology
NPC	—	National Peoples' Congress
NRDC	—	National Research and Development Corporation
NSFC	—	National Science Foundation of China
NSTEDB	—	National Science and Technology Entrepreneurship Development Board
NTP	—	New Technology Policy
OECD	—	Organization for Economic Cooperation and Development
PATSER	—	Programme Aimed at Technological Self Reliance
PCR	—	Polymerase Chain Reaction
PCT	—	Patent Cooperation Treaty
PECC	—	Pacific Economic Cooperation Council
PIS	—	Patent Information System
POA	—	Programme of Action
PSCs	—	Pioneer Service Centers
PTC	—	Plant Tissue Culture
RDDE	—	Research Development Design Engineering
REP	—	Revised Education Policy
SAC	—	Science Advisory Committee

SACC — Scientific Advisory Committee to the Cabinet
SAP — Special Assistance Programme
SAPTA — South Asian Preferential Trade Agreement
SOE — State Owned Enterprises
SEC — State Education Commission
SERC — Science and Engineering Research Council
SIROs — Scientific and Industrial Research Organizations
SPR — Scientific Policy Resolution
SPC — State Planning Commission
SSC — Science Society of China
SSTC — State Science and Technology Commission
STAC — Science and Technology Advisory Committee
STDRC — Science Technology and Development Research Center
STEPs — Science and Technology Entrepreneurship Parks
STG — Science Technology and Globalization
STP — Science and Technology Parks
TDC — Technology Development Centers
TENs — Techno Economic Networks
TIFAC — Technology Information Forecasting and Assessment Council
TIFR — Tata Institute of Fundamental Research
TP — Technology Policy
TPr — Torch Programme
TPS — Technology Policy Statement
TRIMS — Trade Related Investment Measures
TRIPS — Trade Related Intellectual Property Rights
T-R-P — Teaching-Research-Production
TT — Technology Transfer
TVEs — Township and Village Enterprises
UGC — University Grants Commission
UNDP — United Nations Development Programme
UNESCO — United Nations Educational, Scientific & Cultural Organization
UNIDAC — Universities, Industries and Department of Atomic Energy Consortium
UNIDO — United Nations Industrial Development Organization
WTO — World Trade Organization

Contributors

Prof M.G.K. Menon
Ex Chairman, Research Advisory Council, NISTADS
C - 63, Tarang Apartments
19 I.P. Ext., Mother Dairy Road
Patparganj, Delhi-110 092, India

Prof. Gu Shulin
Senior Research Fellow
UNU/INTECH, KEIZER KARELPLEIN 19
6211 TC MAASTRICHT,
THE NETHERLANDS

Dr Ashok Jain
Former Director, NISTADS
DG'S Bungalow, CRRI Residential Colony
C.V. Raman Marg, Maharani Bagh
New Delhi, India

Dr. V.V. Subba Rao
Joint Adviser
Department of Scientific and Industrial Research
Technology Bhavan, New Mehrauli Road
New Delhi — 110016, India

Prof. Chen Wenhua, Director
Scientific, Technology and Development Research Center,
Central South University of Technology,
Changsha — 410083, China

Prof. Fang Xin, Head
Department of Policy Research,
Institute of Policy and Management,
Chinese Academy of Sciences,
P.O.Box 8712, Beijing — 100080, China

Mr. V.P. Kharbanda
Senior Scientist, NISTADS
Dr. K.S. Krishnan Marg
New Delhi —110 012, India

Dr Y.S. Rajan
Senior Advisor, Confederation of Indian Industry
Executive Director, Technology Information
Forecasting and Assessment Council
Department of Science and Technology
New Mehrauli Road
New Delhi —110 016, India

Ms Sunita Wadhwa
Senior Scientific Officer
Technology Information
Forecasting and Assessment Council
Department of Science and Technology
New Mehrauli Road,New Delhi —110 016, India

Prof. Jia Xinmin
Institute of Policy and Management
Chinese Academy of Sciences
P.O.Box 8712, Beijing — 100080, China

Prof. (Mrs) Wen Xikai
Deputy Director General,
Law and Treaty Department,
Chinese Patent Office,
6, Xitucheng Lu, Haidian District, 100088
Beijing, China

Mr. N.R. Subba Ram
Former Head, Patent Unit, CSIR
17, Temple Glade
41 D, Kalakshetra Colony
Chennai — 400 060, India

Mr. M.A. Qureshi
Ex-Deputy Director, NISTADS
P-201, Taj Enclave, Link Road
Geeta Colony, Delhi, India

Prof. Wang Wenyong
Program Officer,
Department of International Cooperation,
State Science and Technology Commission,
Beijing, China

Prof. Chen Hao
Department of Policy Research
Institute of Policy and Management
Chinese Academy of Sciences
P.O.Box 8712, Beijing —100080, China

Dr (Mrs) A.R. Rajeswari
Ex-Adviser, Department of Science and Technology
Flat 202, Second Floor
Nritya, No. 16, 18TH Cross
Malleswaram, Bangalore — 560 055, India

Prof. Wang Qinnan, Director
Division of Genetics, Cell Biology
National Natural Science Foundation of China (NSFC)
Beijing, China

Dr (Mrs) Kavita Mehra
Senior Scientist, NISTADS
Dr. K.S. Krishnan Marg
New Delhi —110 012, India

Prof. Ding Lu, Lecturer
Economics Department
National University of Singapore
10, Kent Ridge Crescent, Singapore 0511

Dr. Partho Banerjee
Senior Scientist, NISTADS
Dr. K.S. Krishnan Marg
New Delhi — 110 012, India

Dr. Sunil Sondhi
Lecturer, Rajdhani College
University of Delhi
Delhi - 110 007, India

Contents

PART I : Conceptual Framework

PART IV : Issues of Technology Transfer and Intellectual Property Rights

PART V : Development and Problems of Science and Technology Manpower

PART VI : Biotechnologies

PART VII : Information Technologies

PART VIII : Regional Cooperation

Introduction

Unlike smaller countries in Asia, both India and China, since independence and liberation respectively, had to deal with heterogeneous development requirements of different regions endowed with characteristic human and natural resources and agro-climatic conditions. S&T inputs were required for rapid industrialization on one hand and for meeting basic minimum needs of large populations on the other. Industrial infrastructure was not developed to the extent it had in the developed countries of the West at the time of launching of their S&T strategies. Therefore, the starting points for science policies were quite different in India and China compared to the Western world.

However, developments in India and China during the last three decades show parallelism. India acquired independence in 1947 and the liberation of China took place in 1949. There after, both the countries had to tackle typically similar problems of population, poverty, illiteracy, disease-bearing environment, and of inherited bureaucratic administrative structures largely controlled by the urban elite. Over the years both India and China have established S&T infrastructures involving strong government support primarily in the context of building strong modern nation states. However, notwithstanding several achievements, there still exist massive poverty and a wide gap between the rich and the poor. The fruits of development have gone largely to a handful of people. China has performed better in widening the base of literacy, but it faces the acute problem of shortage of expert *S&T* manpower in contrast to India. A significant gap continues to exist between the economies of these two countries and those of the developed countries. The gap is actually widening fast.

In relation to science and technology strategies for modernization of industries, while some of the developing countries like Republic of Korea and Singapore from the very beginning relied on technological modernization through transplanting of techniques from outside, India and China did not accept (at least in principle) this mode of technology acquisition as the basis of modernization. Rather, they were inclined towards evolving a hybrid technology policy - a mix of import of technology and development of indigenous technology. The Indian and Chinese leaders had explicit concern for technological over-dependence on advanced countries. India's stated objective of achieving technological self-reliance has hinged around the concept that "In a country of India's size and endowments, self-reliance is inescapable and must be at the very heart of technological development. We must aim at major technological break-throughs in the shortest possible time for the development of indigenous technology appropriate to national priorities and resources." China has expressed its intentions to catch up with the advanced world through an open door policy, mainly by acquiring technologies from advanced countries and adapting them to the local conditions. Since 1980's, both the countries have been promoting a 'liberal, and 'open door policy,' to speed up technological modernization of their industries.

The *New Industrial Policy* Statement placed before the Indian Parliament on 24 July, 1991 clearly recognizes a great need for promoting an industrial environment where the acquisition of technological capability receives priority. In the context of foreign technology agreements, para 13 of the Statement states "foreign investment and technology collaboration will be welcomed to obtain higher technology." We, therefore, see that while both China and India continue to lay emphasis on indigenous science and technology capabilities, a much stronger pull in favour of imported technology has been evident since 1990. Both the countries are trying to cope with the changed international scenario through increased flows of science and technology as well as of investments across national boundaries. In the years to come, adjustment and reorientation of their research and development infrastructures in the wake of open market driven policies is going to be the principal challenge. More specifically, technology transfer is going to be a key issue, because attempts at technology transfer, whether from indigenous sources or from abroad, has not had smooth sailing in both the countries. This, therefore, will continue to occupy the centre place in S&T policies during the 21st century.

It is important for the two countries to understand and assess their experiences of the last 50 years in respect of Science and Technology Strategies adopted by them. In - depth analysis is required as to how these countries, starting in the forties with an economic, technological and R&D base different from that of the industrially advanced nations, should incorporate concepts like market forces, competition, state-intervention, public and private R&D, Intellectual Property Rights, etc. in their science, technology and development policies and programmes. While past experience is useful, it is imperative to take new initiatives, taking into account the increasing trend towards globalization. The resulting institutional changes, such as deregulation and privatization, liberalization of world trade and investments and increased global connectivity of science and technology, including environmental aspects, require a dynamic strategy. Such a strategy should be amenable to continuing readjustment of national and regional priorities and policies in science and technology. Additional demands are thus placed on endogenous S&T capacities of each country to choose its own options and to initiate actions.

This book deals with the above theme and is the direct outcome of a joint bilateral project entitled A "Comparative Study of Science and Technology Development Strategies in India and China" undertaken by the National Institute of Science Technology & Development Studies (NISTADS), New Delhi, India and Central South University of Technology (CSUT), Changsha, China under CSIR-NSFC Agreement, and a Joint Workshop organised on the above subject in 1995.

The book is divided into eight parts. Beginning with the Conceptual Framework relating to close linkages of science and technology with society, M.G.K. Menon, dwells upon scientific creativity and its impact on socio-economic development and the emerging new world order. He has argued that one cannot look at science policy as though it is purely a matter of putting money or providing bricks and mortar or providing education or technology and expecting that somehow it will flower. When we talk of science, science policy, or anything that results from science, we must also look at the interface fabric of the society. In that sense, *science* is embedded in society. Emphasizing the need for greater interaction of S&T with society, Prof. Menon avers that one could learn a lot from the experience of the two countries in S&T as well as their historical past. However, to plan for the future, one must look into the present socio-political needs too. According to him, China has been able to make great achievements with a coherent vision and tremendous self-confidence, which has been less evident in India, although it has been fortunate in having a very supportive political attitude for S&T since Nehru. In the context of the present world scenario of liberalization, Prof. Menon cautions against having excessive reliance on the multinationals which have become global entities and are stronger than the governments. The challenges of the Intellectual Property Rights (IPR) are especially acute, particularly the attempts to put blocking patents in almost every major area. Finally, he stresses the need for greater cooperation among India, China and Japan, not to dominate but to prevent domination by other countries and to provide a more Just World Order.

Gu Shulin, dwelling on National Innovation Systems with particular reference to China, puts forth the view that the concept of a national innovation system is a useful instrument in analyzing endogenous capability building for industrial development. The concept provides perspectives focusing on learning and learning efficiency, and emphasizes that learning is supported and embodied in an institutional context. Furthermore, the concept is sensitive to the unique features of industrially underdeveloped countries in recognizing (1) the indispensability of building and restructuring institutions throughout the modernization process, and (2) the distinctive and complex role of government policy in compensating for inadequately developed market mechanisms, and in inducing institutional transformation. Using this framework, she goes deeper to understand the reasons why particular industrialization efforts in specific countries have been frustrated or have proved successful. The concept of a national innovation system leads us to expect that these internal reasons are directly manifested in the efficiency of technological, institutional and policy learning. Focusing on micro-level events, she points out that it is at lower levels that the delicate mechanisms of interactive technological, institutional and policy learning work. She stresses that a careful identification of critical issues affecting technological capability building is a necessary step if this analytical framework for national innovation systems is to be properly employed.

Part II focusses on a comparative study of the science and technology strategies and their implementation during the various plan periods in India and China. Ashok Jain and V.P.Kharbanda point out that S&T developments in India and China can essentially be talked about in three phases. In India, during the first phase (1947-60s) a sound S&T infrastructure was built. The phase was also characterized by Nehru-Mahalanobis Planning Approach with emphasis on a "Self-Reliance Strategy." During this phase, China also witnessed the building of S&T infrastructure (1949-57). During 1958-65, emphasis was on a self-reliance strategy. In the second phase (1970-80) in India, efforts were directed towards reorientation, import substitution, putting self-reliance into practice and attempting coupling of S&T with socio-economic development. In China, the second phase (1965-76), coinciding with Cultural Revolution (CR), was the period of turmoil which witnessed large scale de-emphasis on scientific research and higher education. Self-reliance was defined as isolation from the world stream of knowledge. This had quite a damaging effect on S&T development in China. In the third phase, starting from 1980s, stress was on accountability of S&T activities. Currently, market orientation is the dominant theme in both India and China.

The distinguishing similarity in these countries has been the initial dominance of conventional Linear Model of Research-Development — Production, which in the later phase has given place to a non-linear feedback model emphasizing the dynamic interlinks between scientific research infrastructure, innovation, production and the economy. Both started with emphasis on building up of heavy industry, particularly steel and nuclear energy. The only difference has been that while India went in for harnessing the nuclear energy for power generation, China's efforts were more defence-oriented. While both started at the invention end of the innovation chain, Chinese efforts at the diffusion end have been more significant. However, during the third phase, efforts in both the countries cover the whole innovation chain. At present, *R&D* intensity is much more in China, which is evidenced by the fact that at present China has almost double the number of *R&D* personnel and the expenditure on R&D is about three times that in India. However, in both the countries, at present there is a declining trend in expenditure on *R&D* as percentage of GNP. This aspect needs particular attention. In the new scenario of increased globalization, finances and production systems with market locale playing a dominant role, it is important to integrate S&T with economic development through 'Techno-Economic Networks' rather than supporting S&T activities *per se.* In this context, Economy-Technology-Science (ETS) model has been discussed.

Part III covers the very important issue of how the developing countries, particularly India and China, should restructure their institutional infrastructure to face the on-going process of globalization. In this context, V.V.Subba Rao focuses on *R&D* efforts of the Indian industry and the challenges it faces in view of liberalization and globalization underway. Linkages and close interaction among various agencies, such as industry, universities and Indian Institutes of Technology, R&D institutions, financial institutions, and the government need to be nurtured and strengthened. He highlights the government policies in regard to building up of technological capabilities conceived in a mix of both domestic know-how and foreign technological assistance. The new policies of the government, thus, are attracting greater foreign investments, improving the quality of our infrastructure and products, increasing competitiveness in industry and boosting our exports in the global markets. These call for higher degree of technological and professional inputs, greater investment for strengthening domestic technological capabilities, and greater commitments for R&D by the industry.

Chen Wenhua, points out that the purpose of reform in China should be to liberate and develop the scientific and technological productive forces, and develop a socialist economy vigorously. It should promote the combination of science, technology and economy in an organic way to suit the needs of socialist market economy. It is pointed out that Technical Innovation is the medium of science, technology and economy. The characteristics of such a Technical Innovation Medium have been described. On this premise, the main aim of the thrust to invigorate the scientific and technological system is to strengthen the new links of technical innovation in the Chinese S&T System, essentially to fill the gap between science, technology and the economic system, characteristic of a planned economy. The motivation behind the reform of the S&T system initiated in 1985 is to strengthen these new links of technological innovations essentially to face the challenges of globalization. It is not essential for all countries to start from basic research. A country can start from technology import and aim at its absorption, digestion and innovation.

Fang Xin dwells upon 'The Progress of Science and Technology Reforms in China.' She points out that the two motives of S&T, i.e., creation of knowledge and meeting practical demands arising from economic and societal development should be activated to the maximum by various mechanisms and managerial methods. This needs much more attention in a country with scarce resources but a huge population. The key decision to be taken in respect of the S&T reform is about how far R&D institutions should go towards market orientation. Other aspects of the reform include enhancing the rights of the research institutions to take independent decisions, promoting R&D setups devoted to tackling problems of the entire country, encouraging mobility of talent and strengthening cooperation between R&D institutions, universities, and enterprises. R&D funds in China are mainly invested by the government with only a small part coming from enterprises. She emphasizes that modernization of the traditional industries and promotion of agricultural technologies is an essential prerequisite for ensuring rapid socio-economic development. The focus of the ongoing reforms is on making S&T undertakings aim at economic construction, meeting of market demand, selectively developing high technologies, and promoting their application in industry.

V.P. Kharbanda, in his contribution on "University -Industry Linkages in India and China" puts forth the view that unless a symbiotic relationship is built up between knowledge generating institutions and industry' it would be difficult for enterprises in the developing countries to compete with the multinational giants of the developed countries in the prevailing world order of globalization and liberalization. Presently, both China and India are generating instruments to build up various forms of university-industry linkages primarily with government support. Development of High Technology Zones, Incubators, Consortia, Science and Technology Entrepreneurship Parks, signing of Memoranda of Understanding between university and industry for joint projects, etc. are some of the welcome developments in these countries. However, industry has to come forward in a big way to assist the knowledge generating institutions by forming regional techno-economic networks and viable National Innovation System.

Technology transfer process is important for industrial growth, and involves a number of critical steps-in both technical aspects and in management. India and China, even after almost half a decade of development, face the critical problem of transfer of technologies, whether they are imported from abroad or have been developed indigenously. This aspect has been dealt with in part IV. The authors primarily stress the need for greater involvement of industry in establishing rapport with research institutions at an early stage of R&D project development. Y.S. Rajan and Sunita Wadhwa concentrate on the problems of absorption and adaptation of technologies in India and highlight the basic problems of technology transfer in so far as the Indian S&T system is concerned. They lay emphasis on the linkages among promotion agencies and their role in technology development and technology transfer. The paper presents case studies on technology development and transfer efforts in different sectors in India, bringing up some issues and problems in technology transfer. They point out that Foreign Collaboration Agreements (FCA) and their equity participation have had positive impact on the development, adaptation and absorption capabilities, but the fact remains that non-FCA units in relative terms spend more on R&D and develop/adapt larger number of products than the FCA units. In India, a number of promising and meaningful technologies have been developed at lab-scale, which can be commercialized. A few agencies are making efforts in this direction through funding programmes, such as, DSIR's *PATSER*, NRDC's schemes, TIFAC's Home Grown Technologies, etc. to develop linkages between R&D groups and industrial production groups. Studies conducted in this respect indicate that the degree of technological dynamism of the firm is the determining factor for its technology transfer and absorption capability.

Jia Xinmin presents a model of mechanism of selecting and shifting the position of 'interface of technology transfer, between the research institutions (the supplier) and the industry (the consumer) . He shows that while the ongoing reforms have laid the foundation in China to finally enter the market economy, the reforms have been one-sided. While R&D institutions have witnessed widespread reforms, the requisite reforms at the enterprise level are lagging behind; this has pushed the interface of technology transfer downwards, where early participation of the enterprise in the R&D project is negligible. This has slackened the pace of the development and building of indigenous technological capabilities for absorption, adaptation and innovation. It has also led to widespread wastage of the institutional resources. The author concludes that if reforms at the enterprise level do not coincide with those at the institution level, the problems being faced will continue and threaten the development of China.

In respect of development of Intellectual Property-Rights System in India and China in line with the international practices and formation of World Trade Organization in January 1995, Wen Xikai in her paper 'Current Status of Patent Protection in China, illustrates the status of patent protection in China, highlights the present Chinese Patent System, related laws and regulations, enforcement of the Chinese Patent law and the plans for the future. She points out that effective patent protection is ensured through both administrative enforcement and judicial mechanisms at various administrative levels. With strict adherence to the reform policy, the Chinese authorities revised their patent law in 1993 in line with the international norms. On the basis of the experience gained so far, China is planning to further amend its patent law in the light of prevailing international practices.

N.R. Subbaram, dwelling on the patent system in India, traces its development since the first legislation brought into force during 1856. He highlights the salient features of the present Patent Law-1970, which came into force in 1972. A well developed Patent Information System was established in India in 1980. Discussing the measures being taken by the Indian government to revise its Patent Law-1970 consequent to the establishment of World Trade Organization in January 1995, the author stresses that India would have to amend the Patents Act, 1970, particularly with respect to allowing *product protection* in all the fields of science & technology, providing protection for micro-organisms and providing a uniform term of 20 years for all patents. Regarding the infrastructural problems faced, he mentions that a major handicap for the patent system in India has been the lack of professionals in this area. Only 150 trained professionals are available in India who can judge the patent system correctly. A comparison of Indian and Chinese Patent System brings to light the weakness of the Indian system. Though China introduced its patent system only in 1985, tremendous development has taken place in a span of 10 years. The number of patents filed and granted in China is much higher in comparison to India. Proper education in patent field and establishment of network institutions has generated awareness about patent culture, which is very much lacking in India. 45 to 50 institutes give diploma courses in patent education, making a good number of professionals available in the country. This is one area where India and China can go hand in hand and learn from each other's experiences.

Modernization of rural traditional technologies is another facet of development crucial for faster economic and technological growth. Both China and India have been facing major problems in their efforts to uplift rural artisans and traditional crafts while implementing various programmes and schemes. While in China, 'SPARK Programme', has been the main instrument since 1986 to carry out such activity, in India a large number of government agencies and voluntary agencies have been involved in an effort to modernize rural technologies. M. A. Qureshi, putting forward a theoretical approach towards choice of appropriate technology in rural areas, points out that attempts to upgrade craft-based technologies through application of modern S&T are constrained by many problems. The main factors responsible for this situation are: an unhealthy orientation of government supported R&D institutions, which favour following of international trends and values; the practice of measuring performance of rural development programmes through physical and financial targets rather than on the basis of delivery of technology to rural people; lack of will to boost technology transfer; and consideration of application of S&T in rural areas as a gesture of benevolence. Two issues related to technology delivery system are crucial. First, in view of growing liberalization of economy and demand for environment—friendly development, the choice of sustainable technology has emerged as a new dimension of the development process. Second, there is need to improve the social relations of different sections of the rural society with the S&T system. After making appropriate choices of technology, the deficiencies faced by the rural society can be made up by evolving an appropriate interaction process. This has necessitated participation of technology receivers, R&D institutions, local authorities, financial institutions, and an interfacing agent. Concerning the situation in China, Wang Wenyong points out that China has made considerable achievements in application of new technologies in rural areas through Spark Programme initiated in 1986, which has helped to remove poverty, increase employment and raise the living standard of the people to a considerable extent. This has been made possible mainly through adoption of reforms and open policy in 1979, which led to the delegation of powers at lower levels. The most important effort has been the establishment of Township and Village Enterprises (TVEs) which were allowed to develop in the highly competitive market environment. However, a number of problems still exist which relate to the disparity of development in rural enterprise in different parts of China, shortage of monetary support, and failure to abide by the regulations, for example, illegal change of ownership or imposing unreasonable charges upon the rural enterprises in some areas. The author emphasizes the need for the national government to incorporate the 'Programme of Introduction of Industrial Technology in Rural Areas' into the national development plans, so that adequate policy and other supportive mechanisms may be built up.

S&T manpower is an important resource and thus its proper development and nurturing is a prerequisite to application of S&T for economic development and building up of technological capabilities. In Part V, A Rajeshwari points out that the rapid advances made in recent years in various fields have necessitated specific manpower development programmes. It is important to bridge the gap between acquired skills and the requirements in job performance. Notable among the emerging technologies are micro-processors, laser, water resource management, biotechnology, ceramics, and semiconductors. To keep pace with the developments in these technologies, inputs of required S&T manpower are important and critical. Over the years, manpower has become an important component in the national educational and S&T policies in India. Tracing India's policies, thrusts, planning and developments relating directly or indirectly to science and technology manpower as a resource, she highlights the mismatch between the higher education system (the supplier) and the industry (the user). She points out that although the supply base of S&T manpower has increased considerably over the years and the estimated stock of S&T manpower by end 1990 was around 4 million, it has failed to deliver the desired results of reducing poverty, unemployment and underemployment to an appreciable extent. Moreover, there has been no real change in the high(er) technology content of India's manufactured goods for exports. There has also been gross under utilization of S&T personnel. This raises doubts as to whether the S&T content of the present educational system suffers from lack of relevance to the needs of industry or whether industry lacks dynamism and professional management, or both. To check the unemployment of S & T manpower, a two-pronged attack is required: streamlining S & T education and increasing industrial growth and exports and, in turn, overall economic growth. In addition, there is a need to improve our educational system to make it more need-based rather than rigid discipline-oriented. This necessitates establishment of a national S&T manpower information system for creating a database on supply and demand of S&T manpower to study and analyse the whole issue of education and employment planning for S&T manpower.

Chen Hao, dealing with the issue of chronic shortage of S&T manpower in China, points out that it is imperative to formulate a rational policy of workforce management in a bid to effectively cultivate, employ and protect the existing manpower resources. The author attempts to delineate a rough picture of S&T manpower management in China with emphasis on higher educational institutions and their interactions with other R&D institutions and enterprises. The irrational distribution of Chinese R&D manpower is epitomized by the fact that a great majority of them (63.7 per cent) are concentrated in universities and R&D institutions, while enterprises account for only 27.5 per cent of them. Discussing the existing problems and issues, the author highlights that internal and external brain drain poses a great threat to development of S&T in China. The root cause of this situation is irrational income structure and unhealthy work conditions in Chinese S&T institutions. Concrete steps are needed to be taken to effectively improve the situation. The current systems of university enrolment and graduate employment have to be improved, while postgraduate training and the system of academic degrees have to be perfected and properly developed. The State should gradually adopt measures to rationalize the deployment of S&T talents across the country in line with market demands. For enhancing the remuneration being currently paid to our S&T workers, some effective steps are needed. Otherwise, the contingent's stability will be affected, further posing a potential threat to China's S&T cause.

Biotechnology is one of the high technology areas which are likely to have considerable impact on the national economies in the 21st century. Both India and China are putting in major efforts in this emerging area, which may have considerable impact in diverse fields of agriculture, health, pharmaceutical industry, food processing, etc. In this context, Kavita Mehra surveys the efforts being made in India to get to the forefront of this high-tech area. Starting with the policy framework, and the organizational structure, the major research programmes being carried out have been identified. Efforts at creating capabilities in indigenous R&D and commercialization of the technologies have been examined. Points out that in India, the existing R & D capabilities have not been fully utilized at the industrial level. Patenting of innovations is also not widely popular among the Indian scientists. Although most of the biotech companies in India are producing a few diagnostics, antibodies, enzymes, vaccines and agricultural products, as yet no company has produced a high-tech biotechnology product. It is concluded that biotechnology research in India is primarily funded by government and the role of private sector in the promotion of biotechnology has been negligible. This aspect needs particular attention. There is also an urgent need to increase research-industry relationship so as to acquire the capability to face international competition in the 21st century.

Wang Qingnan describing China's development programme for biotechnology, points out that Phase 1 of the programme is primarily aimed at basic research and R&D for product development. Some of the achievements made during this phase have been described. Phase II of the biotechnology programme is to concentrate on the national resources in a few well-chosen promising areas which are likely to have a major impact in the 21st century. When selecting major areas, novelty is of paramount importance so as to increase the competitive edge of the Chinese biotechnology industry. In terms of technology, emphasis will be on interdisciplinary aspects. The first biotechnology drug approved by the Chinese government for entry into the market is human interferon. The author concludes that in the 21st century, R&D in biotechnology will be of vital importance for the Chinese economy, for raising the standard of living of the people and for providing adequate healthcare for them. Emphasis in Phase II of the Biotechnology Programmes will be on strengthening the concept of property rights, protecting Chinese patent rights, and at the same time, training a large number of biomedical scientists. To achieve these objectives, it will be important to lay stress on basic and interdisciplinary research, and to encourage international collaborations.

Telecommunications is another high technology area which is likely to have great impact on national economies in the coming century. Countries which excel in this field are likely to have major influence in the international economic scene. Further, in the context of liberalization and globalization, development and modernization of telecommunications industry in the developing countries has attracted significant attention. Both India and China are reorienting their strategies through various reforms in order to accelerate the process of modernization of this sector. Ding Lu highlights some institutional facts about the managerial system of China's telecommunication industry. He points out that State control in this area is being considerably reduced since the adoption of 'Open Policy', and initiation of reforms in the early eighties. Efforts have been geared to make post and telecommunication enterprises financially independent. Telecommunication companies have been separated from the postal service agents. The author holds that China's bid to join the World Trade Organization (WTO) will surely have a profound impact on the country's telecommunication industry. However, WTO's acceptance of China hinges on whether China will agree on terms that mandate opening of its services sector, including the telecommunication market. Dwelling upon the Indian situation, Parthasarathi Banerjee points out that in the present scenario, Information-telecommunication is the key transforming agent for an economy at the cross-roads. Increasing returns on knowledge and information, growth through connectivity, increasing information-intensity in all kinds of activities, new international division of labour, resurgence of macro economic-instruments and freeing of finance are only a few facets of this new manifestations of the economy. The integration of telecommunications with information technology and the media has created a new situation, where the role of the regulator is crucial and thus has to be redefined. This redefining strategy has to take into account a federated domestic political process, the economic rationale of the industry, the global strategic concerns of the state and a resurgence of the importance of social control. He argues, that this redefining of regulations should be carried out at three levels, i.e. global, industrial and social levels. As a result, political and social authorities can appear as regulators at multiple levels of regulations. It is important not to forget the abiding role of political and social institutions and authorities, especially if the country is populous-diverse and poor, as is the case of India, in matters pertaining to regulating and guiding the information-telecommunication-entertainment events.

India and China are the two biggest countries of the world in the range of billion population, and are amongst the largest geographical areas of the world put together. Both the countries over the years, India after 1947 and China after 1949, have built an impressive R&D base and infrastructure in terms of heavy industry as well as production of certain raw materials and consumer products. Therefore, it is very important for the two countries not to remain isolated, but to come together to share experiences and to encourage mutual cooperation. Jointly, they can effectively work towards establishment of a just world order. Touching upon these aspects, Sunil Sondhi discusses pathways to regional cooperation. He points out that bipolarity, which dominated the world system for more than four decades, has resulted in a new multipolarity. The Asian region is economically, culturally and politically more diverse than any other region in the world. By the end of the present trajectory, the Asian region as a group, is likely to grow its economic power to exceed the economic wealth of the present European Economic Commission (EEC) countries. Looking into the need and conditions for successful cooperation in S&T across national boundaries in the Asian region, he points out that the Asian region, particularly India and China are economically, culturally and politically more diverse and rich in natural resources than any other region in the world. Scientific cooperation among these countries could well provide the means for capitalizing on the divided scientific capabilities throughout the region, by nurturing collective economic strength against the new powers of European and North American economic dominance. Nations participating in regional cooperation bring a range of different strengths and weaknesses to a common platform for intensive cooperation in specific areas of research, industrial innovation processes, and commercial exploitation of capabilities in specific industries. There is thus enough reason for India and China to cooperate intensively in areas of S&T not only to mitigate socio-economic problems of development but also to strengthen efforts towards creation of a 'Just World Order.'

V.P. Kharbanda
Ashok Jain

PART I : CONCEPTUAL FRAMEWORK

1. Science in Society–The Emerging Scenario*

M.G.K. Menon

The Social Fabric of Science

Often one makes the mistake of assuming that science is something which operates in a compartment by itself, and is separate from everything else in society. Yes, there are instances when it stands alone; and in this country, perhaps the finest example in recent times is that of Srinivasa Ramanujan, a mathematical genius. In his case, we could say that his work had nothing to do with society, with government, with non-government organizations, with science and technology policies and the like. He was deeply motivated from within and produced the greatest mathematical work through sheer inspiration. But, was even Srinivasa Ramanujan totally divorced from society? He was not a robot. He was not just a computer. What went on in his brain, we do not understand. But surely he was a person who evolved in a certain cultural and social milieu. His creativity arose both from his genes and his socio-cultural roots.

There has to be a deep inner motivational element for science to flower. Such elements may survive in a hostile environment, but have the best chance to fructify in a supportive environment. One can see this in the growth and decay of science in various civilizations of the past such as in China, South East Asia, India, Mesopotamia, Central Asia, Egypt, Greece, Rome and the American Indian empires. What happened to science in all of these? Yes, at a certain stage it flowered in these regions and

---

* Based on the Key Note Address given at the National Institute of Science, Technology and Development Studies (NISTADS) on 25th September 1995.

then largely decayed as the environment turned hostile and barren. They provided some of the building blocks on which later developments took place. But the relevant institutions and cultures vanished. Thus, we use zero today which was discovered in India. Much more than zero is the concept of " Nothingness or Shunya". There is deep philosophical content in 'Shunya' which we still have to understand. There were other great achievements also that one can recount. But what is worth noting is that the growth of science in all these cases was not self-sustaining. Does any body have an answer for why this was so? It lies largely within the framework of society: its flowering and decadence.

In contrast to the earlier situations, we had the Scientific Revolution that took place in parallel with the Industrial Revolution a few hundred years ago in Europe. This has had a self sustaining character. Some assume that there is a direct connection from science to technology to production. While the feedbacks between these different elements are exceedingly strong now, this was not the case when this process got going. The steam engine had little to do with the scientific work on thermodynamics. It was a technical development in its own right. But all of electrical engineering, and production relating to this area, derived from basic scientific discoveries in electricity and magnetism.

What enabled science to develop in Europe; and to continue to do so on a self sustaining basis? This is related to the manner in which society in Europe was transforming at that time. There was continued questioning and a ferment in which established ideas and hierarchy were not allowed to hold sway except on merit, and there was increasing reliance on experimentation and measurement. Theories and hypotheses had to stand the test of experimental verification and observations had to be understood on rigorous and widely applicable theoretical grounds with the potential for prediction. There was an in-built mechanism for feedbacks that kept the process self sustaining.

It was such a ferment in a search for a true understanding of Nature, partly related to social transformations that led to the freedom struggle which gave India, before Independence, almost a similar scientific renaissance. For example, Calcutta University had hardly any facilities in 1920's, but it was one of the great universities of the world. India had a succession of individuals starting from the setting up of the Indian Association for Cultivation of Science by Mahendra Lal Sircar, J.C. Bose, P.C. Ray, C.V. Raman, K.S. Krishnan, S.N. Bose, Meghnad Saha, P.C. Mahalanobis and others who constituted this renaissance. Science suddenly flowered; and there was also a technical renaissance. There was this whole class of people interacting in a vibrant social atmosphere resulting in what one could call the beginning of a coherent scientific community. Thus, when we talk of science and science policy, we must look at the interplay of theory and experiment, interaction within science and the interface it has with the fabric of society. All of this is involved in a self sustaining growth of science.

Fifty Years of the Development of Science in India - An Overview

Let me now look at the manner in which science has developed in India over the past fifty years; and correspondingly what happened in China. India and China are the two biggest countries of the world, each of whose populations lie in the billion range; they are amongst the largest diversified geographical areas of the world, particularly when put together; other than the CIS, USA, Canada and Brazil. Both emerged as independent nations half a century ago. Their comparative experiences are important and of value as one plans for the future.

India and China are great countries, not only in terms of areas and population, but also because of their civilizations, cultures and traditions, deeply rooted in a history of thousands of years. In science and technology also both have had a great productive history and tradition. While the history of Science in China has been documented by Needham for outside readers, no similar effort is available in the case of India. This is a major lacuna that needs to be rectified; an effort needs to be made on a comprehensive analytical history of science in India. Let us make an effort to see what happened in the two countries in the socio-cultural-political context.

Both the countries entered the twentieth century in total disarray. In the case of India, the fabric of its society was in a fragmented state during the nineteenth century. Instead of one great nation, different regions and different strata of society were functioning in bits and pieces. There were internecine conflicts, breakdown of law and order, and science was certainly nowhere on the agenda. It was the experience of British occupation that made us realize that there was one India; it may have diversity of many types but there was also an underlying significant unity. And in the freedom struggle the bits and pieces started to come together. Science in India grew as part of the ethos of the freedom struggle and the social ferment that was related to it. The nineteenth century bequeathed to the twentieth century a profound multi-faceted search for an Indian identity involving deep thinkers like Raja Ram Mohan Roy, Dayanand Saraswati, Debendranath and Rabindranath Tagore, Ramakrishna Paramahamsa, Swami Vivekananda and many others.

India became free in 1947. However, it is important to stress that one emerged into freedom from a comparatively benevolent, colonial past. There were individual brutalities and the trauma of Partition, with enormous loss to life and relationships. There had been severe scarcities, particularly acute in the countryside such as the Bengal famine during the Second World War. There were insidious damaging effects of colonialism, which continue to persist even today in the education system and societal attitudes. There is continuing importance given to white collar jobs and the power of government rather than the entrepreneurial spirit of a free society. Yet, we also had large numbers from the west who came to India in the colonial period with good intentions. The origins of the Indian National Congress and the way in which the Asiatic Society was set up and grown in the best traditions of European liberalism are just two examples. Science and engineering also benefitted from various activities in the colonial period in relation to meteorology, functioning of the Surveys, radio and telegraphy, the railways, irrigation systems and the like.

China suffered much more during their colonial period because it was the battle ground of many foreign powers. It went through the actual experience of war, both during the Second World War and in the Civil War. Loss of life and property was on a much larger scale. The entire framework of government and society was transformed. In 1949, when China emerged as a new nation, it had been badly mauled. It had to recover in a far more intense and different sense than what India had to in 1947.

Economy wise, both the countries, were in a relatively green field situation. If one went to a shop in India in 1947, except for local items that formed part of traditional agricultural economy, everything else that related to industrial production was available only through import. This was also the case in China. Both countries, since they became independent, have built an impressive base and infrastructure of an industrial nation.

In science and technology, both have had great traditions. India had an advantage in the period of three quarters of century prior to independence. Indian science had flowered from the last decades of the nineteenth century inspite of lack of money and of real support, which in any case are not the most important elements for science to flourish; and not particularly important at that phase in the development of science globally. Men like Mahendra Lal Sircar, who created the Indian Association for Cultivation of Science, and Ashutosh Mookherji, Vice Chancellor of Calcutta University made such a difference. These individuals played a role in creating institutions and providing support, so that science could flower. It is in this aspect, and the developments that followed, that the Indian science scene differed from the Chinese science scene in 1950. India had a sound base of high quality science largely in educational institutions. This had resulted in schools and traditions that produced successive generations of scientists from Calcutta, Bangalore, Allahabad, Lahore and so on. There was work done of Nobel prize quality. Quite clearly, while work in Indian science was at the cutting edge, there was really no technology, or large scale manufacturing needed for a modern industrial state.

In the case of China, there was very little in this regard when it became free in 1949. There were some distinguished Chinese studying or working abroad, who could make a difference to China later; but very little in China itself. China and the Soviet

Union were very close in the immediate post World War II era. The Soviet Union was functioning on Marxist ideology and theories; so was China. There was initially an affinity based on a common ideology. They were also neighbours. Until 1956, this led to a close cooperation between the two countries, including in science. This can be seen in the defence equipment used by the Chinese armed forces; and in science the pattern, adopted for the highly centralized Chinese Academy of Sciences and its constituent laboratories, was based on the Soviet Academy of Sciences. In 1956, the two countries broke apart. China decided "We can and will stand on our own"; and they have indeed done so with remarkable success. Dependency is not good. The resolve of China to stand on its own strengthened it immensely. It led to self confidence. In any case, China has always had a strong characteristic of national pride and faith in itself as a great nation and a great race. They embarked on plans to become a world super power. China was isolated from a large part of the industrialized world, particularly the West - because of its Marxist ideology and its immediate post World War II links with the Soviet Union. Later, in areas of confrontation such as in Korea and Viet Nam they were clearly in a camp opposite to the USA. Then size, their pride in themselves and isolation led them on a self confident pathway of development as a nation. There were prices to pay at various stages; but they paid these.

India emerged into independence with a sense of self confidence that was born out of the freedom struggle - but in the spirit of Mahatma Gandhi: a meek little man outwardly, but an individual prepared to take on the might of the British Empire and say " do what you like, we will not bow". The entire Indian society was galvanized by his "satyagraha" and stood behind him. But that spirit does not exist in the same way in India today. There are abundant human and material resources. There is a great history and tradition to be proud of. There is a free democratic society. Yet the coming together of the people to make a great society and a great nation does not manifest itself.

What did China do after 1956? Let us take just one area. China decided that it would not allow itself to be in a position where a large country such as itself remained weak militarily in a world of super power confrontation. Atomic bombing of Japan by the USA in 1945 was followed by an increasing spiral of the nuclear arms race. The making of hydrogen bomb by USA, corresponding developments in the Soviet Union and the coming into existence of increasingly sophisticated delivery systems made China feel insecure. So China asked: "how can we, as a large country, which has an enormous border with the Soviet Union with whom we do not have friendly relations and with a super power, the USA, which is opposed to our ideology, not develop our own capabilities in the field of nuclear weapons and delivery systems. China proceeded to fulfill the objectives of making the atomic bomb, then the hydrogen bomb, and along with it delivery systems upto ICBM and submarine launched ballistic missiles. Recently, China has been able to test ICBMs with solid propellants which can be mobile. This policy decision arose from the spirit of self confidence, and to feel secure; not, as far as I can see, aimed at anybody as a first strike, to conquer, exploit or colonize; its aim was to ensure that nobody could dominate China.

India approached the same area very differently. We had decided right from the beginning, from Bhaba's vision, (which predates independence and indeed the first atomic bomb test in 1945 at Alamogordo) that India should have nuclear power. Bhaba was in the category of those great scientists who understood the possibility of the atomic bomb. From the discovery of the neutron to that of fission, and then a controlled chain reaction would enable a million fold increase in energy compared to the electron volts that characterize chemical reactions. He realized that abundant power was essential for India's development. Thus, India embarked on a nuclear power programme to produce nuclear electricity for the country. When China exploded its first atomic bomb in 1964, Bhaba understood the implications that this had for India going along that path. It was, however, only in 1974 that India conducted the Pokharan underground nuclear test. This was a restrained statement that India could be a nuclear weapon power. Since then, India's nuclear power programme has gone ahead, limited largely by financial allocations; but not its nuclear weaponization until May 11, 1998 when underground nuclear tests were conducted at Pokharan followed by the statement of the Prime Minister of India, that India was now a modern weapon state.

India was particularly fortunate in one respect. When India became independent, it had as its first Prime Minister, Jawahar Lal Nehru, who understood the connection between the growth of science and technology and its implications for national development; its value for society, not only as a motive force in the production process and for economic development, but as importantly a part of the cultural base for objective rational thinking. He supported science very strongly. He was succeeded for a brief period by Lal Bahadur Shastri who was very respected Prime Minister but did not live long enough. Thereafter, Prime Ministers Indira Gandhi and Rajiv Gandhi gave strong support to science. One can, therefore, say that India found strong support for science for a significant part of the half century since Independence, at the highest level of the Prime Minister who was also Minister in Charge of the various science departments and ministries.

In China also there was strong support for science based on the Marxist - Leninist ideology and the recognized importance of self reliance. I gave the example of the nuclear area; there are related aspects of delivery systems too. China went ahead with single minded dedication in this regard. Rocketry and missile development were emphasized. Their great scientists settled abroad, like Chou Peiyuan and others, were facilitated to come back to China. The programs were dynamic, time-targeted and accomplished well.

But China had a major disruption, the period of the Cultural Revolution (CR). The entire scientific fabric collapsed during that period, when vast number of scientists and professors were asked to sweep the streets, go to work in communes, equipment were thrown out of the laboratories, etc. But that was in the late sixties and early seventies. I have been to China many times, soon after the CR, but not during, and many other times through the eighties. I have seen their development, and a variety of institutions. What has always impressed me is hard headedness and an approach based on self confidence.

Both countries emphasized strategic 'import substitution' in the early years of their development. This was but natural. Neither country could export on any significant basis other than low value added commodities. The exports were needed to finance essential imports. The rest had to be covered by import substitution. Also imports in areas of high technology or of a strategic nature were not easy to arrange. How could China import what was needed for their nuclear weapon programme, missiles etc? How could India? The spirit of import substitution was wide spread in the early days of freedom. The situation has since changed more rapidly in recent past: as the rest of the world begins to appreciate the enormous market potential in these two populous countries; and they have moved up the ladder to be able to export significantly through high value added products.

Science, Technology and Globalization

Globalization is largely viewed in terms of financial transfers, and economic aspects which are the end manifestations of a process engineered through developments in science and technology.

First, large scale rapid physical transportation has been made possible by science and technology. With the invention of the jet engine and wide bodied aircraft, there is large volume global movement of people and freight. Components can be flown in from different parts of the world where they are best manufactured, to assembly plants and finished products flown out to markets globally. Entire aircraft fly fruits and flowers across the globe into countries where they do not grow. Then, there is the movement of enormous tankers. We never had oil moved on the scale it is carried today. If one looks at a satellite picture of the Indian ocean, you can see, almost like the way ants move one after the other, a conveyor belt of oil tankers moving from the Persian Gulf to Japan, a great consumer of imported oil. Not only are there big tankers but containerization for carrying goods of all types too. The concept of a container is not new; it is just a box as we all use. But the concept of "containerization" where for a certain destination various types of goods are put into giant containers that go into vessels and aircraft is relatively new. Then, there are pipelines not only across land areas but below the ocean. With these pipelines, oil and natural gas is moved and distributed across thousands of miles. These are engineering and technological developments which have made the transport of people and goods possible on a scale never realized before. As important are the other technological developments which have made possible the movement of ideas and information. This revolution started with the transistor invented in 1947. I have seen days when transistor did not exist. I built my electronic circuits without transistors, based on valves. Then came the first integrated circuit in the late 1950s moving across small scale, medium scale, now to large scale and very large scale integrated circuits where one can have as many as a million transistors on a chip. And it is not going to end there. We know now, with developments in solid state physics, with new quantum dots and so on, that one will be able to achieve a packing density of 10 billion devices. That is the scale factor on which one has moved over the past half a century. What has this done? It, essentially, has made electronic equipment very light, very low in power consumption, very small in volume. One sees this most dramatically in the revolution that has taken place in the field of computers where one has enormous computing power in desk and lap-top systems. One has been able to make satellites highly complex. All complexity can now be in space, with no complexity on the decentralized ground systems which can therefore reach out everywhere. Meantime, a laser revolution has taken place too. Fiber optics having enormous band width are strung under the Atlantic and Pacific, and across wide land areas. Today the world is completely inter connected through space, ground based terrestrial fibre optic, coaxial, line of sight and wireless systems. Telecommunications has moved from the analog to digital and this has resulted in coming together of computers and telecounters. It is this which has led to the modern world of information technology, enabling for example, 'money' to flow on a scale much bigger and more rapidly than the central banks of most countries can handle.

It is these in which are underpinning foundations of globalization. Some of this is the result of demands from defence and space programmes.

Looking Ahead

With these developments, obviously there have to be changes on the economic scene. There was a time, when if you wanted to buy a technology, you discussed it with a firm that had it, and you bought it - like buying in a super market. In many cases multinational firms established offshore operations because of availability of cheap labour at these locations. That is how electronics industry in Singapore, Hong Kong etc. started several decades ago. This offshore processing involved, for example in electronics production, a large number of girls sitting in lines doing jobs like soldering, on an assembly line. But from those days, there has been a shift: from offshore operation to subsidiaries and joint ventures. Today, a multinational company is almost a stand alone system, belonging to but not really connected with its own country. They look for factor advantages that are not natural resources or low cost labour alone; one has to look taxation laws, stability of governments, infrastructure such as power and transportation and all aspects involved in ensuring low cost quality production. A major factor is the availability of local markets. It is the sum total of all these economic factors that lead to production advantages. Since these change from time to time, and from country to country, there is an increasing degree of mobility in movement of investments and production centres.

In addition, scientific advances are such that it has now become possible, which was not so before, to have production of very high reliability, high quality, and in very large volumes. The Industrial Revolution was characterized by a transformation exactly of this nature, when you could produce through a machine, nuts, bolts, screws, and various parts identical to one another, with none of the differences that arise when they are made by hand; and also in large volumes. That was the change from hand made to machine made production. What we have today with computer based designs, computer aided production, and computer controlled machines, is that the scales and flexibility are totally different. A thousand, a million or ten billion transistors can not be put on a chip, except by these new automated techniques; and these techniques that are backed up by mathematics, physics, chemistry and material sciences of the highest order. One has to look at the efficiency involved, because that defines ultimately in economic terms, the unit cost of the product. And the unit cost is brought down through efficiency in production. Thus, one now demands reliability of a product, quality of a product, as also efficiency in the production process which results in low costs. Previously high value products are now greatly reduced in price and affordable to large numbers; while low cost items have achieved the highest levels of quality and reliability, now one wishes that one could transform society similarly.

Factor endowments are thus totally different today compared to what they were earlier. It is no longer a matter of where the resources like coal, copper, iron, wood etc. are available nor of a site that offers a large labour force at low costs, or even skilled labour; but it is the 'totality' of the system. As a result of all this, mergers are increasingly taking place on the economic scene to achieve the size needed. Large multinational companies of the past - say of ten years ago - have begun to wonder whether they are not already sub critical - too small to face the densely populated world. Therefore, we are now in a new regime which is a result of all these scientific and technological changes. We are moving increasingly to a knowledge based society and a science and technology driven economy. It is towards the emergence of such a regime that science and technology policy will have to be directed in future.

Conclusions

In conclusion, one can certainly see major political changes on the horizon, like a tremendous change that took place after the Second World War. After the Second World War, the old form of colonialism which had been established through military conquest disappeared. Vast numbers of countries became free.

The world was inching towards a democratic international set up. India was one of the leaders in this effort, being among the earliest countries to emerge into freedom. But the nuclear arms race, the cold war and superpower confrontation dominated the world scene. Large numbers of small wars have and continue to be fought with continued human misery. A major transformation happened with the collapse of the Soviet Union. Today we have the situation of an apparently single super power. But there is a limit to the use of military power. The superpower capabilities are now increasingly being exercised in matters which relate to economic assertion. This is where science and technology comes in as it is the main engine of economic advancement today. New aspects have come to the fore: globalization and its implications; the World Trade Organization and intellectual property rights; the revolution in information technology and biotechnology - one establishing a global connectivity among humans never seen before, and the other probing the very meaning of life and transforming living systems and the environment. In this continuing realignment can two of the largest countries be left out? Both have a long way to go to ensure a good quality of life for their people in the long term. Energy needs, sustainable development, protection of the environment and moving into the new age of information technology and of modern biology are all challenges as we move into the next millennium. It cannot be a matter of the two countries being separate on these issues, and allow once again, because they are divided, a new world order which represents the high degree of injustice one sees in the world today. The two countries have to cooperate; and the most significant area to establish cooperation is science and technology. Chinese science policy has progressively shifted from the post 1949, post 1956, post 1978 policies to new levels of openness and cooperation. India has moved in somewhat similar fashion. We now need to work towards policies that relate to a cooperative, equitable, inter-connected global system - and cooperate in this so that with one combined might this can lead to a Just World Order.

2. Towards an Analytic Framework for National Innovation Systems

Gu Shulin

Abstract

The author reports on the preliminary development of a conceptual framework for national innovation systems, with special reference to problems and difficulties contemporarily faced by developing countries in their industrial development, in the circumstances of economic globalization and liberalization. The study is divided into two sections. Section I presents an outline conceptualization of national innovation systems, emphasizing that learning and learning efficiency, and the institutional context of learning constitute a coherent core of the approach of national innovation system. It is argued that the concept with these core elements is sensitive to the unique features of industrially underdeveloped countries in recognizing (1) the indispensability in building and restructuring institutions throughout the modernization process, and (2) the distinctive and complex role of government policy in compensating for inadequately developed market mechanisms, and in inducing institutional transformation. Section II highlights some essential facets of research approach suggested by the concept of national innovation system, through illustrating issues and methods confronted, under the headings of (i) distinguishing between development stages; (ii) industrial structure and learning; (iii) institutional setup and learning; (iv) the market mechanism, opening to the world and learning; (v) economic reform, path-shifting and policy learning; and (vi) institutional and technological analysis. Demonstrates that while the concept reminds about national competitiveness and national specificity, the focus for analysis using this concept is on micro-level events. It is at lower levels that the delicate mechanisms of interactive technological, institutional and policy learning work. It points to the importance of a careful identification of critical issues that consist of a necessary step for a proper application of the analytical approach of national innovation system.

Introduction

The study reports on the preliminary development of an analytic framework for national innovation systems. It was prepared as part of the current stock-taking and synthesis of INTECH's recent research, and as a step towards making further studies on innovation systems in developing countries. Research at INTECH centres on the question of how a developing economy can enhance the efficiency of its development. One key theme highlighted in INTECH's studies has been the importance of learning capabilities in relation to 'high technological development', or 'high productivity development', (Cooper, 1995; Cooper and Turner, 1995). A conceptual framework for national innovation systems is clearly of interest in relation to this theme. While the paper is intended to develop a conceptual framework, this cannot be achieved without observing and responding to specific problems and difficulties currently arising in industrial development in developing countries. As has been widely recognized, contemporary industrial development is characterized by economic liberalization and globalization. Because my studies at INTECH have concentrated on the transformation of the national innovation system in China, this is also the starting point for the theoretical framework. The concepts developed here will, therefore, be applicable primarily to larger and technologically capable developing countries which have begun the transition from a closed system under tight administrative control towards a system which is more open to market competition and international exchange.

The study is divided into two sections: Section I presents an outline conceptualization of national innovation systems, based on a survey of literature and highlighting any differences or unique features which might be evident in industrially underdeveloped countries. This section is confined to an abstract delineation, to ensure that the conceptual issues are as clear as possible. The conceptual issues are elaborated in section II, which addresses some issues that are highlighted by the conceptual framework and some of the analytical methods required in studying the institutional and technological dimensions of a national innovation system. This section is intended to illustrate how this conceptualization of a national innovation system can provide new perspectives on development issues which are currently seen as being important and urgent, and how an analytical approach can be developed in accordance with this conceptualization.

I. THE CONCEPT OF A NATIONAL INNOVATION SYSTEM: Why national, and why a system?

Since the 1980s, there has been growing interest in national innovation systems, partly as a result of current economic trends. As Nelson and Rosenberg put it:

> "The slowdown of growth since the early 1970s in all of the advanced industrial nations, the rise of Japan as a major economic and technological power, the relative decline of the United States, and widespread concerns in Europe about being behind both, have led to a rash of writing and policy concerned with supporting the technical innovative prowess of national firms. At the same time, the enhanced technical sophistication of South Korea, Taiwan province of ,China and other NICs (Newly Industrialized Countries) has broadened the range of nations whose firms are competitive players in fields that used to be the preserve of a few and has led other nations who today have a weak manufacturing sector to wonder how they might emulate the performance of the successful NICs. There clearly is a new spirit of what might be called "technonationalism" in the air, combining a strong belief that the technological capabilities of a nation's firms are a key source of their competitive prowess, with a belief that these capabilities are in a sense national, and can be built by national action." (Nelson and Rosenberg, in Nelson (ed.) 1993, p.3)

Previous works on this topic[1] have concentrated mostly on advanced industrial nations, or at least are organized from this point of view, even if a few cases from other countries are included. Furthermore, although some efforts to understand national innovation systems have been made, the situation remains unsatisfactory, mainly because of the lack of a coherent analytical framework. Nelson puts it in this way:

> "[T] he organizers ... were bothered by the tendency of various analysts to argue, on the basis of very little evidence or analysis, that this feature, or that one, was a major factor behind country performance differences, a tendency checked neither by strong conceptual understanding of what is likely and what is not a causal factor, nor by the requirement that the proposed causal connections be consistent with a wide range of country observations." (Nelson, in Nelson (ed.) 1993, p.505)

The importance of developing a conceptual framework of national innovation systems for studies relating to developing countries has been clear in discussions taking place at the Institute for New Technologies. One of the most crucial issues discussed is the relationship between trade liberalization and government intervention. There is a deep concern that trade liberalization alone is inadequate to guide the economic reforms taking place in most developing countries. 'Getting the price right', which is the main purpose of current reforms, is perceived as not being adequate, in itself, to explain the success or failure of the ongoing reform efforts. Such concerns are not confined to discussions at INTECH, as can be seen from some extracts from Lall's work on the determinants of industrial success in the developing world (1991, pp. 118-121):

> "Developing countries vary greatly in their experience in industrialization....

---

[1] Notable examples include works by Chris Freeman (1987), Bengt-Aake Lundvall *et al.* (Lundvall (ed.) 1992), Richard R. Nelson *et al.* (Nelson (ed.) 1993) , and a series of papers published in the *Cambridge Journal of Economics*, 1996, 19.

> What explains this enormous diversity of experiences? Much of the recent literature on industrialization in the Third World has focused on one set of explanations for industrial success: trade policy. It has been argued that export-oriented strategies are more conducive to rapid, efficient and sustainable industrial development than inward-looking, import substituting strategies....
>
> The superiority of outward-orientation has been traced to the neutrality of government interventions in the incentive structure facing manufacturing enterprises. The most neutral of regimes is free trade, but others also qualify for neutrality if biases in favour of domestic market are offset by compensating measures to make exports attractive....
>
> This general case for trade strategy as the main determinant of industrial success is based on the assumption that incentives are the most important determinant of performance. It is assumed, in other words, that market prices provide correct signals for economic activity and that the ability to respond to those signals is present: stated simply, product and factor markets are supposed to be efficient in developing countries....
>
> There is no doubt that the neoclassical analysis of industrialization has a great deal of validity...but...other factors... are necessary for successful industrialization to proceed. Incentives are a necessary but not sufficient condition for industrial development. "Correct incentives may, moreover, well deviate from prices thrown up by free markets, and incentives by themselves may not produce the 'supply-side' factors needed for industrialization."

Obviously, a more coherent conceptual understanding for studies focusing on developing countries is required, if the growing interest in national innovation systems is to bear fruit in effective analysis. Such a framework is particularly critical for studies relating to building national endogenous technological capability as a means of promoting industrial development and competitiveness in the context of economic liberalization and globalization.

The conceptualization suggested here begins with the work of a group of economists at Aalborg University (Lundvall, 1992), whose theoretical framework is particularly helpful, although the empirical observations which support it are mainly from Nordic countries. The framework developed by the Aalborg experts highlights two fundamental components as constituting a coherent core for a national system of innovation: *learning and the institutional set-up.* These are adopted as fundamental components of this conceptual framework, and are used below to outline the concept of a 'national innovation system'.

I.1 Learning and Learning Efficiency

The first of these components, which the Aalborg group calls the learning assumption, is explained as follows:

> "First it is assumed that the most fundamental resource in the modern economy is knowledge and, accordingly, that the most important process is learning. The fact that knowledge differs in crucial respects from other resources in the economy makes standard economics less relevant ...."(Lundvall, in Lundvall (ed.) 1992, P.1)

This assumption suggests that one key element in the concept of a national innovation system, is knowledge and learning. The assumption can be supposed to be basically valid for developing countries, because it is widely accepted that (1) technological changes are accelerating, (2) competition in both international and domestic markets is becoming more intense because of trade liberalization and (3) ultimately, industrialization for countries means a strategic choice to join the club of modern economies which rely heavily on knowledge and learning.

Such a learning-centred approach implies that *learning efficiency* is a major explanatory factor in successful long-term economic development. This contrasts with classical development economics, which has tended to identify efficient factor allocation as a key to economic development, with the factors concerned being capital, labour and land. Effective land, labour and capital allocation do not logically presuppose effective learning, although the market mechanisms which serve as an important medium for effective factor allocation have in many cases proved to provide effective incentives for learning and transmission of the information which is necessary for learning. It is conceptually useful to distinguish between learning efficiency and allocative efficiency (and also between allocative efficiency and distributive efficiency).

Assuming that the concept of a national innovation system focuses on learning, an analysis of such a system must focus on *process*. This focus is in effect distinguished from the capability argument which has been popular in development studies since the 1980s and which leads to a focus on the consequences of learning. There is certainly a close causal relationship between the process of learning and its consequences in capabilities, and this relationship can be taken into account in a learning-centred approach. The capability-centred approach, on the other hand, has proven to be weaker in exploring the origins of capabilities.

I.2 The Institutional Context of Learning

The second assumption is that support from the institutional set-up is crucial to the learning process. As Lundvall says:

> "Second, it is assumed that learning is predominantly an interactive and, therefore, a socially embedded process which cannot be understood without taking into consideration its institutional and cultural context. Specifically, it is assumed that the historical establishment and development of the modern nation state was a necessary prerequisite for the acceleration of the process of learning which propelled the process of industrialization in the last centuries." (Lundvall, in Lundvall (ed.) 1992, p.1)

The proposed conceptual framework will, therefore, seek to identify elements which influence the capacity and efficiency of a national innovation system - that is, the institutional set-up within which innovation occurs - and how this supports learning. This assumption is based on the social nature of the learning process. Learning, as an accumulative process, is largely incorporated in the daily operations of institutions, and these operations are activated, motivated and directed by internal tensions and imbalances, and by external demands and disruptions. The framework of institutional establishment also serves as a medium in which knowledge, which may be tacit or expressed, codified or non - codified, is combined into routines, norms, habits, and mutual understandings so as to reduce communication costs and increase effectiveness of the learning process. The various interactions, ranged in a system, determine the dynamics of the learning process.

Given the assumption that the institutional support for learning is critical, those *cultural and social elements* which are crucial to an innovation system should be considered as intrinsic rather than extrinsic parameters. This is because development processes are characterized by widely 'divergent initial conditions,' which are profoundly shaped by national cultural and social backgrounds. An analytical instrument which did not appreciate such cultural and societal elements could have little analytic power in dealing with issues relating to the developing world.

The *nation state* is considered as a special kind of institutional context, so that it is included in the suggested framework. The nation state is heavily involved in establishing the institutions necessary for industrial development, by means such as industrial strategies, laws and regulations, and direct government expenditure. The nation state is also heavily involved in economic reforms, by reformulating the industrial strategy, altering laws and regulations, and re-organizing the policy institutions. While recognition of the nation state's role has been based on the experiences of the industrialized Nordic countries, the state's role is equally significant and relevant in countries now beginning their industrialization.

The nation state, as a distinctive institutional context, has a great influence on the pattern of the whole innovation system, and on the learning process and degree of learning efficiency which takes place within such a system. Interactive learning is, therefore, assumed to embrace the feedback between policy formulation and the responses of economic actors, which leads to a process of *policy learning* and *institutional learning*. These processes affect and are affected by the efficiency of *technological learning*.

To sum up, a national innovation system is a system constituted by various actors, most of them organized, who influence technological innovation. It operates through various interactions taking place within the context of the institutional establishment of the system, and it is significantly influenced by national policies. A national innovation system is "national" to the extent that it has identifiable national and societal specificities. It is a "system" in as much as we can identify the importance of institutional support for the learning and innovation process. The concept of a 'national innovation system' thus focuses attention on (1) learning efficiency, and (2) the capacity and efficiency of the institutional set-up to support learning. The latter is assumed to be a critical causal factor in the former.

Some key aspects of a national innovation system may be different for less developed countries, as compared to the industrially advanced countries. From various cases and experiences in less developed countries, three general features of national innovation systems in developing countries can be suggested: (1) the institutions needed for industrialization (e.g. institutions of the market, technology, science, and engineering education) have initially to be implanted from outside, and have to develop from a very rudimental basis, (2) therefore, factor management in less developed countries relies more on a framework which combines organizational hierarchies and quasi-market mechanisms, with the degree and form of the mixture differing country by country, and altering with time, and (3) because of (1) and (2), the capacity of the system to become functional in local 'initial conditions' and to retain its flexibility for further evolution is crucial.

II DEVELOPMENT OF THE CONCEPTUAL FRAMEWORK: Issues and methods

This section will elaborate on some issues that are highlighted by the conceptual framework, and on the analytical methods required in applying it to a national innovation system, under the headings: (1) distinguishing between development stages; (2) industrial structure and learning; (3) institutional set-up and learning; (4) the market mechanism, opening to the world and learning; (5) economic reform, path shifting and policy learning; and (6) institutional and technological analysis. The intention is to illustrate how this conceptualization of a national innovation system can provide new perspectives on important development issues, and how an analytical approach can be developed in accordance with this conceptualization. The approach is illustrative rather than systematic, to call attention to issues and methods which will be important in the further development of this approach to national innovation systems. As will be evident, this elaboration of the concept of national innovation systems is in effect a creative application of insights developed in transaction costs economics and the economics of technological change, among others in that it incorporates an evolutionary theory of technological change, institutional theory, and concepts such as learning and information flows, and technological capability building, to name just a few examples.

II.1. Distinguishing between development stages

A classification scheme suggested by a conceptual framework may be regarded as both a test and an application of the concept. Classifying the subject under investigation is also an important part of a research methodology. If searching is to be informative and the findings relevant and valid, the questions need to be specified in appropriate terms.

Assuming that the institutional set-up makes a critical contribution to supporting learning and innovation, and that the institutions needed for industrialization in developing countries have largely to be imported rather than endogenously developed, it may be useful in studying national innovation systems to distinguish between degrees of development of the institutional set - ups in various developing countries. Such a classification could be based on a number of criteria including the following:

(1) Whether a country has established its own engineering education system to generate a labour force meeting the requirements of modern industrial development.
(2) Whether there is an R&D infrastructure which is sufficient to provide a domestic basis for the necessary generation

of scientific and technological knowledge for industrial production, and for other social needs, given that contemporary industrialization is a much more science-intensive and knowledge-intensive activity than it was, for instance, fifty years ago.

(3) Whether there are sufficient industrial enterprises in 'modern' production, that is production with modern means and using modern management.[2]

(4) Whether a developing economy has a certain capacity to produce capital goods. This capacity represents a kind of learning potential in adapting technologies which offer opportunities to local demands, and a forum for interactive learning between suppliers and users based upon dynamic applications of the means of production.

(5) Whether there is an institutional framework for the formulation of governmental policy (given the importance of such policies in shaping the institutional set-up and the operation of relevant institutions, and thus in affecting the learning efficiency of an economy).

By combining these criteria, two groups of developing countries can be distinguished in a preliminary analysis. The first consists of countries in which industrial development is still at the stage of 'institutionalization'. At this stage, engineering education may have started, but it still plays only a trivial role in generating human capital, the modern industrial firms in the country are largely subsidiaries of multinational companies, there is little or

[2]The industrial revolution did not create an industrial sector where none had been before. It transformed an industrial sector that already existed by introducing new ways of making the same old things. (Lewis 1978, p. 10) . In this sense, the 'informal industrial sector', which is pervasive in Southern Asian and African countries, is not modern. It is necessary in development studies to distinguish between pre-industrial and post-industrial organizations; the latter are expanding in industrial society, and take forms which are to some extent 'informal'; but informal in a way which is quite different from the informal relationships which may exist in a pre-industrial context. It will also be necessary, if economies in which the pre-industrial and modern forms of organization co-exist are to be positioned within the suggested framework of development stages, to identify which form of industrial organization is dominant.

no local ability to produce capital goods, and the R&D infrastructure and policy-making institutions have not begun to function properly. Many African countries are at this stage[3]. Considering these characteristics, it is unlikely that countries in this group will actively respond to trade liberalization by harnessing the 'industrial rent,' more efficiently than they did when incentives for better allocative efficiency were weaker. Research with respect to countries in this group should be formulated so as to reach an understanding of the conditions for effective and healthy institutionalization which can provide a domestic basis for learning. Without the institutional structures, there is no basis for technological investments. The necessary institutionalization may be facilitated by taking account of the lessons and experiences of countries which have largely completed the establishment of a learning context, while bearing in mind that technology has changed very rapidly and that the circumstances in which international trade occurs now may be rather different from those which applied when these countries were designing their policies.[4]

---

[3]To take the final two criteria as examples. The infrastructure for scientific and technological R&D was institutionalized in China and India in the 1950s, while in most African countries this occurred in the 1970s and 1980s or later. It is reported that by 1987 only about one third of African countries had 'institutionalized' their scientific and technological system, according to the criteria of having established a government body responsible for S&T policy making, and of getting such a body to function properly, at least to some extent. Kenya and Tanzania are examples of countries in this group (see Forje, 1988). Applying the fourth criterion, African countries would stand out as having generally weak capacities to produce capital goods or make local adaptations to capital goods technologies (see section II.2 below) .

[4] The relationship between trade and industrialization has been aptly expressed by Lewis: "The engine of growth should be technological change, with international trade serving as a lubricating oil and not as a fuel. The gateway to technological change is through agricultural and industrial revolutions, which are mutually dependent" (Lewis 1978, p.74). To this one might add that if the agricultural and industrial revolutions are the gateway to change, institutionalization implies another kind of revolution, but one which is imported since the institutions for industrial development are not a natural outgrowth of the traditions of today's less developed countries. Such an institutional revolution requires a revolution in social structures and networks.

The second group consists of countries which have largely achieved the creation of institutions but now face the need to adjust their innovation systems in accordance with dramatically altered economic and technological circumstances. This group includes China, India and some other countries in Asia and Latin America. The system adjustments have to be rather dramatic for many reasons, not the least of which is the radical change in their economic regimes, that is, in the basic pattern of government intervention in economic activities. China had a centrally planned regime, while India had a tight regulatory regime exerting physical control over economic activities. Both were very closed to international exchanges. The reforms currently taking place in countries of this group are characterized by extensive withdrawal of direct control by the governments, and expansion of market rule to govern various domestic and international economic activities. For this group, the most pressing issues at present seem to centre on transforming the institutional structures for learning, which will enable their entire systems to adapt to changing demands and the new incentive structure during a transition period when they face rather radical change in the economic and technological environments. Studies of innovation systems in such countries should then be approached in a different way, aiming at finding out how the country in question can redeploy the technological strengths which it has accumulated during the previous phase of its development in order to improve and promote learning efficiency in the further development process.

The classification above provides one example of the utility of this conceptual framework. This topology contrasts with classifications currently applied to developing countries. One of the most popular of these is based on average per capita GNP, a static measure of the output of an economy at a given time. According to this measure, China and India are placed in the low income group. Another classification distinguishes between newly industrializing countries' (NICs) and other non-NICs. This distinction, which relies heavily on the competitiveness of countries in manufacturing exports (a measure of the strength of an economy at a given time), is commonly applied in OECD studies.[5] A handful of countries are often treated as *NICs* in such

[5] See. for example, OECD: *The Newly Industrializing Countries, Challenge and Opportunity for OECD Industries*. Paris; OECD, 1988.

studies: usually those which are thought to provide serious competition for the industrial countries. China and India have not usually been included as NICs, but this is now becoming more frequent. Using either of these accepted classifications, the international competitiveness of China and India appears to have improved suddenly. However, these classifications provide no indication of the potential of an existing national innovation system to serve as a springboard for accelerated learning dynamism.

II.2. Industrial Structure and Learning

Given the importance of learning and innovation in the dynamics of economic development, one question which arises is how learning effectiveness varies in different industrial sectors. Is learning promoted equally by all economic activities? If it is not, which sectors seem to be more significant, and why? This is a critical question for understanding the long-term technological upgrading of an economy.

Learning is embodied in, and backed up by, an industrial structure. Different industrial sectors have different potentials, because they require more or less intense and sophisticated learning on the part of the agents involved. It will be argued below that the 'capital goods' sector is of central importance. The heart of this sector has traditionally been the machinery industry, but nowadays microelectronics and telecommunications are becoming increasingly vital.

Table 1 shows the structure of the manufacturing sectors in four developing countries: China, India, Brazil and South Korea, in the mid-1980s. It can be seen that China's machinery industry (ISIC 38) accounts for about 30 per cent of the value added in manufacturing as a whole, which is approximately the same as in India, Brazil and the Republic of Korea. Table 2 shows the same data for Japan, the U.S. and West Germany, illustrating the position attained by these typical industrially developed countries at the same time. In these three countries, ISIC 38 industry accounts for about 45 per cent of the value added in manufacturing as a whole, which is apparent evidence of a higher degree of development, despite some differences in the ISIC 38 structure. For most other developing countries (not shown in the tables), 20 per cent or less of the value added in manufacturing comes from ISIC 38 industry. Thus, they lag behind the four developing countries shown in Table 1 in this respect. Although the comparison remains rather intuitive, these figures do appear to show an interesting correlation between development of the capital goods sector and the degree of industrialization.

The technology of modern industrial capitalism has developed through mastering increasingly complicated means of production in various economic activities, and in various sectors. The capital goods sector plays a central role in the innovative strength of an economy. One recent study (Dalum, 1992), which covers 21 OECD countries and a quarter of a century (from 1961-1987), concludes that an internationally specialized engineering sector has been a major, though not sufficient, condition for successful long-term economic strength of the OECD countries.[6]

To understand why this sector should be so important in lending learning dynamism to the system in which it is embedded, it will be useful to compare the interactive learning between users and producers in the sectors of capital goods and bulk intermediates. The capital goods sector is characterized by a high degree of complexity in the underpinning knowledge and skills which are required for its own operations, which implies strong pressures to learn. Yet, from the point of view of the extent of 'linkages,' the bulk intermediates sector would seem to be more important in interactive learning between users and producers, since this sector has more extensive backward

[6] It is interesting that a few of the sample countries have extraordinary strong intermediate goods sectors, compared to other OECD countries. It is from such evidence that this study concludes that 'an internationally specialized engineering sector' is a major precondition for long-term economic strength (Dalum, 1992) . This relates to the role of engineering sector and intermediate sector in learning, which will be discussed further below. Dalum's emphasis on the 'internationally specialized' character of these engineering sectors warrants consideration in studies focusing on developing countries, since the economic reforms being pursued at present entail opening to international trade, which implies greater international specialization.

and forward links. Moreover, the bulk intermediates, such as industrial chemicals, have been among the most science-intensive sectors, and have heavily relied on fundamental R&D, which is a factor promoting complicated learning. But if we consider the character of the interfaces between users and producers, it is apparent that there must be more intensive and enduring interactions between capital goods producers and their users, because capital goods are more user-specific than bulk intermediates. Greater user-specificity entails higher levels of tacitness and uncertainty, which will promote more intensive interactions and thereby spread learning dynamics, not only within the producer sector but also broadly throughout various user sectors. Bulk intermediates, in contrast, are generally traded anonymously on a 'pure' market, so that enduring interactions may not be established. While linkages are important in estimating the strength of dynamic learning effects, we have to distinguish between the *commodity flows* and the *information flows* between firms or industrial sectors. It is the extent and strength of information flows which are more intimately related to *interactive learning*.

This is not to deny the importance of other sectors in the system as a whole. The most that can be asserted is that the capital goods sector not only provides productive means which increase physical productivity in other sectors (and in this respect is qualitatively different from other sectors), but also promotes particularly intensive information flows across the boundaries of firms and sectors, which disseminate learning and the knowledge of how to learn.

There is other evidence that the capital goods sector demands more intensive user-producer interactive learning. According to Pavitt's categorization of firm-based technological trajectories, the machinery industry is typically in the category of 'specialized suppliers'. In this category, innovations are characterized by close interaction with customers (Pavitt, 1984). Moreover, the particular significance of economies of specialization for the capital goods sector has been recognized. Rosenberg has shown that economies of specialization mean that as an economy expands in size and complexity, it achieves greater productive efficiency through increasing specialization among machine tool producers, in close interaction with users (Rosenberg, 1976).[7]

Given the importance of the intensive learning which takes place in the capital goods sector, the capacity to introduce commercially successful capital goods can be taken as one indicator of the strength of an innovation system. In developing countries, this highlights two policy areas for further study: policies to induce entry to the capital goods sector and policies to guide the transition of an existing sector. The capital goods sector does not automatically emerge in a developing economy. Those in South Korea, India, China and Brazil were all strongly 'policy induced' to develop to the stage of making adaptive innovations for industry (Pack and Westphal, 1986; Chudnovsky and Nagao, 1983). A 'pure evolutionary approach,' based on the insights of classical economics which discourage any radical policy inducement, may not be effective in producing this structural dynamism because of the increasingly high entry barriers. The weakness of neoclassical prescriptions may partly explain why most developing countries do not even attempt to

[7] The significance of several works by Rosenberg for development studies has been seriously neglected, with a few exceptions. One point made in these works is that the dynamism which results from specialization in the machinery industry is due "in large part to the fact that certain technical processes were common to many industries" (technological convergence) . Furthermore, a specialized machinery industry "may be regarded as the centre for the acquisition and diffusion of new skills and techniques in a machinofacture type of economy. Its chief importance, therefore, lay in its strategic role in the learning process associated with industrialization. This role,... is a dual one: (i) new skills and techniques were developed or perfected here in response to the demands of specific customers; and (2) once they were acquired, the machine tool industry was the main transmission centre for the transfer of new skills and techniques to the entire machine-using sector of the economy. "See Rosenberg: Technological Change in the Machine Tool Industry, 1940-1970"; "Neglected Dimensions in the Analysis of Economic Change"; and "Capital Goods, Technology, and Economic Growth." All these works are in Rosenberg (1976), pp. 9-31, 85-107, and 141-150.

develop capabilities in the capital goods sector.[8] This has significantly hampered development in many countries.

The second area for policy studies in developing countries relates to policies for the transitional period for an established capital goods sector. It is still not clear what is happening to established capital goods sectors under the current reforms, although debates are extensive. The sector seems to be at the centre of transformation and restructuring in terms of technological acquisition, competition, and the relationship between users and suppliers. Transformation and restructuring of the sector are vitally important for the subsequent effectiveness of a national innovation system. Once again, it must be asked whether a 'neutral' policy regime is an effective condition for a sector undergoing such a transformation.

In addition to the traditional capital goods industry based on mechanical technology, which is still important for developing economies, the new technology-based capital goods industry which has developed around electronics and telecommunication (broadly defined) warrants attention, not least because of its increasing economic importance. The very high degree of tacitness and the uncoded nature of user specificity in computer-

---

[8] Pack and Westphal have strongly contested the applicability of a pure evolutionary approach, based on the experience of East Asian *NICs* in the area of governments' role in industrialization as part of an industrial strategy.' They find that "a neutral policy regime is not a necessary condition for successful industrialization. The other new element is a ... conceptualization that puts indigenous dynamic phenomena involving technological change ... at the center of industrialization. Industrial strategy therefore "should be seen as a matter of managing technological change to achieve dynamically efficient industrialization." Selective infant industry protection is thus crucial in the two - pronged industrial strategy which would accompany what they call 'a dual policy structure.' The two components here are investments in established industries in which there is a static comparative advantage and investments in infant industries in which there is a potential for dynamically achieving a comparative advantage. But it is not clear from their argument how infant industries are to be defined. Implicitly, infant industries would be new or emerging industries which are technologically more sophisticated and thus are potentially able to yield greater externalities in both pecuniary and non-pecuniary terms (See Pack and Westphal 1986).

aided telecommunication technology have impeded applications in most developing countries due to the lack of services to support the user-producer interface (James, 1991, Ernst and O'Connor, 1992). Thus, the emergence of a computer (application) software industry, which must be locally based, given the necessity of close communication with users, may be seen as a predictive indicator of how rapidly a particular developing economy will be able to adopt the revolutionary new production technologies. India, China and Brazil are again among the very few developing countries which have had a reasonably strong computer software industry, according to the literature. The Chinese software industry has benefited quite dramatically from economic liberalization. In fact, China's software industry is the youngest of the three, since the central planning system had particularly serious disadvantages when dealing with the numerous and rapidly-changing producer-user (horizontal) interactions in this sector.[9]

11.3. Institutional Set-up and Learning

While the structure of particular industries may provide particular opportunities to foster more dynamic interactive learning in an economy, the institutional set-up provides the forum in which learning takes place, and can determine how practically effective it is. Because learning is incorporated in doing, in using, and in interactions within and among various economic agents, it must largely be incorporated in operational routines. It is the institutional set-up that sustains the organizational and operational routines, which in turn, shape the incentives for learning, guide the flow and provision of information for both doing and

---

[9] A previous study has outlined the emergence of the locally based computer software industry in China since the 1980s. This industry is concentrated on user application services. The local specificity of applications, such as the complexity and special norms in Chinese writing and spelling in general, and in tacit user parameters in particular, have favoured the development of a locally-based industry. The industry's growth has been linked with the creation of new 'non-state enterprises,' operating outside the planned distribution of such services, because intimate interactions between suppliers, users and foreign producers of computers is indispensable. See Gu, 1994 and Gu, 1995.

learning, and reinforce the locus where learning is accumulated. The importance of the institutional set-up, which is assumed to be a fundamental element of a national innovation system, is a specially relevant issue in countries of the second group mentioned above, that is, in countries such as China and India which have largely established the primary institutional elements but are now undergoing radical institutional transformation as they shift to market-based economic regimes. As has been mentioned, the term 'economic regime' is used here to refer to the basic patterns of government interventions aimed at promoting industrialization. Economic regimes, once chosen, become deeply embedded in the entire institutional set-up: not just in the structures for policy formulation and implementation, but also in the ordinary economic activities taking place in the system. The success of reforms is, therefore, largely determined by the success of the institutional transformation. This point has largely been neglected by researchers and decision-makers concerned with the current reforms.[10] This section deals with the institutions for ordinary economic activities, leaving policy institutions and policy learning for Section II.5.

Two types of institutional context need to be examined closely, in particular to identify the critical features of the current transformation. These are industrial R&D establishments and the structures of industrial firms. Where there have been tight administrative controls on industrial technology investments, as in China from the 1950s to 1970s and in other countries with

---

[10] Westphal's (1990) argument is pertinent to policy institutions. He has noted political differences in addressing how and why Korea's government has succeeded where many other governments have failed "... political difference" he says, "pertain to political processes that have conditioned the use of policy." " ... attempts to institute effective industrial policies elsewhere in the Third World would generally require drastic change in domestic political processes before they could succeed." This aspect is incorporated in our research framework under the heading policy learning, which is supported by, and conducted in the policy institution of a nation state (see Section II.5). More and more empirical evidence has shown that drastic changes in domestic political processes are underway in a variety of countries in the Third World in which it has been possible to reach a domestic consensus on the reform agenda. A key motivation has been the attempt to learn from the development experiences of some successful East Asian cases.

similar institutional establishments, both R&D establishments and industrial firms inherit an institutional legacy from these controls. One clear consequence is the serious separation between R&D in the field of industrial technology and the industrial users of this technology. Table 3 shows the distribution, in the years 1988 - 1990, of R&D scientists and engineers across the sectors of production, higher education and 'general services', in China, India, Argentina (as a representative case of Latin American developing countries), Japan and South Korea. The latter two countries provide particularly interesting comparisons with China and India. In China and India, more than half of the R&D scientists and engineers are in the 'general service' sector, while the production sector provides employment for only 23 per cent in China and 25 per cent in India. In contrast, the production sector accounts for 60 per cent of Japanese R&D scientists and engineers, and 50 per cent of their Korean colleagues. This decisive difference indicates some fundamental disparity between China and India (and many other developing countries) on the one hand, and Japan and South Korea on the other. It might be relevant to examine this difference further, especially if the successful industrialization in Japan and South Korea is considered as instructive for the current reforms in the developing world.

These simple statistics have profound implications, given that the acquisition, application, adaptation and upgrading of industrial production technology is a costly process requiring ongoing investments and the energies of scientific and technical manpower, and that technological learning in industrial production is incorporated in routine activities and accumulates continuously within the framework of the productive institution. The contrasting distributions of R&D manpower, as a critical resource for innovation, can be seen, in retrospect, to reflect institutional factors which have hampered technological learning under the previous economic regimes in China and in some other countries. Looking to the future, these distribution patterns presage one significant complication which will inevitably be associated with the transformation process, and which must be carefully monitored if the current reform is to be as successful as hoped.

Since a large part of the manpower and other assets relating to industrial technology are tied up in the general services sector, that is, in industrial R&D institutions which are not directly integrated with production facilities, making more extensive and efficient use of this accumulated strength is very important for the future of the entire innovation system in these countries. China's reform programme for R&D institutions was officially launched in 1985. Initially, hopes focused on using market mechanisms to mediate transactions in industrial technology, so as to overcome the inherited organizational separation. Later, this was supplemented with more wide-ranging processes entailing various forms of institutional restructuring. India, according to recent reports, has also started to see the necessity of reforming its industrial R&D institutions.

Various aspects of this institutional legacy are incorporated in the structures of industrial firms and their linkages with suppliers, users and other agents, such as the technological infrastructure and government. Planned allocation and distribution in China, and regulatory restrictions and monopolistic market as in non-centrally planned but tightly administratively controlled developing economies like India, meant that firms were not under pressure to compete. They, therefore, did not need to invest in marketing, in technological searching, and in communication with users. The very limited development of interfirm subcontracting relationships and excessive vertical integration within firms, which are largely common features in the two countries, although they derive from different policy structures, severely erode the potential productivity gains from certain investments. A study of the machinery industry in China which is underway at INTECH, reveals that the whole institutional set-up which developed under the old regime locked the industry in a path which made it proficient at quantitative expansion, but inept in technological upgrading. The innovation system of the Chinese machinery industry apparently worked satisfactorily in the 1950s and 1960s, but an examination of the long-term dynamics shows that it was the institutional forms of this pattern that obstructed further institutional evolution. The productivity of the economy as a whole, and of the machinery industry in particular, has stagnated since the 1960s. A parallel situation can be observed in India.

The strength of industrial firms is central to the strength of a national innovation system because of the importance of efficient and effective learning for progress in industrial technology. Manufacturing, marketing, procurement, and technological searching for problem-solving purposes and for identifying new technological opportunities are all integral parts of commercial success in industrial production. More rapid technological learning and innovation is fuelled by more efficient information flows between these integral parts.[11]

Arm's length communications, which resulted from the distance between the functions of product design, problem-solving, opportunity- searching and manufacturing under the previous regimes, are workable only where technological change is slow, there is little pressure for technological change, and the incentives for change provide only weak motivation. Research regarding this aspect, therefore, needs to give high priority to:

(a) The transformation of 'independent' R&D institutes for industrial technology, and of other elements of the technological infrastructure which were developed to fit the previous economic regime. How were the technological activities which were previously integrated by administrative coordination re-oriented and re-allocated in both institutional and functional terms? This question is pertinent to technologically capable developing countries, but not to those which have not yet built such institutions, or to developed countries. For this reason it has not been seriously addressed in the literature.

(b) The restructuring of industrial firms and institutions which support firms, technological efforts, which were also largely developed to fit the previous environments. For instance,

[11]Studies of the success of Japanese firms have enriched our understanding of the importance of close communication and mutual learning among the various functions necessary for commercial production (Aoki, 1988 and 1990, among other works). Aoki's work also shows convincingly how peculiar local conditions can be incorporated in the institutional terms to be analyzed.

how are industrial firms moving to respond to stronger competition through restructuring their internal work organization? And how are they restructuring their external links where the previous technology supply and support framework is also being reorganized? The institutional restructuring may well assume different forms in large and leading firms, and in small rural enterprises. In the case of the small enterprises, more attention needs to be given to the supporting framework, to explore how their supporting infrastructure is changing under radical market reforms.

II.4. The Market Mechanism, Opening to the World, and Learning

The discussion of the institutional set-up in Section II.3 compared two kinds of economic regime. The outline of the first can be drawn from the situation in China before the current reforms, supplemented in some respects by reference to India in its pre-reform era. The other regime is typified by Japan and South Korea. What are the major differences between the patterns of government intervention to promote industrialization in these two stylized regimes? For many Chinese scholars, whose experiences with development efforts in the pre-reform era have been discouraging, the issues are clear: the key issues are whether or not the basic goal of governmental intervention is the development of market mechanisms, and whether it relies as much as possible on market power. In other words, the basic difference between the two regimes is between intervention to complement market mechanisms and intervention to substitute for such mechanisms.[12] Industrial firms were placed in different positions under the two kinds of regime: in one, the promotion of competition is a goal, and industrial firms are, therefore, expected to be the key players in learning and innovation; in the other, industrial firms' activities were controlled by planning or regulation.

Now that market-oriented reform programmes are in effect

[12]See, for instance, various debates appeared in *jingji shehui tizhi bi jiao* (*The Journal of Comparative Economic and Social System*), especially in the second half of the 1980s.

throughout the third world, it is necessary to re-assess the *effects of market mechanisms* with respect to learning and innovation, and to gauge what difference *economic liberalization* makes to the functioning of a national innovation system. This reassessment would appear to be particularly critical when considering a government industrial strategy based on 'selective intervention.' The success of selective intervention policies in newly industrializing countries such as South Korea has been widely recognized, and it is also explicitly supported by the concept of the national innovation system developed in this paper. Yet it should be clear that a sound market-based regime has been a prerequisite for the success of 'selective intervention' policies in these countries.

Effects of Market Mechanisms

It is generally argued that market mechanisms favour learning and innovation because of the flexibility of a market regime, which allows the various economic agents to adapt readily to changing opportunities and prices. Businesses thus achieve allocative efficiency by seeking to maximize profits (discounted for risk and over time). Learning and innovation are largely incorporated in the process of effective factor allocation. Where the specific industrial innovation process is examined, it has been argued that market mechanisms imply greater incentives and pluralism, so favouring an experimental approach by industrial firms to cope with the technological and economic uncertainties which are intrinsic to technological innovation (Rosenberg and Birdzell, 1986; Hanson and Pavitt 1987). A slightly different argument is that the market favours natural selection of technological progress (Nelson and Winter, 1982), which is much more effective than other known regimes, and this effectiveness is again attributed to the flexibility and plurality of market mechanisms.

Observations of China's reforms conform the general acceptance that conforming to market mechanisms will normally produce benefits. Two observed effects of market reforms are closely related to learning and institutional evolution. One is that market mechanisms favour closer user-supplier relationships. When producers move to marketing their outputs directly, communications with users and suppliers are significantly intensified, which, in turn, provides much stronger and more precise guidance to actors' technological efforts in areas such as product diversification, production scheduling, testing, R&D, and standardization and codification of their tacit knowledge. Not only is learning more intense due to closer interaction, it also takes place in a more appropriate locus, since industrial firms are the institutional structures in charge of technological innovation and they should be fully enabled to learn by doing, by using, and through these interactions.

A second effect of market mechanisms is that they enable economic agents to be more open to institutional evolution. Market mechanisms imply the freedom or autonomy to decide what kind of institutional arrangement is suitable for managing innovation resources, and for better management of their relationships with users and suppliers, which provide an opportunity for interactive learning. Institutional restructuring is, therefore, possible only under market mechanisms, as part of the move towards more effective allocation of resources to maximize competitiveness. Plural approaches to restructuring industrial R&D institutes have been created under the market reform in China, including spinning off technology-intensive business, merging R&D institutes into existing enterprises to be the in-house R&D and design department of the host enterprise, and wholly transforming the functional objectives of an R&D institute to make it an engineering company. These have been documented in an INTECH study (Gu, 1995). Other observed forms of enterprise restructuring that have already started include merging enterprises, splitting up into smaller units, and the development of subcontracting relationships. As the restructuring goes deeper, the ambiguity of rights under public ownership has emerged as the most constricting factor in China. Nevertheless, preliminary institutional restructuring could not have been achieved without a certain degree of freedom to choose, and this freedom comes from the introduction of market mechanisms. It is useful to add here that the current distribution of R&D engineers and scientists in Korea and Japan (which has been mentioned in strong contrast to the distribution in China and India) was a result of flexible institutional evolution in a market environment. By expanding their in-house technological R&D, industrial enterprises replaced the government as the major setting for R&D in Japan in the 1960s, and in Korea in the 1970s.

Effects of Trade Liberalization

International trade liberalization, which is a central part of the current reform programmes, offers in principle all of the merits of market mechanisms. Firms' incentives to learn come from international competition. The information to be learnt comes from international suppliers and users, who are often much more experienced than the domestic suppliers and users. The interactive communications entailed in international trade relationships thus turn out to be much more informative. Besides, an example of technological dynamism, and of the advanced management and entrepreneurship in industrially advanced economies, is very important for technological and institutional learning. All these non-pecuniary benefits help to transmit learning dynamism to an underdeveloped system. Export promotion is in this sense indispensable if a development strategy calls for high learning efficiency. National innovation systems which are not open to international exchanges, regardless of their political appellation, have seemed to inevitably fall into a vicious cycle in which administrative control and technological pessimism[13] dominate, market mechanisms are curbed or inhibited in their development, and it is not possible to keep government intervention selective. A selective intervention strategy is only applicable in combination with an open national system.[14]

---

[13] There is ample evidence, in closed systems, of technological pessimism in relation to industrial activity which co-exists with extraordinary enthusiasm for technology and science in non-industrial applications, such as academic or military uses.

[14] After a detailed analysis of the South Korean industrial policy of selective intervention, Westphal concludes with the remarks that: "Korea's industrial performance owes much to the governments' reliance on free market institutions to provide for flexibility in resource allocation...." ".... If nothing else, the policies towards exports have created an atmosphere - rare in the Third World - in which businessmen could be certain that the economic system would respond to and adequately reward their efforts aimed at expanding and upgrading exports." (Westphal, 1990)

Market reform, in brief, means the opening of an innovation system to much more abundant challenges and opportunities. In China, before the current reform, for instance, systems for innovative activities were often closed, even to international exchanges. Closure was a pervasive reality: industrial firms were substantially closed to information from users and suppliers and to information on new technological opportunities; industrial sectors were also significantly closed to each other, with the limited communications between them passing mainly through coordination by the planning bodies. Thus, system opening, which proceeds from international and domestic trade liberalization, does not invalidate any of the national attributes of an innovation system. Trade liberalization changes only the external conditions for learning; it cannot be a substitute for learning, which is backed by an institutional set-up and is largely embodied in tacit knowledge and incorporated in routine operations. A preoccupation with 'globalization' can obscure the importance of another side of the current trends, which is 'regional specification' (Lundvall, 1992). In fact, the argument for selective protection, which is advanced in relation to an export oriented policy, is in essence saying that the development of endogenous capabilities falls within the non-tradable sphere.

Opening to international trade, however, has changed some things crucially. The supplier structure for industrial technology has changed, to rely largely on the international technology market. A significant portion of domestic efforts, which were to a large extent organized in government-run R&D institutes and were devoted mainly to the re-innovation of production technology, often wastefully and with the effect of raising costs, have, therefore, had to discover new orientations for their services. Industrial firms have had to adapt to the change in the supplier structure. This implies that the changed interface between endogenous ability building and external technological opportunity should be examined closely. It is at this interface that the major difference between a relatively closed system and an open one is manifest.[15] Under an open system, the goal of a national innovation system has to be reformulated to pursue and promote learning efficiency and innovative capability through approaches combining choices in the dimensions of 'make/buy' 'search/research,' and domestic/foreign, in new ways.[16]

II.5. Economic Reform, Path-Shifting and Policy Learning

Researchers working on technological change have been greatly influenced by the concept of 'path-dependent' development. But radical technological innovation implies that the technological trajectory shifts from one path to another, and economic reform in principle requires that the institutional set-up shifts to another path, as has become evident in China. At first, people in China generally expected that remedies for 'government failures' would automatically begin to work well once the economic orientation had turned to 'the market.' However, the process of reform in China has proved to be far more complicated. It entails first of all institutional restructuring, i.e. institutional path-shifting. The potential efficiency gains under a market regime may then follow gradually, along with the evolution of applicable institutions. In this regard, China's experience is an example of the importance of what Williamson (1994) has called 'getting the institutions right,' if a reform programme is to be workable.

---

[15]A study (Gu, 1996) of the emergence of the commercial computer software industry in China reveals that getting access to an 'international technology infrastructure' was an important factor enabling the capabilities accumulated in a closed system to be translated into some degree of international competitiveness. A 'technology infrastructure' here refers to particular technological knowledge derived from the commercial efforts of leading international companies, and which embodies information on the technological trajectory of the commercial development of the telecommunication industry in the form of technological designs, norms, specifications and so on (see Steinmueller 1996) . This case shows that mechanisms at the interface are message-rich, but also that we know too little about them.

[16]This expression is from Richard P. Suttmeier, personal communication, January 1996.

Radical technological innovation takes place when some new key element of technology is inserted into a combination of factors, with many complements already available or able to be incorporated into the new combination with some degree of modification. The learning-backed path-dependent concept, therefore, does not deny the possibility of path-shifting. Rather, it should be understood as pointing to the complex evolutionary process entailed in innovation. Path-shifting for the institutional structure implies innovation in institutional terms. The transformation of a national innovation system, which is required by the economic reforms, entails decisive path-shifting in terms of organizational patterns, operational routines, and relationships among the various agents who constitute the large system which, as a whole, supports technological changes. Like a shift to a new technological path, such a shift in organizational patterns starts with the introduction of a new element (new rules of operation) , which is introduced because of change in the policy and international exchange environment. Successful institutional innovation will be selected out, only through a process of experimentation by which institutional evolution is tested for suitability in the changed and still changing environment.

Two characteristics are necessary for the process, if it is to be workable. One is the gradual or evolutionary nature of institutional path-shifting. This is because of the high degree of social uncertainty which results from radical institutional innovation. The necessary probing to find proper solutions, and passing through numerous failures, takes time. Even where preliminary experience has been promising, enormous adjustments are required if a substantial part of the learning which has accumulated under the old order is to be passed on to the new and transformed institutions.

Another aspect of the process is that it is very much policy-induced. Radical and widespread institutional restructuring is an indispensable concomitant of radical economic reform. At the outset, the reform policy functions as a general signal of the reform programme, stimulating the beginning of institutional change. Policy then has to evolve in response to the reactions of various actors, with endless additions, modifications, elaborations and revisions. Both evolutionary and policy-induced characteristics are evident in the evolutionary formulation of policy for reform of the scientific and technological management system in China during 1985-1995 (Gu, 1995). This experience shows that institutional innovation involves intensive policy learning, which occurs through the interactions between policy makers and economic actors. There is a parallel between the need for policy learning when instituting institutional innovations, which entail high levels of social uncertainty, and the technological learning process involved in technological innovation, which entails high levels of technological and economic uncertainty.

It is reasonable to hypothesize that the gradualism and experimentalism associated with policy learning and institutional learning are essential attributes of the policy process for successful reform. In fact, they might be more generally regarded as essential attributes of any successful policy process. For instance, effective implementation of the industrial strategy in South Korea has been interpreted in a similar way.[17] In other words, a reform programme would not be likely to be effective unless interactive policy learning is achieved in the society in which the programme is to be enacted. Furthermore, policy learning needs institutional support, and an important ingredient of institutional support is a certain degree of social consensus. Social consensus is necessary to provide the conditions for interactive policy learning and institutional learning. Once there is a consensus on the need for reform, as in the case of Chinese and Indian reforms (Rosen, 1992), the reform process itself flows mainly through an evolutionary cycle characterized by continuing accumulation and

---

[17]"There has necessarily been a great deal of uncertainty in selecting infant industries for promotion and in setting export targets for all industries. There is ample evidence that this uncertainty has been progressively resolved by using information gained during implementation to evaluate and, where deemed appropriate, revise intentions .... The evidence suggests that the monitoring process has assured that exports are both privately profitable and internationally competitive, at least in the medium run." (Westphal, 1990)

elaboration of policy capacity, and realized through interaction (interactive learning) between policy makers and economic actors. Other factors, such as conflicts regarding the re-distribution of power and benefits among interest groups, can apparently be outweighed by the mainstream process of societal learning. Unfortunately, this is not the case in a number of other developing countries which are embarking on reform programmes.[18]

II.6. Institutional and Technological Analysis

If institutional structures and their transformation are regarded as intrinsic factors which explain the efficiency of technological learning from a given investment, it follows that research in this field must combine institutional analysis with technological and economic analysis. The institutional analysis required by the research approach may not, by itself, provide direct information about economic performance, since the transformation of institutional structures does not directly tell us what positive or negative consequences this will have for performance. It is thus necessary to incorporate all appropriate economic analyses, covering performance measures such as productivity, competitiveness, product diversification and upgrading, and indicators of international trade performance, such as export and import structures and volumes.

It is worth emphasizing that the transformation of a national innovation system entails extraordinarily intensive learning in relation to technology institutions and policy areas in which the system is in fact in a state of 'low equilibrium.' Thus, the information which may be highlighted by an analysis in technological and institutional terms may go far beyond that offered by a merely classical analysis. For instance, the ongoing INTECH study of China's machinery industry has revealed that a number of R&D institutes which had the best staff and equipment under the planned economy are moving to provide

[18]For instance in Chile, where the current reform initiatives have mainly come from outside (Weiss 1988).

"custom-made' processing technology,[19] resembling the activities of a Western engineering company. This kind of restructuring would appear to be a positive development, since a move in this direction indicates the possibility of considerably enhancing the system's capability in adapting appropriate software and hardware to more comprehensive local uses, which has been an area of enduring weakness in the past. Knowledge of this kind may not be well reflected in price and profit terms, but it is this kind of knowledge that indicates more precisely in what directions the various kinds of restructuring are leading, and the problems and difficulties entailed in restructuring. Policy implications are also more precisely illuminated by this knowledge, which helps to identify problems and difficulties which make further adjustments to policy measures desirable.

Many published works that address policy issues of transitional period were apparently of reduced informative value because of a lack of institutional and technological analysis, that was not carefully pursued to be incorporated with ordinary economic analysis. One pointer of this lack may be the study titled "Electronics Policy and the Television Manufacturing Industry: Lessons from India's Liberalization Efforts" by Guhathakurta (1994). After a clear description of liberalization policies, and performance evaluation of firms drawing upon a number of economic indicators, the "micro-level empirical analysis" of firms' strategy remains limited to show "the role of market and the state institution" that is the emphasis of the work. Because the strategic behaviour of firms is not explained in a clear institutional and technological context, without knowing what the big, the medium-sized, and the small firms are doing technologically, and what kind of subcontracting or infrastructure links are developing for specific technological and managerial functions in relation to the changes due to liberalization, it is hardly possible to estimate what the learning consequences or the policy

[19] 'Custom-made' processing technology refers to the specially tailored development of manufacturing technology for a certain user's processing, usually with a high degree of integration of various kinds of technology, including hardware and software.

implications of the strategic movements may be. The paper concludes with the statement that liberalization, together with innovative state institutions, can achieve the twin objectives of growth and equity. This conclusion is, in fact, no more than an echo of the general policy faith long enunciated by the Indian (and Chinese) government, both before and after the current reforms.

What is urgently needed is to know in precise terms how the markets have been deregulated, how state institutions are working with enterprises in the deregulated market, what institutions have been created in response to the reforms, and what firm's strategic responses have been correspondingly. In short, we need to develop an understanding of the reforms based on a more detailed picture, which includes institutional and technological analysis, to serve as a knowledge basis upon which policy learning can accumulate.[20]

More generally, that technology and institutions are important subjects that should be given more attention has recently been recognized widely by researchers working on development studies. As Knight expresses it (Knight, 1991, 18-19):

> "There are certain issues which development economists tended to neglect in the past, and which deserve more attention in the future .... The common theme is that economic analysis should be extended into areas unfamiliar or suspect to market oriented economists-broader welfare objectives, technology, institution and the state."

---

[20]The urgent need for policy learning has been expressed by Mohan (Mohan, 1992) as follows:

"Government intervention in industry has to be far more knowledge-based than it has been .... Although [India's] self-image is one of the great technological achievements, there is little industrial technology development that is not imported. The control system which has aimed to restrict foreign technology imports seems to have been counter-productive. Focused attention, therefore, has to be given to the mode of government intervention which promotes greater indigenous technological absorption, adoption and development. This calls for a national effort at finding and creating new institutional mechanisms for the development of technological competence."

(First, broader welfare objectives)

"A second issue concerns technology,.... Development economists have tended to neglect technology, ... partly because the conventional tools of analysis, concentrating on prices, tend to push it out of view. Yet, economic agents in less developed countries find it difficult to learn: much learning is by doing, and the ability to learn is itself partly learned. There are important externalities in the learning process which cannot easily be internalised, and the outcome can be a low technology equilibrium .... Moreover, ...it (technology) raises fascinating analytical issues concerning externalities, market failure and imperfect information; and it offers scope for cross - disciplinary work with technologists.

A third issue ... is that of institutions. The analysis of institutions is now more common than it used to be, but there is a lot of interesting work to be done in analysing the extent to which particular institutions are functional or dysfunctional. Endogenous theories of institutions have been produced by both the transaction cost school and the imperfect information school: institutions can be seen as ways of dealing with high transaction costs and information problems. Having plenty of both, less developed economies are hospitable territories for institutional analysis... "

For doing institutional and technological analysis effectively, a substantial development of methodology is desired. Insights developed in the schools of such aspects as transaction cost, imperfect information, as Knight mentioned, and technological innovation economics, together with necessary engineering knowledge are possibly valuable inputs. This topic is, however, beyond the scope of the current paper.

Conclusions

This study has shown that the concept of a national innovation system is a useful instrument for analysing endogenous capability building for industrial development. The concept provides perspectives focusing on learning and learning efficiency, and emphasizes that learning is supported and embodied in an institutional context. Furthermore, the concept is sensitive to the unique features of industrially underdeveloped countries in recognizing (1) the indispensability of building and restructuring institutions throughout the modernization process, and (2) the distinctive and complex role of government policy in compensating for inadequately developed market mechanisms, and in inducing institutional transformation. This analytic framework offers a focussing device that enables our observation to go deeper, to discover the reasons why particular industrialization efforts have been frustrated or successful. The concept of a national innovation system leads us to expect that these internal reasons will be directly manifested in the efficiency of technological, institutional and policy learning processes (other conditions permitting).

While the concept requires us to remember the 'national' element as a pointer to national or local specialities, the focus for analysis using this concept is on micro-level events. It is at lower levels that the delicate mechanisms of interactive technological, institutional and policy learning work. Some elaborations of the concept, as developed in Section II, also demonstrate that although the concept relates to national competitiveness in general, the key issues faced by countries in different circumstances may be very different. The key issues discussed in Section II should, therefore, be seen only as illustrations, which may be more pertinent to countries in which certain institutions and technological capabilities have been established, but which are now undertaking radical market reform programmes requiring dramatic transformations in such institutions. The elaboration in Section II also shows that a careful identification of critical issues is a necessary step if this analytical framework for national innovation systems is to be properly employed.

Table 1 : Structure of the Manufacturing Sector in Developing Countries, by Share of Selected Subsectors in Total Value Added, 1985-1987

Per cent

Branch (ISIC code)	China	India	Brazil	Korea, Republic of
Food products (311/2) Beverages (313) Tobacco (314)	11.4	11.6	14.3	12.6
Textiles (321) Wearing apparel (322) Leather and fur products (323) Footwear (324)	14.1	15.2	10.4	16.6
Wood and cork products (331) Furniture, fixtures excluding metal (332)	1.4	0.5	4.3	1.4
Paper (341)	2.0	1.6	3.6	2.5
Printing and publishing (342)	1.2	1.9	2.4	2.3
Industrial chemicals (351)	7.4	7.3	6.6	3.9
Other chemicals (352)	3.1	7.5	5.5	4.6
Petroleum refineries (353)	4.6	3.5	4.3	3.0
Rubber products (355)	1.9	2.4	1.3	3.1
Plastic products (356)	1.7	0.9	2.8	2.4
Pottery, chinaware earthenware (361)	0.6	0.2	0.2	0.3
Glass (362)	1.0	0.5	0.7	1.0
Other nonmetal. min. prods. (369)	6.1	4.6	5.1	3.3
Iron and steel (371)	7.8	10.8	5.7	6.2
Non-ferrous metals (372)	2.2	0.9	2.5	1.1
Metal products (381)	3.3	2.7	4.7	4.3
Non-electrical machinery (382)	14.0	9.1	8.4	5.4
Electrical machinery (383)	7.7	8.1	6.3	13.2
Transport equipment (384)	4.7	8.1	5.9	8.7
Prof. and scient. equipment (385)	1.2	0.8	0.6	1.1
Total, Machinery. Industry (38)	30.9	28.8	25.9	32.7

Source: UNIDO : Handbook of Industrial Statistics 1990, pp. 94, 96,105,115

***Table* 2 : Structure of the Manufacturing Sector in industrially Developed Countries, by Share of Selected Subsectors in Total Value Added, 1985-1987**

Percent

Branch (ISIC code)	Japan	U.S.	Germany, Federal Republic of
Food products (311/2) Beverages (313) Tobacco (314)	9.6	11.8	9.6
Textiles (321) Wearing apparel (322) Leather and fur products (323) Footwear (324)	5.4	5.3	4.1
Wood and cork products (331) Furniture, fixtures excluding metal (332)	2.7	3.0	2.5
Paper (341)	2.5	4.2	2.4
Printing and publishing (342)	5.3	7.4	1.9
Industrial chemicals (351)	4.3	4.4	7.4
Other chemicals (352)	5.2	5.5	5.2
Petroleum refineries (353)	1.1	1.4	4.3
Rubber products (355)	1.3	1.1	1.1
Plastic products (356)	3.4	2.5	2.6
Pottery, chinaware, earthenware (361)	0.4	0.1	0.3
Glass (362)	1.0	0.8	0.9
Other nonmetal. min. prods. (369)	3.1	2.0	2.2
Iron and steel (371)	5.5	2.4	4.1
Non-ferrous metals (372)	1.2	1.2	1.5
Metal products (381)	6.5	6.1	6.4
Non-electrical machinery (382)	12.6	11.4	15.1
Electrical machinery (383)	15.1	11.1	12.8
Transport equipment (384)	10.4	12.8	12.8
Prof. and scient. equipment (385)	1.6	4.0	1.5
Total, Machinery Industry (38)	46.2	45.4	48.6

Source: UNIDO: Handbook of Industrial Statistics 1990, pp. 102, 107, 120

Table 3 : Scientists and Engineers Engaged in R&D, and Distribution by Sector of Employment, in Selected Countries

Numbers

Country	Year	Scientists and engineers (total)	Production sector	Higher education	General services
China	1990	653,000	147,961	116,000	389,181
India	1988	119,027	31,145	22,100	65,782
Argentina	1988	11,088	196 2,098 (Non-Integrated)*	5,602	3,192
Korea, Republic of	1988	56,545	28,299	18,665	9,581
Japan	1989	636,817	376,151	211,723	48,943

Source: 1. For countries other than China: *Statistical Yearbook* 1992, UNESCO, pp. 5-29, 5-31.

2. For China: *China Statistical Yearbook and Technology*, 1991, pp. 3, 231. The number of R&D scientists and engineers in higher education in China is measured in full-time equivalents.

* 'Non-integrated' refers to activities aimed for productive sector, but conducted in organisations separate from productive enterprises.

Selected References

AOKI Masahiko: *Information, Incentives, and Bargaining in the Japanese Economy.* Cambridge etc. ; Cambridge University Press, 1988.

AOKI Masahiko: Toward an Economic Model of the Japanese Firm. *Journal of Economic Literature* 1990, Vol. XXVIII (March) p1-27.

CHUDNOVSKY Daniel, NAGAO Masafumi: *Capital Goods Production in the Third World: An economic study of technology acquisition.* London; Pinter, 1983.

COOPER Charles: *Technology, Manufactured Exports and Competitiveness,* paper for the Global Forum on Industry - Perspective for 2000 and Beyond. New Delhi; UNIDO, 1995.

COOPER Charles, TURNER Roy: *Technological Upgrading, Manufactured Exports and Optimal Accumulation,* Maastricht, UNU/INTECH discussion paper 9506, 1995.

DALUM Bent: Export Specification, Structural Competitiveness and National System of Innovation, in Bengt-Aake Lundvall (Ed): *National Systems of Innovation: Towards a theory of innovation and interactive learning.* London; Pinter, 1992. p 192-208.

ERNST Dieter, O'CONNOR David: *Technology and Global Competition: The challenge for newly industrialising economies.* Paris; OECD, 1989.

ERNST Dieter, O'CONNOR David: *Competing in the Electronics Industry: the experience in newly industrialising economies.* Paris; OECD Development Centre, 1992.

FORJE John W.: In Search of a Strategy for a National Science and Technology Policy in Africa. In Atul Wad (Ed.): *Science, Technology, and Development.* Boulder Co. ; Westview, 1988. p229-257.

FREEMAN Chris: *Technology Policy and Economic Performance: Lessons from Japan.* London; Pinter, 1987.

GU Shulin: Spin-off Enterprises in China: *Channelling the components of R&D institutes into innovative businesses,* Maastrich, UNU/Intech working paper No. 16, 1994.

GU Shulin: *A Review of Reform Policy for the S&T System in China; From paid transactions for technology to organizational restructuring,* Maastricht, UNU/ Intech Working Paper No. 17, 1995.

GU Shulin: The Emergence of New Technology Enterprise in China: A study of endogenous capability building via restructuring. (Forthcoming) *Journal of Development Studies* 1996, 32 (4).

GUHATHAKURTA Subhrajit: Electronics Policy and the Television Manufacturing Industry: Lessons from India's liberalization efforts. *Economic Development and Cultural Change* 1994, p 845-868.

HANSON Philip, PAVITT Keith: *The Comparative Economics of Research Development and Innovation in East and West: A survey.* Chur etc.; Harwood Academic, 1987.

HIPPLE Eric von: *The Source of Innovation.* New York and Oxford; Oxford University Press, 1988.

JAMES Jeffrey: *Microelectronics and the Third World.* Maastricht; UNU/INTECH Working Paper No. 2, 1991.

KLINE Stephen J., ROSENBERG Nathan: An Overview of Innovation. In Ralph Landau and Nathan Rosenberg (Eds): *The Positive Sum Strategy-Harnessing technology for economic growth.* Washington DC; National Academic Press, 1986.

KNIGHT J.B.: The Evolution of Development Economics. In V. N. Balasubramanyam and S.Lall (Eds): *Current Issues in Development Economics.* Basingstoke etc.; Macmillan, 1991.

LALL Sanjaya: Explaining Industrial Success in the Developing World. In V. N. Balasubramanyam and S. Lall (Eds) : *Current Issues in Development Economics.* Basingstoke etc.; Macmillan, 1991.

LEWIS W Arthur: *The Evolution of the International Economic Order.* Princeton N.J.; Princeton University Press, 1978.

LUNDVALL Bengt-Aake: User-Producer Relationships, National System of Innovation and Internationalisation. In Bengt-Aake Lundvall (Ed): *National System of Innovation.* London; Pinter, 1992. p45-67.

LUNDVALL Bengt-Aake: *National Systems of Innovation.* London; Pinter, 1992.

MOHAN Rakesh: Industrial Policy and Control. In Bimal Jalan (Ed): *The Indian Economy-Problems and Prospects.* New Delhi and N.Y.; Viking, 1992. p. 85-115.

NELSON Richard R., (Ed): *National Innovation Systems-A comparative analysis.* New York, Oxford; Oxford University Press, 1993.

NELSON Richard R, WINTER Sidney G: *An Evolutionary Theory of Economic Change.* Cambridge, Mass and London; The Belknap Press of Harvard University Press, 1982.

PACK Howard, WESTPHAL Larry E: Industrial Strategy and Technological Change. *Journal of Development Economics* 1986, 22, p. 87-128.

PAVITT K.: Sectoral Patterns of Technical Change: Towards a taxonomy and a theory. *Research Policy* 1984, 13, p342-373.

ROSEN George: Contrasting Styles of Industrial Reform: *China and India in the 1980s.* Chicago and London; University of Chicago Press, 1992.

ROSENBERG Nathan: *Perspectives on Technology.* Cambridge and New York; Cambridge University Press, 1976.

ROSENBERG Nathan, BIRDZELL L. E.: *How the West Grew Rich: The Economic transformation of the industrial world.* New York; Basic Books, 1986.

STEINMUELLER W.E.: Technology Infrastructure in Information Technology. In M Teubal, D Foray, M Justman and E Zuscovitch (Eds): *Technological Infrastructure Policy-An international perspective,* Dordrecht; Kluwer Press, (Forthcoming) 1996.

WEISS John: *Industry in Developing Countries-Theory, Policy and Evidence.* Crown Helm, 1988. (Reprint, London and New York; Routledge, 1990)

WESTPHAL Larry E.: Industrial Policy in an Export Propelled Economy: Lessons from South Korea's experience. *Journal of Economic Perspectives* 1990, 4 (3) p41-59.

WILLIAMSON Oliver E: *The Economic Institutions of Capitalism-firms, markets, relational contracting.* New York; Free Press, 1985.

WILLIAMSON Oliver E.: The Institutions and Governance of Economic Development and Reform. In *Proceedings of the World Bank Annual Conference on Development Economics 1994,* p171-205.

PART II : SCIENCE AND TECHNOLOGY STRATEGIES AND THEIR IMPLEMENTATION DURING VARIOUS PLAN PERIODS IN INDIA AND CHINA

3. Science and Technology Strategies and Their Implementation During Various Plan Periods Since Independence in India and China - A Comparative Study

Ashok Jain and V.P. Kharbanda

Abstract

The study compares the experiences of India and China in managing science and technology strategies for development over the last 40 years. The paper assesses how government support has been instrumental in formulation of science and technology policies and their implementation during the different Five Year Plan periods in the two countries. The study reveals that in the initial years, while both the countries started at the invention end of the innovation chain, over the years, both have moved closer to the diffusion end of the innovation chain. The capabilities built up have been discussed. In the new scenario of increased globalization, finances and production systems with market locale playing a dominant role, it is important to integrate S&T with economic development through 'Techno-Economic Networks', rather than supporting S&T activities *per se*.

The Backdrop

India (Before Independence, 1947)

Before independence, India had only a limited capacity for introducing minor process innovations in sectors like railways, telegraphy, mining of minerals and exports, processing of grains, oilseeds, sugar, tea, coffee, tobacco, jute, cotton hides, etc. This was the situation almost two centuries after the Industrial Revolution in Europe. However, some Indians had made pioneering efforts at learning by doing and could set up industries for coke, iron and steel, cotton spinning, jute processing and chemicals.

There were some individual scientists who had made important research contributions and had even acquired international acclaim. However, no organized effort was made by the British government to support basic research. The research capabilities of individual scientists of the time were known to the national leaders, who perceived their role in the building up of an independent nation, but the rulers did not make any attempt at encouraging linkages between research and production activities. By and large, the Indian industry depended on foreign technologies and did not develop research programmes of its own. In 1916, Indian Industrial Commission (IIC) was set up for coordinating efforts to link research activities with production, under the chairmanship of Sir Thomas Holland. However, the report of the commission was not implemented. According to Shiv Vishvanathan, "the seven volumes of the IIC report along with the journal *Science and Culture,* constitute the two most important texts in the archives of Science Policy in India since 1900. Strangely, the reports of the Commission have never been systematically considered in the histories of science and technology"[1]

The next important step taken by the then government was the constitution of the Industrial Intelligence and Research Bureau (IIRB) in 1934, which started functioning in April 1935. The 'Intelligence' activities of IIRB were confined to information collection and dissemination, publication of bulletins and looking after organisation of exhibitions. The creation of IIRB with insufficient funding and many projects under its charge proved a mere sop to the demands of Indian scientists for paying better attention to industrial research in India.[2]

Then in 1938, the National Planning Committee was set up under the chairmanship of Jawaharlal Nehru for framing schemes for the economic development of India. This step constituted a land-mark in the history of the country for bringing about

---

[1]VISHVANATHAN Shiv: *Organizing for Science: the making of an industrial research laboratory,* Delhi; Oxford University Press, 1985 275P.

[2]RAJAGOPAL N.R., QURESHI M.A. and BALDEV SINGH: *The CSIR* Saga. New Delhi; CSIR, 1991. 200 P.

the desired nexus between political thought and scientific research. During World War II, the importance of scientific and industrial research received considerable attention. Need was felt for indigenous production of a large number of industrial products which were being imported. In these circumstances, Government of India constituted the Board of Scientific and Industrial Research in 1940. The Council of Scientific and Industrial Research (CSIR) was established in 1942. This helped in tackling a number of problems related to military needs. Processes were also-developed for producing essential articals from indigenous materials. Since independence in 1947, science and technology policy, planning and building up of an indigenous R&D system became a part of the overall economic policies.[3]

China (Before Liberation, 1949)

Though China's encounter with modern western science and technology began around the seventeenth century in a small way, it was not until the mid-nineteenth century that Chinese were introduced to western mathematics and astronomy to a considerable extent by the Jesuits.[4] Compared to the Chinese superiority in scientific and technical pursuits of the pre-sixteenth century era, the Chinese realized their declining position in a large measure by the mid-nineteenth century, particularly during the Opium War (1839-1842) . Apart from the monumental works of Needham, a number of important studies (see, for instance, Moore, 1966; Einstadt, 1987; Fairbank *et al.*, 1973; and Sivin, 1980,

[3]JAIN Ashok and OBEROY Asha: *Science and Technology Policies: A Comparative Study of the Japanese and Indian Experience.* Journal of Scientific and Industrial Research 1993 (52) p717-727.

[4]According to Needham's monumental volumes on China, it was exactly 1601 - the year in which the Italian Jesuit Matteo Ricci obtained stay permit in Peking from the Emperor. He is known to have transferred western mathematics, astronomy and also rendered help in reforming the Calendar. (Yang An Xian et al: Twists and Turns in the Formation of Chinese Scientific Community. In *Sciencific Communities in the Developing World edited by J. Gaillard etal. New Delhi; sage, 1997.*

1982)[5], lend considerable support to the view that till the beginning of the nineteenth century, China was not able to rapidly institutionalize and operationalise modern science due to specific social and structural factors.

However, the situation changed gradually after the Opium War and with rapid Western intrusion from the last quarter of the nineteenth century. During this period, Zhang Zhidong, a regional reformer, coined the theory of "zhongxue weiti xixue weiyong" (Chinese learning for the fundamental principles, Western learning for the practical application). Even though the political elites realized the benefits of promoting reforms and Western science and technology, they also foresaw the threat to their own legitimacy, which prevented China from undertaking any major reforms till the onset of the Republican Era (1912-1949).

The return of a small group of Chinese students from Europe, Japan and North America during this time greatly contributed to the prevailing local intellectual efforts for promoting the values of science, democracy and autonomy. These efforts culminated in what is known as the May Fourth Movement (MFM) during 1915 and 1920's. This led to the creation of several intellectual and cultural organizations from the neutral political groups, intellectuals and other sympathizers for Chinese modernization and reception of modern science and technical institutions. The most significant event during this period was the constitution of Science Society of China (SSC) in 1914 at the Cornell University

---

[5]MOORE, Barrington Jr.: *Social Origins of Dictatorship and Democracy: Lord and Peasant in the Making of the Modern World.*, Cambridge, Mass; Harvard University Press,1966.

EINSTADT S. N. (Ed.) : *Patterns of Modernity, Volume II: Beyond the West.*, London: Frances Pinter, 1987.

FAIRBANK J. K., EDWIN 0. R. and CRAIG A. M.: *East Asia: Tradition and Transformations.* Boston; Houghton Miffilin, 1973.

SIVIN N.: 'Science in China's Past' in Leo A.Orleans (ed.) *Science in Contemporary China.* Stanford, California; Standord University Press, 1980.

SIVIN N.: 'Why the Scientific Revo lution did not take place in China or didn't it?, *Chinese Science* 1982, 5 (June).

by Chinese students of natural sciences which was later transferred to China in 1918. By 1920's, the SSC became an important first professional body of Chinese scientists in modern sciences. The most significant achievement of SSC was its radiating influence in the creation of several important scientific societies, such as the Chinese Medical Society (1915), the Agricultural Society, (1917), the Geological Society of China (1922), the Societe Astronomique de China (1924), the Chinese Meteorological Society (1925), the Physiological Society of China (1926), the Chinese Chemical Society (1932), the Chinese Physics Society (1932), the Chinese Botanical Society (1933), the Chinese Zoological Society (1934), the Chinese Geographical Society (1934), and the Mathematical Society of China (1935). By the end of 1930's there were about 65 periodicals and journals in various scientific disciplines. In the higher education sector, there were 200 institutes, including universities and colleges with an enrolment of 110,000 students by late 1940's[6].

The government created the Academia Sinica in Nanjing in 1928, and the Peking Academy in 1929. By 1940, there were about 13 research institutions under the umbrella of Academia Sinica and nine institutions under Peking Academy. These institutions and scientific societies sustained their activities due to the individual scientific efforts, most of whom were educated in USA, Europe and Japan.

The nascent stage of China's S&T infrastructure in 1940's was not in a position to aid industrial development and promote manufacturing activity. The Chinese industries depended basically on buying readymade goods and equipment from foreign sources. Though the consecutive civil wars and the Japanese aggression awakened the political elites, corresponding encouragement to the strengthening of research and development activities did not come about till the communist regime took over the government in 1949. On the whole, the S&T infrastructure was very weak in

---

[6]UNESCO: *Statistical Year Book, Paris;* UNESCO, 1985.

[7]LUO WEI: Interflow of Research Achievements, Knowledge and S&T Personnel. In Rustam Lalkaka and Wu Mingyu (Eds.) : *Managing Science Policy and Technology Acquisition - Strategies for China and a Changing World.* New York; UNFSSTD, 1984.

1940's[7]. There were altogether 40 research institutions with less than 50,000 S&T personnel, only 500 of whom were involved in research[8,9].

After independence (1947) in India and liberation (1949) in China, the development proces could be divided into three phases as under:

(i) Infrastructure Phase (1947-1960s)
(ii) Re-orientation Phase (1960s-1970s)
(iii) Market orientation Phase (1980s)

(i) The Infrastructure Phase (1947-1960s)

India

In India, after independence, emphasis was laid on rapid industrialization, increasing agricultural productivity and establishment of a strong base for scientific and technological infrastructure[10]. The strategy of self - reliance under Nehru - Mahalanobes planning approach was pursued during the period, which witnessed three Five Year Plans (FYPs) upto 1966. The strategy emphasized the role of investment in the development process, centering on the production of capital goods. Accordingly, resource allocation and foreign trade policies were geared to the promotion of industrialization with emphasis on heavy industry. Imports of foreign technologies were encouraged to set up core industries in the state sector, particularly steel, electricity, oil exploration and mining. This was done due to inability of the private sector to make large scale investments and the risk factors involved. In general, the strategy of import

---

[8]Chinese Academy of Sciences: *History of CAS*. Beijing; Contemporary China Books, 1989.

[9]CHEN Wenhua et al: *A Comparative Study on the Strategies for the Development of Science and Technology of China and India*-Working Paper on China. Changsha(China); Central South University of Technology, 1995, 52p.

[10]JAIN Ashok and KHARBANDA V. P.: *Science and Technology Policies during different FYP periods since Independence in India*. New Delhi; NISTADS, 1995. 68p. (Working Paper).

substitution for industrialization was emphasized in consumer goods industries (First FYP years) and also in capital goods and intermediate goods industries (Second FYP years)[11]. Scientific research was given a big push because of the immense faith the planners had in its automatic transfer to industrial development. Hence, the basic infrastructure for scientific research was established. The university system and engineering education were expanded to generate scientific, engineering and technical manpower. By the end of the Third FYP, a total of 26 research laboratories were established under CSIR, besides 81 other research institutes, 80 universities and 87 medical colleges. With this, the strength of S&T manpower rose from 0.188 m in 1950 to 0.732 m in 1965. However, expenditure on R&D incurred by the industry was only 3.5 per cent of the total expenditure. As industry was not in a position to invest in R&D, the government took the direct responsibility to carry out industrial R&D. To link research institutions with industry, the National Research Development Corporation (NRDC) was established in 1953. In 1956, a conventional forum Scientific Advisory Committee to Cabinet (SACC) was set up to deliberate on and advise the government on matters related to science policy. Scientific Policy Resolution, was adopted in 1958 articulating the government's commitment to science and technology as an instrument for national development[12]. While SPR emphasized scientific and technological self-reliance, little effort was made at establishing the link between research activity and the production system. The S&T infrastructure was established and nurtured for building up research capabilities on a broad basis covering academic, industrial, agricultural and defence research. However, it was considered too premature to put specific demands on the S&T infrastructure to contribute to problems of industry, agriculture, etc. Investments in R&D laboratories

---

[11]DESAI, Padma: *Alternative Measures of Import Substitutions*. Oxford Economic Papers (1969). Qouted in "The Economy of India by V.N. Balasubramanyam. London; Weidenfeld, 1984.

[12]Government of India: *Scientific Policy Resolution*- 1958. New Delhi; 1958, the 4th March, 2p.

as well as in university education were viewed essentially for creating assets in the form of manpower and infrastructure that would, in due course, become available for tackling the economic and other developmental needs.

The Scientific Policy Resolution, with accent on the spirit of free enquiry, supported the growth of research capabilities at the invention and innovation ends *(Fig. 1)*. An extensive R&D infrastructure was established outside the production units within a short span of 20 years, almost undisturbed by the successes or failures on the economic front. The period was characterized by the simple 'Linear Model' approach to economic development, i.e. Science-Technology-Economy (S-T-E) *(Fig. 2)*. The priorities of research were determined by the leaders of science in whom the policy leadership committed to science had trust and faith. The Third Review Committee of CSIR set up in 1963 stressed the need for applied research and industrial association with R&D institutes, but this could not be achieved satisfactorily.

While the R&D infrastructure was set up essentially for the purpose of invention and innovation, experience of technology came through industry and services set - up during this period as a part of the planned development programme and through learning by doing process. A policy for industrial development was announced (Industrial Policy Resolution) in 1948[13]. Planning exercise got under way that allowed considerable flexibility for industrial development. Investments by foreign companies were welcomed with minor restrictions placed on equity holding or terms of technology transfer. Under these liberal policies for foreign investments, technology got introduced in several areas. Some examples are:

Introduction of foreign ownership companies in the fields of petroleum refining, pharmaceuticals and fertilizers; and establishment of joint ventures in the sectors of polyethylene, nylon, automobiles, detergents, cement plants, etc.

The Sterling balance accumulated during the war enabled the government to import capital equipment without restrictions for setting up industries.

[13]*Industrial Policy Resolution*, New Delhi, 6 April, 1948.

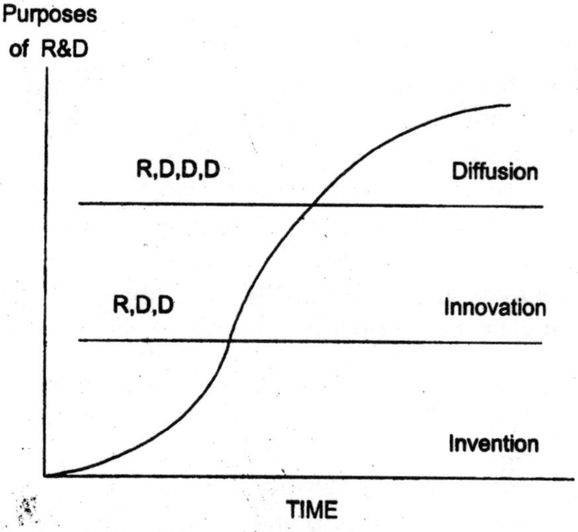

R, D, D, D = Research, design, development, and diffusion

Fig. 1 : Innovation Chain and Purposes of R&D

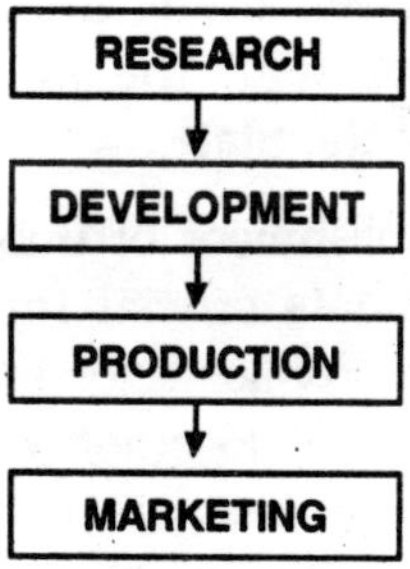

Fig. 2 : The Conventional Linear Model of Technological Innovation

The favourable climate for foreign investments, however, did not receive response from investors in areas like exploration of oil and gas, steel, major size capital goods, heavy machinery, and basic organic chemicals. Perhaps, investments in these areas were not considered as viable due to lack of confidence in the available technical manpower, future growth prospects and economic stability. Under the circumstances, the first Industrial Policy Resolution, announced in 1948, was sharpened and revised in 1956[14]. It classified industries into three categories.

(i) Schedule A : Industries whose development was to be exclusive responsibility of the State, i.e. public, e.g. mining and quarrying (94 per cent), electricity, gas and water supply (91 per cent), railways (100 per cent), communication (99.9 per cent), banking and insurance (82 per cent), and public administration and defence (100 per cent).

(ii) Schedule B : Industries to be progressively owned by the State. The State would set up new undertakings with cooperation from the private sector.

(iii) Schedule C : All remaining industries were to be set up by private sector with the development of necessary infrastructure by the State.

It thus required establishment of public sector corporations in areas of large investments crucial for establishing a modern industrial infrastructure. Public sector industries were set up, for example, for steel, fertilizers, locomotives, petrochemicals, oil and gas exploration and refining, heavy machinery, etc.

In the agriculture sector, immediate food requirements were met through imports. The research system was strengthened for meeting the future needs.

China

After liberation in China, under moderate policies advocated by Liu Shaoqi and Chou Enlai, radical changes were brought about in S & T organizations as compared to the pre-revolutionary period. In

[14]*Industrial Policy Resolution*, New Delhi, 30 April 1956.

November 1949, the Chinese Academy of Sciences (CAS) came into existence with 21 research institutes by merging of Academia Sinica and Peking Academy. The main goals of the CAS were to: engage in basic studies and applied research activities on a planned basis; strengthen liaison among different disciplines; promote interaction between science and application; and absorb advanced scientific developments from other countries.

By the end of 1950's, the research system in China was organised in terms of five major sectors with CAS as the major academic body for scientific research. These sectors included: (1) CAS and its subsidiary academies; (2) Institutes of higher education with limited research activities; (3) Academies and institutes under the government ministries in health, agriculture, railways, mechanical engineering, metal industry, etc; (4) Provincial and municipality controlled research institutes; and (5) Defence research institutes. The overall structure of research system in China in 1950's actually followed the Soviet model by distinguishing between basic and applied research reflecting broadly the division of labour between CAS and the industrial sectors.

The total number of research institutes under CAS increased from 21 in 1950 to 120 in 1965. The total S&T manpower increased from 316 to 21937 during the same period[8]. The scientific societies also gained support from the government, as their number increased from 12 in the pre-1949 period to 30 in 1965 under the newly set up apex body called the National Federation of Scientific Societies.

As a part of Soviet influence, an important event in this period was the launching of the first long-term 12 Year Plan (1956–67) for S&T development in 1956. The Plan was drawn up with the assistance of 200 scientists and advice of Soviet experts. The main objectives set included the achievement of scientific and technological excellence to catch up with the advanced countries in a time of 20 years. It prioritized four sectors, viz. computers, semiconductor technology, electronics and automation, atomic energy and jet technology. This Plan is said to have ended in 1962, five years ahead of schedule. Defence – related space and nuclear technology programmes were also initiated during this period. Together with these high technology areas, biological sciences drew considerable attention. In mid 1960's S&T personnel in biology made up 30 per cent of the total- occupying the first rank amongst the natural sciences. By the time Prime Minister Zhou Enlai announced the Second Five-Year Plan in 1958, there had been a massive Soviet transfer of technology to China, including organization and management of science and technology. Even though Soviet assistance was abruptly withdrawn in 1960, the foundation of industrial and technological infrastructure had been already established.

During late 50's, scientific and technological growth was interrupted by the onset of the Great Leap Forward (GLF) (1958–60). It called for emphasis on agriculture and small scale industry with a mass-oriented approach. As steel was considered as the backbone of the economy, a large number of small coal mines and backyard steel furnaces sprang up. The S&T strategies adopted during GLF were oriented to meet the immediate technological demands of the economy through 'Workers' Innovations' at the shop floor and Mass Movements. It advocated self-reliance, which meant total isolation from the main stream of the world pool of scientific knowledge. Emphasis was only on technology diffusion without taking into consideration the invention end of the Innovation Chain.

By the end of the First Phase, while both the countries had built up the base for heavy industries and S&T infrastructure, industry was largely divorced from the research institutions. It is observed that the priority purpose of R&D, visualized in the policies adopted for S&T in India and China during this period, was invention and innovation, unlike in Japan, where R&D originated at the production and diffusion ends[15] *(Fig. 1).* Further, whereas in Japan the engineers played a leading role in shaping the direction of industry with accent on technical capabilities,

---

[15]The Chinese attempts at the diffusion end through policies of mass innovation 'Three in One Combinations' and 'Walking on Two Legs' during GLF and CR did not meet with success. For details see Kharbanda V.P., and Qureshi M.A.: Science, Technology and Economic Development in China. New Delhi; Navrang Publishers, 1986.

in India the civil servants and economists in the government shaped the industrial policy on the basis of a planned economy approach on one hand and science and technology policy shaped by scientists coming out of academic research on the other. In contrast, the Chinese Science Policy was highly politically motivated. The Indian and Chinese efforts at formulating policy instruments and plans to couple S&T capabilities with the requirements of agriculture, industry and other sectors of the economy remained weak.

Table 1

Production of Principal Commodities in India and China

	1950-51		1964-65		1984-85		1994-95	
	India	China	India	China	India	China	India	China
Coal (mt.)	33	32.43	67.2	232	145	872	263	1298
Oil (mt)	0.2	0.12	3.0	10.8	29	125	32	149
Electricity (bkwh)	5.3	4.3	29.5	42	167	408	351	1000
Cement (mt)	2.7	1.4	11	16.34	30	143	58	450
Steel (mt.)	1.0	0.61	5	9.4	8	47	13	94
Grains (mt.)	50.8	113.2	90	194	145	407	182	465

Note: mt - million tonnes; b kwh - billion kwh.

Starting at almost the same level of economic development in 1950, China had already outpaced India at the end of this phase (Table 1). During this phase, one peculiar feature of similarity in both the countries has been that both emphasized establishment of heavy industry with steel as the priority sector by importing turnkey large scale plants initially. In India, investment in industry increased from 7.6 per cent in the Ist Plan to 18.5 per cent in the 2nd Plan, 50 per cent of which was invested for steel production. While India continued to emphasize on large scale plants, the Chinese went in for small scale and micro plants (to the extent of backyard steel furnaces) particularly during GLF and thereafter in line with their strategy of decentralized approach. While India imported turnkey plants from USSR, UK and West Germany for its Bhilai, Durgapur and Rourkela plants respectively with 1 million tonnes capacity each (1954-56), China imported turnkey plants from USSR for its Wuhan and Baotou plants with 2 million tonnes and 1.5 m tonnes capacity respectively (1953). As USSR stopped aid during late fifties, this led the Chinese to make their indigenization efforts and expanded capacity to 8m tonnes. Comparatively, indigenization efforts in India have been at a low ebb. From setting up of Bhilai plant in 1956 to Bokaro steel plant in late 1960s the import content was reduced as under:

	Bhilai	Bokaro
Equipment	87%	40%
Structures	78%	6%
Refractories	94%	39%

With the notable efforts of M.N. Dastur a non-resident Indian, a design and engineering organization was set up, later named as Metallurgical and Engineering Consultants (MECON). This organisation basically helped in respect of techno-economic feasibility studies, study of locations, overall design and preparation of detailed projects[16].

(ii) Re-orientation Phase (1960s-1970s)

India

In the two FYPs during this phase, the work done on the application of S&T to development was revieved. In accordance with the recommendations of the Third Review Committee of

[16]ETIENNE Gilbert: *Asian Crucible - The Steel Industry in China and India.* New Delhi; Sage, 1992. 304 P.

CSIR (1963), and those of the 1965 'Industry R&D Get-together', the expectations that research institutions should contribute to economic development started finding expression in the mass media, Parliament and other forums. Several factors led to this situation. Despite the open-door policy for foreign investments and technology imports, the extent of such investments in certain sectors had remained low. For example, areas like exploration of oil and gas, steel, major sized capital goods, heavy machinery, basic organic chemicals, etc. were not considered viable by foreign investors, perhaps due to lack of confidence in the available technical manpower, future growth prospects and economic stability. In addition, food imports for a rising population, conflict with China in 1962 and with Pakistan in 1965 led to depletion of the Sterling balance. It was also realized that modalities of technology imports adopted thus far had contributed little to the enhancement of indigenous technological capabilities. The foreign exchange crisis leading to a balance of payments problem caused concern in 1964-65. A country-wide appeal was made for import substitution efforts and for bringing together the country's research facilities and industry's know-how requirements. As a result, planning priorities for research were reshuffled. CSIR was asked by the Vice-President, M.C. Chagla, to give top priority to those research activities which aimed at finding ways and means, of conserving foreign exchange potential for the purpose of making the country self-reliant. The concern for applied research was such that even a laboratory like National Chemical Laboratory (NCL) with a balance of 50:50 between basic and applied research was asked to alter it to 20:80[17]. In general, investment decisions came under critical review and in this context it was felt that the system had to be strengthened.

During the 1970s, therefore, science policy and plans came in for re-examination and reorientation, which can be traced to the following key issues:

---

[17]SANDHYA G. D., JAIN Ashok and MATHUR Purnima: *S&T Planning, Policy Directions and CSIR. Economic and Political Weekly 1992* (Dec. 22) p 2801.

(1) Foreign technologies introduced so far had contributed little to the enhancement of 'Technological Capacity.' The importation had been by and large in packaged form, with little or no scope for unpackaging.
(2) Interaction between the research institutions and industry had been poor.
(3) Research institutions had mostly confined their activities to the invention end of the innovation chain without taking any interest and initiative in the activities required for moving up to innovation and diffusion.

Several steps were taken to remedy the above situation.

At the highest level, SACC was replaced by the Committee on Science and Technology (COST) in 1968. This committee had a broader purview than SACC, where the issue of utilization of scientific research got included and hence the emphasis on technology. In 1970, Patent Law was enacted to reiterate the government's will to encourage indigenous R&D with its utilization in the production system.

The first S&T Plan (1974-79) prepared by the National Committee on Science and Technology (NCST) (established in 1973) adopted the sectoral approach, identifying 24 major sectors in order to take a systems view of science and technology. The sectoral approach was considered important, as it was recognized that S&T could be applied effectively in different sectors of the economy and the infrastructure for S&T could be utilized in the best possible manner. Later during this phase, it was noted that application of S&T demands, in the first instance, a strong foundation in basic sciences. For promoting this, the formation of Science and Engineering Research Council (SERC) was recommended to support and foster basic scientific and industrial research in new and emerging areas. Particular attention was to be paid to the fuller use of existing facilities, avoiding of unplanned duplication of research by different agencies, minimizing sub-critical funding by spreading resources over too many projects, and closer monitoring of research programmes right up to the stage of their application in the field.

On the industrial front, important policy changes were introduced that had a strong influence on the utilization of

Table 2

EXPLOITATION OF INDIGENOUS TECHNOLOGIES THROUGH NRDC

Year	Technologies Received by NRDC (Cumulative)	Technologies Transferred to Industry	%
1962	648	64	9.8
1965	746	96	12.8
1970	1214	187	15.4
1980	2147	408	19.00
1984	2251	433	19.23
1990	2513	NA	—
1994	2634	NA	—

indigenous R&D capabilities. Import of technology came under modified regulations through the Foreign Investment Board (set up in 1969), the Industrial Licensing Policy Instruments and Foreign Exchange Regulation Act (1973), and Technology Development Fund Scheme (1976) . Industries were classified into categories where foreign collaboration was permitted or not permitted. This enabled the research institutions to direct their efforts in areas where foreign technology was not permitted. The Indian Patent Act allowed only processes to be patented in sectors like food, medicine, drugs and chemical substances. In 1973, a package of extraordinary incentives by DST to promote in-house R&D in industry was introduced in the form of tax incentives. Write-offs of 100 per cent on capital investments for R&D, and 133 per cent expenditure on sponsored research were made available to industry.

The above changes introduced during the 1970s had a strong influence on the S&T structure in India. S&T plans became an integral part of the five year planning exercise. In several areas, indigenous research moved up to production, for example, in electronics, drugs and pharmaceutical, pesticides, chemicals, power generating equipment, agriculture and food processing. A large number of in-house R&D units were established in industry. Private

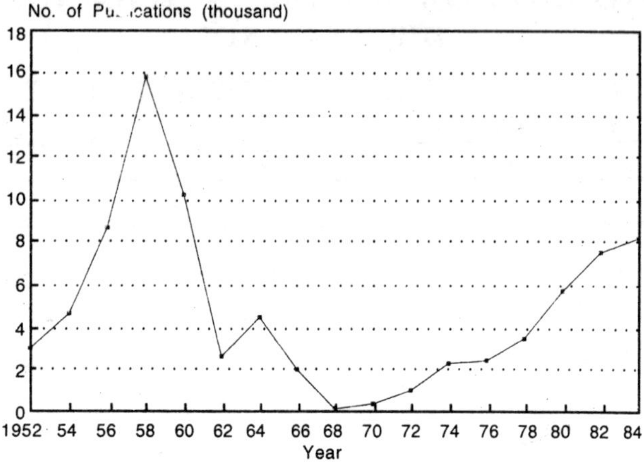

Fig. 3 : Number of Natural Science Publications/Books

sector industrial R&D expenditure increased from 3.5 per cent in 1965 to 16 per cent in 1980. Exploitation of indigenous technologies through NRDC also increased from 9.8 per cent in 1962 to 19 per cent in 1980 *(Table 2).* Capabilities in new areas like non-conventional energy sources, ocean resources, environment, new materials, communication and informatics, space applications, remote sensing, fast breeder reactor technology and biotechnology were established. Sophisticated equipment for front-line research became available to universities at a fairly extensive scale. Self-sufficiency in food grains had been achieved. It became possible to bargain for reduced payments for technology imports and also endogenously develop parts of technology in the form of detailed engineering and introduce procurement inspection and construction through Indian engineering consultancy organizations.

China

In China, Phase II period (late 1960's - early 70's) was the period of turmoil. The Cultural Revolution (CR) (1965-76), which downgraded the scientists and questioned the use of scientific research for the masses, played havoc with the scientific institutions. The highest scientific body, the Scientific and Technology Commission, was abolished. In 1967, the CAS budget was cut to the extent of 84 per cent to that of the 1965 level and was further reduced by 1976[8] . CAS also lost more than 60 per cent of its major research institutes and by 1976, there were only 36 of them, which were directly linked to defence and military technologies[18]. The basic research activities virtually ceased until about 1973. Higher technical education was also disrupted. The number of universities was drastically reduced and China lost a whole generation of educated personnel during the period 1965–1976[19]. The period witnessed all round retardation of scientific research activities, as examplified by drastic decrease in the number of scientific publications *(Fig. 3)* as well as in the number of scientific achievements *(Fig. 4)*. While the *Second FYP (1958-62)* was overtaken by the GLF, the *Third FYP (1965-70)* and the *Fourth FYP(1970-75)* were abandoned because of CR. This gave a setback to the national economy and the development of science and technology. Some important technological areas, once having reached the international advanced level, lagged behind by the end of 1970's. The narrowed gap once again widened. For example, China started the experiments on oxygen blasting furnace and continuous ingot casting techniques almost at the same time as Japan started the same experiment. However, by 1978, Japan had transformed its all open hearth furnaces into revolving furnaces and continuous ingot casting was widely applied.

---

[18]YAN Jici: *Development of Chinese S&T in Last Three Decades.* (Mimeo) 1988.

[19]KHARBANDA V. P. and QURESHI M. A.: Science, Technology and Economic Development in China, *Op cit.*

However, the political leadership guarded the high technology and military–related research organizations from major upheavals. Even though a large segment of scientific personnel in these areas worked under tremendous pressure and military directives, the projects in electronics, lasers, semiconductors, advanced computer technology, nuclear and space programmes continued to draw considerable financial and political support. For example, China was able to explode atomic bomb and hydrogen bomb in 1964 and 1967 respectively, to carryout artificial synthesis of bovine insulin in 1965, to develop the first hybrid rice in 1973, and to launch several satellites between 1970 and 1976.

The death of Mao in September 1976 changed the whole political scene in favour of moderates, which once again emphasized the crucial role of scientific research in the economy, which set the development stage to enter the third phase.

The *Fifth FYP (1976-1980)* was not worked out. It only came out in the form of a draft "The Outline of the Ten Year Plan for the Development of National Economy 1976-1985." However, the "Outline" included many high and unrealistic targets. In December 1978, the Third Full Session of the llth Conference of Chinese Communist Party shifted emphasis to socialist modernization construction, and pointed out that the key to realization of four modernizations lay in the modernization of science and technology. The new leadership in 1978 convened the National Science Conference, which was the first after 1956 and was participated by 6000 S&T personnel and administrators.

"1978-1985 National Development Plan for Science and Technology" was worked out according to the principles of "Adjustment, Renovation, Strengthening and Improvement." The Plan concentrated on the eight crucial fields of science, newly emerging technologies, such as in agriculture, energy, materials, electronic computers, laser, space, high energy physics and genetic engineering. The plan also embodied 14 specific measures, such as going all out to develop college education and train technically talented people; speeding up applications of scientific achievements; studying and absorbing imported advanced foreign technology; and promoting international academic co-operation.

In 1979, CAS Scientific Council was revamped with the creation of five academic departments in place of four. These were physical mathematical, chemical, biological, geological and technological sciences. From the beginning of 1980, post-graduate studies began in the universities and CAS. By the early 1980's, the new leadership restored the whole spectrum of S&T system that was dismantled during the CR to enter into the third phase.

(iii) Market Orientation Phase (1980s -)

India

In *India* the reorientation of policies and plans initiated during the 1970s, started putting additional demands on S&T. During the 1980s, the capabilities in S&T, built over four decades of independence, matured to a stage where R&D programmes could be conceived and executed covering the entire innovation chain. Under this ethos, technology acquired a stronger focus. Accountability and questions relating to returns on investments in S&T became important. The Technology Policy Statement of 1983, announced 25 years after the Scientific Policy Resolution of 1958, reflected this change. Various other S&T policy statements announced from time to time are given in *Table 3*.

The *six Technology Missions* launched in a short period of three years (1985-88) for meeting the immediate needs in the areas of drinking water, literacy, edible oils, immunization, telecommunication and dairy products have been manifestations of the accent on accountability and performance.

Table 3 : Various S&T Policy Statements (India)

Year	S&T policy statements	
1894	National Forest Policy	
1927	Industrial Policy Resolution (revised in 1956, 1973, 1977, 1980, 1990)	
1958	Scientific Policy Resolution	
1968	Policy on Education	
1970	The Patent Law (revised 1999)	
1972	Archival Policy Resolution	
1978	Health for All (Population Policy)	
1980	The Forest (Conservation and Control of Pollution Act	
1982	National Health Policy	
1983	Technology Policy Statement	
1984	Ocean Policy Statement	
	Computer Policy	
	Telecommunication Report	
1985	Textile Policy	
	Policy Measures in Electronics	
1986	Drug Policy	
	National Policy on Education	
	Computer Software Report	
	Environment Protection Act	
	R&D Cess Act	
	Consumer Protection Act	
1987	National Water Policy	
	The Air (Preservation and Control of Pollution 1981) Act (revised)	
1988	National Forest Policy of 1952 (revised)	
	National Housing Policy	
	The Water (Preservation and Control of Pollution 1974) Act (revised)	
1989	Motor and Vehicle Act	
1990	New Comprehensive Guidelines for Steel	
	National Mineral Policy	
	New Industrial Policy	
	Agriculture Policy	(Expected)
	Environment Policy	(Expected)
	Comprehensive National Housing Policy	(Expected)
	Review of National Gas Utilization Policy	(Expected)
1991	New Industrial Policy	
1992	New Fertilizer Pricing Policy	
	New Technology Policy	(Expected)

Source : J Sci Ind Res, 50 (3) (1991) P. 212-221.

In 1987, a Technology Information, Forecasting and Assessment Council (TIFAC) was established to examine and evaluate the existing State of Art of Technology and directions of future technological developments in various sectors; and prepare Technology Forecasting Reports covering 10 or more years.

The setting up of Technology Development Fund and risk investment mechanisms within the financial institutions and of Science and Technology Advisory Committees (STACs) in economic ministries are also steps towards forging stronger linkages between R&D and economic activities.

A mandate has been given to publicly funded R&D laboratories to supplement their financial resources by 30 per cent or more of what they directly receive from the government by undertaking sponsored or consultancy research. Thus, user (market) oriented research gained importance. The associations of industry, commerce and trade are getting increasingly involved in S&T policy debates, reflecting a change in the perception of S&T towards economic issues.

The Industrial Policy Statement issued in 1991 was a major shift in giving a business approach to Indian R&D in the wake of adoption of liberalization and globalization policies. At the beginning of the *Eighth FYP (1992-97)*, globalization of business and of R&D emerged as major developments. Meaningful partnerships on the international plane in the areas of mutual global interests became necessary. Thus, the S&T strategies were to be suitably moulded to meet these realities. The resource crunch, the adverse balance of payments position and the shortage of foreign exchange were also hard realities. Liberalization in technology transfer and trade intensified.

The major thrust areas of S&T in the Eighth FYP were:

(I) Basic research in frontline areas;

(II) Innovative research in exploitable areas of S&T with emphasis on R&D activities in emerging technologies;

(III) Diffusion of appropriate technologies and technology support for ancillaries of large units.

(IV) Integration of S&T in socio-economic and rural sector to fulfil the basic needs of water, food, nutrition, health and sanitation, shelter, education, energy, clothing, employment etc.

Various measures were adopted to meet the objectives of the Eighth Plan. Using SERC and other mechanisms, certain national R&D programmes involving a number of institutions in multi-disciplinary fields were being taken up. The disciplines included: super conductivity, instrument development, molecular electronics, astronomy, seismicity and deep continental studies. Under the technology utilization scheme, the National Register of Foreign Collaborations provides an opportunity for the national laboratory system to tune its activities to meet the current demands and possible future demands of technology by the industry and for preparing the technology status reports of several industries in the country. The Technology Absorption and Adaptation Scheme provides a mechanism of interaction for R&D by the companies which have imported the technology and will take advantage of the existing national infrastructure.

The policy of economic liberalization accepted by India during the last couple of years has accelerated change towards economic accountability of R&D. The new industrial and economic policies adopted by Government of India during 1991-92 have weakened the protection earlier available to the Indian R&D system as well as to the industry through instruments such as licensing, restrictions on foreign investments, subsidies, etc. Marketing of R&D and generation of technological capabilities to face open international competition have become the major concerns of all those involved in either generating or exploiting science and technology capabilities. A draft paper on New Technology Policy (NTP) prepared in February 1993 was designed to fulfill these objectives in the context of Industrial Policy Statement of 1991[20]. However, this NTP Statement is still to be implemented. At the operational level, a new Cabinet Committee on Science and Technology was constituted in early 1994 under the chairmanship of the Prime Minister to provide policy directions and approve integrated programmes. It includes Ministers of Human Resource

---

[20]*A Draft Paper for a New Technology Policy*. New Delhi; Department of Science and Technology, 1993. 10p.

Development(HRD), Agriculture, Finance and Defence, besides Deputy Chairman of the Planning Commission, Ministers of State for Industry, Rural Development and Science and Technology[21]. The present S&T strategies are being oriented to have a business approach for research, to create wealth from knowledge by breaking the boundaries of various disciplines and those of research and market institutions[22]. In this context, there is an urgent need to strengthen Intellectual Property Rights (IPR), which would not only protect the interests of the researchers but also attract global R&D[23].

China

In China too, "An Outline of the Report of the Development Policies for Science and Technology by the State Science and Technology Commission," (February 23, 1981) summed up the past experiences and lessons, further clarified the principles and policies for the development of science and technology for the coming years, which can be generalized as follows:

(1) Development of science and technology should be well coordinated with economic and social development;
(2) Emphasis on research to strengthen productive technology;
(3) Choosing the appropriate technologies so as to form a rational technical structure;
(4) Work of technology development and application must be strengthened;
(5) Learning, absorbing and digesting foreign technology should be strengthened.

These points were reflected in the subsequent Sixth FYP (1981-85). It pursued the policy of "Adjustment, Reform, Correction and Improvement." The Seventh FYP *(1986-1990)*

---

[21]Panel for Science and Technology. *University News*, January 17, 1994.

[22]MASHELKAR R. A.: India's Emergence as a Global R&D Platform- The new challenges and opportunities. *Current Science* 1995, 68 (12) p 1170-77.

[23]BIJLANI S. K. Indian Innovators Need Protection. *Times of India*, 1995 (June 17) p10.

covered a critical period for overall reform of the economic, scientific and technological systems. Major reforms were carried out in favour of "market economy" and liberalization in technology acquisition which began in the early 1980's. Together with the major reforms which came in the 1980's, particularly after 1985, the efforts of the post-Mao leadership have been in favour of building of the scientific and technical infrastructure in line with modernization of the industrial sector.

In the first half of 1980's, all formal R&D work and most of the product and process commercialization activities have been carried out in the public sector and directly or indirectly funded by government. In fact, state control had been all-pervasive in allocating resources to specific projects and programmes until the reforms of 1985. After initiation of the reforms in 1985, one aspect of the funding system that has received major attention is the policy of reducing direct state allocations of operating funds, thereby putting pressure on the R&D sector across the board to find alternative sources of funds through contract work, transfer fees and sale of technological know-how. For the first time, a new Patent Law was implemented in 1985; and the Technology Contract Law was promulgated in 1987.

The S&T policies in China during the 80's called for a liberal environment giving full play to market forces through establishment of 'Contract Management System'; 'Technology Markets'; establishment of 'Spin-off Companies, ; and 'Open Laboratories, for facilitating mobility of scientific personnel. A number of S&T programmes like 'SPARK Programme' (1986) for application of technologies to rural areas; '863 Programme' (1986) for basic research in high technologies; and 'TORCH Programme' (1988) for application of high technologies have been initiated. On the whole, the policies were directed to forging of better linkages between research institutes and enterprises through market orientations[9]. The market-oriented liberal approach adopted during this phase, till now, has had considerable impact on economic growth, as depicted in *Table 4*, so much so that the production of basic industrial and agricultural products like coal, cement, cotton and grains reached the highest level in the world and that of steel and electricity rose from 5th and 7th position respectively in 1978 to the 3rd position in 1993.

***Table 4* : Output of Major Industrial Products in China and Ranking in the world**

	Output (mt)			Rank		
	1950	1978	1993	1950	1978	1993
Steel	0.61	31.78	88.68	26	5	3
Coal	43	618	1141	9	3	1
Crude Oil	0.2	104.05	144	27	8	5
Electricity (b kwh)	4.6	256.6	820	25	7	3
Cement	1.4	65.24	360	8	4	1
Cotton	0.69	2.17	3.76	4	3	1
Grain	112.3	290.85	409.56	2	2	1

Note : mt - million tonnes; b kwh - billion kwh

Experience and Existing Problems

Looking at the various plan periods since independence, generally the changes are in the direction of moving from infrastructure and capability building phase to emphasis on performance and accountability of S&T activities in relation to goals of social and economic development, and to formulation of policies and structures in line with prevailing worldwide trend of liberalization and globalization. In other words, both India and China launched their science and technology activities from a policy regime closer to the inventive end of the technological capability spectrum *(Fig. 1)* and then, over the years, attempted reorientation so as to cover the entire spectrum, from invention to diffusion or production. What has been the outcome of these attempts? A few indicators of the S&T capabilities thus developed are given in *Table 5*. From the table we find that although India and China had adopted similar approaches of initially concentrating on the invention end to establish S&T capabilities as an essential infrastructure and then over the years moved closer to the diffusion or production end, the resulting capabilities in India and China show interesting differences. From the data presented, the following key contrastive features are evident :

TABLE 5 : Broad Indicators of Science and Technology in India and China

Indicator	India	China
1. Total S&T manpower (including technicians)	6.31m (1996)	19.13m (1996)
2. Total Scientists and Engineers,	3.8m (1990)	1.48m (1993)
per million of population	4500	1529
3. R&D Personnel:	314,489 (1994)	917,000 (1993)
per million of population	370	774
4. No. of Institutions of Higher Learning	8421 (1994)	1065 (1993)
5. Total Enrolment	5.007m (1994)	2.065m (1990)
6. Total No. of Teachers	0.286m (1994)	0.4m(1990)
7. Student/Teacher Ratio	17.5:1	5:1
8. Enrolment in Post graduation & Research	531000 (1994)	107000 (1993)
Total No. of		
Postgraduates	476,000 (1994)	89000 (1993)
PhD's	55,000 (1994)	18,000 (1993)
9. Total Expenditure on R&D (billion US $)	2.43 (1994)	2.70 (1994)
As % of GNP	0.81%	0.5%
In it share of		
Government	73.5%	70%
Public Industry	10.1%	30%
Private Industry	16.4%	–
10. Expenditure on R&D		
per scientist (1990)	US$ 614	US$ 1873
per capita	US$ 2.19	US$ 2.54
11. Patents (1989-93)		
Filed	18313	268825
Granted	8075	157935
Indigenous	2142	141155
Foreign	5933	16780
12. Publications		
All scientific Research Papers		
Total (1989-93)	48661	27172
Share in World Output	1.87%	1.04%
Relative Citation Rate	0.55%	0.54%
13. Per Capita GNP (1993)	US$ 300	US$ 490

Sources: See reference 36.

(a) India has developed a much stronger research base than China. For example, India has three times more scientists and engineers per million population than China. India has three times as many PhDs and twice the research publications compared to China. Its expenditure on R&D as percentage of GNP is more than that of China.

(b) Compared to India, the Chinese effort of moving closer to the diffusion or production end has been more successful. For example, China filed about fifteen times more patents than India during the period 1989-93. Another indication of China's R&D being closer to the production end comes through the pattern of funding. The industry's contribution to R&D funds in China being 30 per cent compared to 26.5 per cent in India.

(c) Whereas the Indian S&T capabilities are extensive, the Chinese tend to intensify these. For example, they have three times more teachers per student compared to India and their R&D expenditure per scientist too is three times higher than in India.

Till the mid-seventies, China did not have technology policy instruments for providing economic incentives to individuals and for protection of patents. There was no provision for ownership of an invention by a private individual. The invention was owned by the 'entity,' or 'unit'[24]. On the other hand, in India, although policy instruments and patent laws existed during the same period, these, to a large extent, restricted the import of technologies[25-27].

In late seventies, China promoted liberal imports of technologies and foreign investments in contrast to Indian

---

[24]KHARBANDA V. P.: China's Technology Policy and Modernization. *China Report* 1986, 12(1), p. 17-27.

[25]The Industries (Development and Regulation) Act, 1951. 31st October 1951.

[26]The Patent Act, 1970. 19th September, 1970. *Gazette of India*, Part II, 21st Sept. 1990 p. 421-493.

[27]Guidelines for Administering Section 29 of Foreign Exchange Regulation Act 1973, 20th December 1973.

situation, which continued to be selective and restricted till very recently, when a liberalized industrial policy was announced in 1991. A comparison of the provisions of Policy Instruments for Import of Technology in India and China is made in *Table 6*, which clearly indicates that China is able to provide a more liberal environment than India.

Technology absorption and adaptation in both the countries has been at a low level and very little attention has been paid to the R&D activities pertaining to either incremental or radical innovations. There are three reasons for this. *First*, Investments for industrial R&D as percentage of sales turnover are extremely low as compared to advanced countries. These are only 0.82 per cent in India and 1 - 2 per cent in China. In Japan, USA and Germany, it is more than 7 per cent. *Second*, industries and research institutions do not work in consonance in efforts to absorb and update imported technologies. Technology transfer, whether from indigenous source or from outside has not been a smooth sail in both the countries. There in very little investment in reverse engineering or R&D to update the technologies. Various administrative or financial problems come in the way, which need to be solved by coordination of activities in industries and research institutes[28]. Only 5-10 per cent of the research results go to industry in both the countries. *Third*, the resources invested in industry are too scattered, particularly in China. For example, although one factory in Beijing has the capacity to make enough carburetors to fit 300,000 cars, China has 32 more factories producing the same components. Automobile industry in China has 4000 units producing vehicle parts. According to People's Construction Bank of China, if investments are concentrated only in 49 enterprises, it would help improve the quality as well as indigenization of the product[29].

---

[28]JAIN Ashok, KHARBANDA V. P., QIU Chengben and HUANG Ailan: *Industrialization in India and China*. New Delhi; Har Anand, 1995. (In English and Chinese).

[29]Planning Key for auto parts sector. *China Daily*, 7 February, 1991.p.4.

The main underlying reason for the contrast between China and India and the developed countries is the lack of fund input directly affecting economic, scientific and technological developments in these countries. A comparison of the expenditure on R&D in India and China is made in *Tables 7 and 8.* In developed countries, the proportion of research and development funds to GNP in the 1980s was 2 - 4 per cent. However, in China, this proportion has fallen from 1.56 per cent in 1979 to 0.50 per cent in 1994. In India, it has fallen from 0.96 per cent in 1988 to 0.81 per cent in 1994. The intermediate of the transfer from research to production has also been too weak, which keeps a large number of research achievements in laboratory rather than tapping their economic efficiency.

Table 6 : Comparison of provisions of Policy Instruments for Import of Technology in China and India

Instrument	China	India
1. General policy	Since liberation in 1949, technology import policy has shifted in favour of imports or no imports in consonance with economic policy during different periods. After 1979, highly liberalized.	Since independence in 1947, has adopted a continuous and sustained policy of selective technology imports.
2. Forms of technology transfer	– License Trade – Joint Venture – Cooperative Production – Wholly owned enterprises.	– License Trade – Joint Venture – Wholly owned enterprises (only for 100% exports).
3. Royality for technology acquisition	Not more than 8% (taxable, but may be exempted in case of advanced technologies)	Not more than 5% (taxable) In case of royality+ lumpsum payments, it is not more than 8%.
Period of contract	Not more than 10 years.	Not more than 8 years
4. Foreign equity Participation	Not less than 25%	Not more than 40% (51% according to new industrial policy, 1991 in high priority industries),

			exceptions are con sidered on merit in case of NRI's 74%.
		Allow 100% foreign equity but all or majority of products to be exported.	Allow 100% equity only in case of 100% exports.
	Terms of the joint venture projects	10-30 years (May be more for specific projects).	Not more than 8 years.
5.	Exemption from income tax for joint ventures	For first two to three profit making years. For next third to fifth year, 50% exemption	For first five years on profit.
6.	Approving authority	Highly decentralised. Apart from Ministry of Foreign Economic Relations and Trade (MFERT), various departments, commissions and administrative organs of provinces, autonomous regions and municipalities, etc. In case of joint ventures, applications are to be disposed of within 90 days. In case of technology import, contract applications must be disposed of within 60 days by the approving authority. If not, it is considered to be approved.	Highly centralised. Only Secretariat for Industrial Approval (SIA) and relevant ministries authorised to apporve. Applications to be disposed of within 60 days.
7.	Absorption and adaptation of imported technology.	Indigenise some percentage of imported technology every year, with complete indigenization in 4 - 5 years.	Obligatory on enterprises whose technology payments are more than Rs. 20 million, to put up R&D facilities for indigenization

As the absorption of imported technologies in both the countries has not been very effective, this has led to the problem of repetitive imports and continued technological gap between them and the advanced countries. Indigenous R&D in India has been utilised largely by small and medium scale companies[30]. In China too, indigenous technologies have been mainly adopted by small and medium scale enterprises[31]. While there are sufficient policy instruments to encourage indigenous technology development and transfer in both the countries, the problem remains acute. It will require not only structural changes in the R&D organizations[32] but, more importantly, an attitudinal change among actors involved in technology generation and exploitation. While both the countries have competence to do so, lack of general consensus and continuous political commitment may interfere with the present ongoing reforms in both[33]. Particularly in China, political interference has played a crucial role in accelerating or retarding the building up of endogenous scientific and technological capabilities. Thus, the success of the reforms and restructuring of the R&D institutions in line with socioeconomic needs basically depends on integration of S&T strategies with national economic policies supported by the stability of the political system. This will remain a crucial issue confronting the two nations in the realization of their aspirations of achieving scientific and technological self-reliance in the twenty first century.

---

[30]QURESHI M. A. et al: *The role of financial institutions and commercial banks in Science and Technology development in India. A case of commercialization of the CSIR know-how.* New Delhi; NISTADS, 1987. (Preliminary Report).

[31]LUO WEI: *Development and Reform-Ten years of reform in China's science and technology system.* Paper presented at symposium on Scientific Temper and National Development, 6-8, March, 1989, New Delhi.

[32]GU Shulin: *Reform Policy for S&T System in China.* Paper presented at the Joint Seminar between the Institute of Economic Growth(IEG) University of Delhi, and the United Nations University Institute for New Technologies. (UNU/TECH), Maastricht,Netherlands; 1995, (April 3) 61p.

[33]DEEPAK LAL: India and China: Contrasts in Economic Liberalization? *World Development,* 1995, 23(9) p 1475-1494.

In order to achieve this, Science, Technology and Economy have to interact in a definite way. According to Niwa Fujio[34] invention, innovation and diffusion are interconnected with feed backs and loops (Fig. 5). In other words, scientific infrastructure, technological innovation and manufacturing production are dynamically interconnected. This requires several structures, institutions and actors which together constitute 'Techno-Economic Networks' (TENS)[35]. These 'TENS' are "Coordinated set of heterogenous actors, laboratories, technical research centres, financial organisations, users, and public authorities which participate actively in the development and diffusion of innovations, and which via numerous interactions organize relationships between scientific-technical research and market place". These Networks interconnect diverse skills and expertize into communities to innovate, produce and use distinctive product and process technologies. Networks are composed of both internal (e.g. within an organization) and external (e.g. among organizations) linkages. Different national settings, cultures and institutions, as well as different public policies make real difference in building and sustaining innovation networks. In continuous innovation of complex technologies the organization of such 'Techno Economic Networks' between diverse organisations both public and private has become important.

[34]NIWA F, TOMIZAWAH and HIRAHARA F (1993): Recommendation of the Japanese Science and Technology indicators. the Proceedings of NISTEP the Third International Conference on Science and Technology Policy Research, 23-24 Mita Press, Tokyo.

[35]JAIN Ashok : Innovation Networks in Social System for Development. Paper presented at the ASCA Seminar on Science, Technology and Regional Innovation, Ohmiya City, Japan, March 16-18, 1994

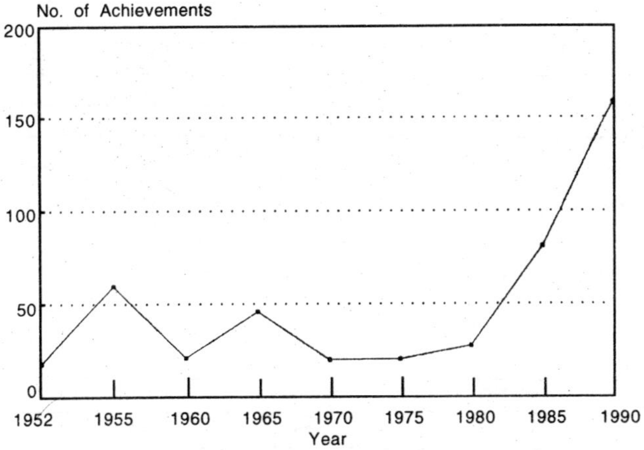

Fig. 4 : Major S&T Achievements in China
(excluding patents)

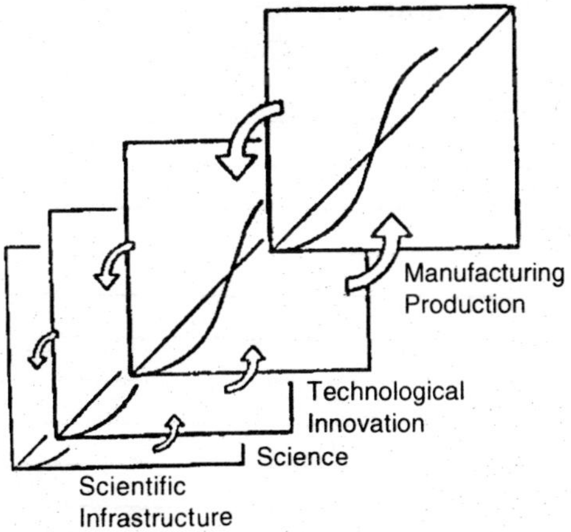

Fig. 5 : Dynamics of the Science & Technology/Development Cycle.

Table 7: Total Expenditure on R&D in India and China

(billions)

	India (Rupees)	US $	China (Yuan)	US $
1958	0.229	—	0.298	—
1965	0.684	—	2.717	—
1970	1.396	—	2.996	—
1975	3.567	—	3.764	—
1980	7.605	0.89	5.332	3.58
1985	20.687	1.38	8.383	2.85
1990	39.257	2.25	12.543	2.65
1991	45.128	2.50	14.230	2.85
1992	50.046	1.99	16.9	3.13
1993	60.730	1.94	19.6	3.06
1994	68.210	2.16	22.2	2.6

(**Source:** Research and Development Statistics, 1994-95. New Delhi; DST. 1997. Data for China compiled by the authors from various sources)

Table 8: Expenditure on R&D as % of GNP in India and China

	India	China
1978	–	1.5
1979	–	1.6
1980	0.62	1.5
1981	–	1.3
1982	–	1.3
1983	–	1.4
1984	–	1.4
1985	0.89	–
1987	–	0.65
1988	0.96	0.64
1989	0.93	0.70
1990	0.85	0.71
1991	0.84	0.72
1992	0.83	0.70
1993	0.86	0.62
1994	0.81	0.50

(**Source :** Research and Development Statistics, 1994-95. New Delhi; DST, 1997. Data for China compiled by the authors from various sources.)

Conclusions

In the new scenario of increased globalization of trade, finances and production system, with the market locale playing a dominant role, it is important to stress the new role of the government in planning and promoting integration of S&T with economic development. The following broad conclusions can be drawn:

(i) Generation of S&T manpower remains a priority. However, careful thought needs to be given to a proper mix between highly qualified scientists, engineers and well trained technicians and workers.

(ii) Initiative has to be taken for preparing a vision for techno-economic development involving all the stake holders. The job cannot be entrusted to either the scientists or the economists. Extensive surveys have proved very effective in generating a large-based consensus on the direction of techno-economic change. This approach has several advantages over the conventional approach of white papers being prepared by 'expert' committees.

(iii) 'Learning by doing, and factory as laboratory, are key concepts. Incremental innovations and improvement in efficiencies of production are all legitimate targets for R&D. The community of scientists, technologists and decision makers has to give high recognition to the attainment of these targets rather than applauding only the major breakthroughs in science or technology.

(iv) Articulating the purposes of R&D, i.e. for diffusion, innovation or invention, is a better strategy for resource allocation than going by the conventional categories of basic, applied or development research (S-T-E). It is only by committing to a particular purpose that one can plan, organize and manage an R&D programme. This calls for a strategy advocating (E-T-S) approach.

(v) Importation of technology must be coupled with an immense local effort for its absorption and adaptation. This leads to technological capabilities at several levels, for example:

- Choice of technology
- Procurement and selection of equipment
- Maintenance and repair
- Material substitution

- Minor improvements, and
- Major improvements

In this concern, the oft-quoted phrase of 'repetitive, imports is misleading. Import devoid of any of the above capability building elements can be wasteful. On the other hand, if the first time import leads to building up of these capabilities then the second import will definitely start with a higher base of Indigenous Technological Capability. In other words, technological capability building is to be viewed as a cumulative exercise and the investments in import of technology as a price to be paid for accumulated learning rather than only for providing goods and services.

(vi) In the changing global scenario, the term science and technology policy is getting replaced by innovation policy.' The aim of innovation policy becomes establishment and strengthening of 'Techno-Economic Network' rather than supporting science and technology activities per se.

(vii) There is a need to strengthen development research greatly, particularly in technical creation and diffusion, so as to promote the harmonious and consistent development of economy and society in a down-to-earth manner. In this context, policies of research funds must be adjusted. The single source structure of research funds from government allocation must be changed to the double source, i.e. from both government and the enterprises. The enterprises must provide more than 60 per cent of the total input funds for R&D, and increase the input of R&D funds to more than 1 per cent of the GNP.

(*viii*) Rules of competition must be reformed and *people who have made outstanding contribution to the development of production technology must be well rewarded.* Social and economic efficiency or the actual achievement itself should be taken as the main criterion of evaluation.

(*ix*) The priority of the present time is to *transform the traditional industries through the application of new high technologies.* Agriculture in India and China is still dominated by manual labour, with low degree of mechanization. In industry, there are islands of modernity in backwardness.

Therefore, the transformation of our traditional industries by way of new high technologies ought to be taken as an important part of the strategy and policy of our country's scientific and technological development.

[36]State Statistical Bureau of the People's Republic China : Statistical Communique on National Economic and Social Development in 1991. *China Daily* (March 8-19) 1992. p.4; More Funds for Basic Research. *China Daily* (Sept.5) 1991. p.3; Cai Yan: Putting Science in Perspective. *China Daily* (Aug.15) 1991. p.4; State Council Research Office: Bridging the Economic Gap. *Beijing Review*, 1989 (Jan 30-Feb 5) p.16-23; Huang Wei: Burgeoning Patent Industry in China. *Beijing Review*, 1990 (Dec. 3-9) p.15-16; Zhang Lin: Degree Holders Increase. *China Daily* (March 6) 1990. p.3; Zhang Lin: Education Generation Gap Closes. *China Daily* (Feb. 1) 1991. p.4; China Develops Education for the 21st Century. *China Daily* (Feb. 1) 1991. p.4; *Research and Development Statistics* 1994-95. New Delhi; Department of Science & Technology, 1997. p. 196; World Development Report 1995. Washington; World Bank, 1995. 225p. China's S&T Expenditure Statistics, China Science and Technology Newsletter, State Science and Technology Commission(SSTC) (Oct. 15 1993, Nov. 25,1994) p.4; China's Science and Technology Indicators 1992. Beijing; SSTC, 1992. (In Chinese) For S&T Manpower & Industrial contribution to R&D. XIAO Wu: Technology Work Yield Success. *China Daily Bussiness Weekly* 1996 (June 16-22, 1996, p8. *Scientiometrics* 1995, Vol. 33 (3).

PART III : GLOBALIZATION, S&T REFORMS AND CHALLENGES TO R&D

4. Globalization and Challenges to R&D in Indian Industry

V.V. Subba Rao

Abstract

Indian industry over the years has grown substantially covering the manufacture of a host of items ranging from common consumer goods to hi-tech products both for domestic needs and exports. It has also set up a sound R&D base. Under the changed scenario, technology imports and foreign investments have been discussed. The liberalised policies have posed a challenge and an opportunity to Indian industry and R&D system to become globally competitive. Towards achieving this, government has provided several support measures and fiscal incentives to the industry. Several government departments have launched major programmes for supporting competitive technology development. Strategies for making Indian industry technologically competitive include strengthening government-R&D-industry cooperation, increasing industrial investments in R&D, human resources development, improved information systems, and providing greater technological support to small scale industries. Proper planning and efficient management of technology development and R&D needs attention.

Introduction

Industry is the backbone of developing economies. Indian industry, practically non-existent at the time of independence, has now grown substantially covering the manufacture of a host of items ranging from common consumer products to hi-tech products, for both domestic needs and export. However, the Indian industry is largely based on the inputs it had received in the past by way of foreign investments, technical collaborations, import of industrial machinery, capital goods and other items. It is estimated that nearly 90 per cent of the industrial production is based on technologies imported and/or indigenised and the remaining is based on technologies generated within the country. Investments in the manufacturing industry alone are estimated to be over Rs. 1000 billion by the end of 1991-92. Industrial production is estimated to be of the order of Rs. 6000 billion during 1993-94. Of this, it is estimated that small scale industries contributed to 40-45 per cent, with the remaining coming from medium and large scale industrial units.

With a view to making Indian industry globally competitive, in 1991 government announced several policy changes in the fiscal, industrial and trade policies. The new policies of the government have direct and indirect bearing on indigenous technology development and utilization of domestic technological capabilities by the industry. It is important to note that indigenous technologies are available generally on relatively lower spectrum of investments and industrial R&D is still not strong enough to meet the emerging needs. It is also important to note that although the licensing policy has been liberalized and the industry is free to import technologies in a number of areas, foreign technologies are very expensive and also the latest technologies are not generally available for import; what is normally available is relatively older technologies. This paper presents the status of R&D in Indian industry, changed scenario and issues of relevance to industry and R&D in view of globalisation, foreign technology imports and foreign investments in the pre-and post-liberalisation periods and suggests initiatives for making Indian industry technologically competitive to meet the challenge.

Research and Development in Indian Industry

Government has taken several measures towards promoting industrial research in industry itself and to establish workable linkages between the national laboratory system, educational institutions and the industry. Several incentives have been provided which encourage and make it financially attractive for private, joint and public sector industrial units to establish their own in-house R&D centres. A scheme for granting recognition to in-house R&D units in industry was initiated in 1973. Department of Scientific and Industrial Research (DSIR) is the nodal administrative department for Research and Development by Industry.

There are 1240 in-house R&D units in industry duly recognised by DSIR. A broad sector-wise break-up of the in-house R&D units is as under

Chemical and Allied Industries	400
Electrical and Electronics Industries	350
Mechanical Engineering Industries	250
Processing Industries including Metallurgical, Refractories, Ceramics, Cement, Textile, Paper, Leather and others	160
Agro Industries	80

The R&D expenditure incurred by these 1240 in-house R&D units is estimated to be of the order of Rs. 14 billion (1994-95). The share of public sector is 45 per cent and that of private sector, 55 per cent. The R&D expenditure incurred by around 300 small scale units is estimated to be around Rs. 500 million. R&D expenditure incurred by the industry as percentage of their annual turnover as a whole is 0.57 per cent; the ratio for the private sector is 0.64 per cent and for the public sector it is 0.51 per cent.

The present estimated manpower employed by 1240 recognised in-house R&D units is around 50,000, of which around 17,500 R&D personnel were employed in the public sector in-house R&D units and around 32,500 in private sector in-house R&D units. Of the total 50,000 R&D personnel, there are around 2600 PhDs, 8400 post graduates, 17,000 graduates and 22,000 other qualified personnel.

The R&D centres in industry have made a number of significant achievements, including process development, product development, technology absorption, cost reduction and commercialisation of indigenous R&D. Value of goods produced based on in-house R&D efforts is estimated to be Rs. 350 billion per year with an average commercialisation factor of 1:25 (ratio of R&D expenditure to goods produced based on R&D) . During the past four years, the in-house R&D centres have also filed around 700 patents. These achievements indicate that R&D in Indian industry is mature and can take challenges.

Changed Scenario and Issues of Relevance

The liberalized policies of the government announced in 1991 provided *inter-alia* greater freedom to industry to select technologies of their choice and encouragement to foreign investments. As a consequence, technology imports in a number of areas have increased in the past four years, as also foreign equity participation. Some of the issues of relevance arising out of the new liberalised policies are:

* Licensing liberalization
* Removal of investment limits for large companies
* Automatic foreign investments
* Reduced import duties
* Changed status of small scale industries
* Energy and environment considerations
* Increased quality standards
* Growth of infrastructure and service sector
* Need for competitive technologies
* Information services
* Domestic competition
* Export competitiveness
* International policy changes
* HRD and efficient management systems

The past four years have seen significant changes in international policies and programmes which have a bearing on domestic technological capabilities. These policies include: initiatives of the United Nations System with regard to General Agreement on Trade and Tariffs (GATT), environmental concerns, including the Montreal Protocol and restrictions on export of technologies, products and services from some of the industrially advanced countries. The new GATT agreement regarding Intellectual Property Rights (IPR) implies modifications in the Indian Patent System. Innovative technology development assumes greater importance. The General Agreement on Trade in Services (GATS) also calls for greater technological inputs in the services sector, including consulting engineering services, to take advantage of the export opportunities which would become available. The global environmental concern places greater emphasis on the need for development and application of cleaner technologies to minimize pollution and conserve natural resources. Several instances of restrictions on technology export by advanced countries due to political and other considerations also invite greater thrust on strengthening domestic technological capabilities in crucial and sensitive areas and increased utilization of locally available R&D options by the industry.

Technology Imports and Foreign Investments

Government policies in regard to building up of technological capabilities conceived a mix of both domestic know-how and foreign technological assistance. Towards this, the government has carefully regulated the inflow of technologies and financial participation in industrial development for about four decades. Fig. 1 presents the year-wise foreign technology collaborations (FC) approved, including foreign collaborations associated with foreign equity participation. As can be seen, the foreign collaborations approved were in the range 130-520 per year during the period 1961-1980. However, from 1981 onwards, there was a steady increase to as high as about 1050 collaborations during 1985; the next three years showed a downward trend with a steep fall during the years 1989 and 1990, a period of import restrictions. However, following the liberalization in 1991, there has been a substantial increase in the number of collaborations, reaching 2337 in 1995. The number of collaborations associated with foreign equity, which remained within 300 per year upto 1990 increased steeply from 1991 onwards with 1335 during 1995. It is seen that foreign investments are generally associated with technology imports.

Fig. 2 presents the foreign investment approvals during the period 1981-1995. As can be seen, the foreign investments during the decade 1981-1990 remained within Rs. 3 billion per year. It increased to Rs. 5.29 billion during 1991 and thereafter steeply increased several fold to Rs. 326.13 billion during 1995.

Sector-wise distribution of technology imports during the period 1981-90 and the following four years, 1991, 1992, 1993 and 1994, indicates that during the period 1981-90, on an average, electrical and electronics sector accounted for 23.8 per cent of the total FC approvals followed by 18.4 per cent in industrial machinery, 11.9 per cent in mechanical engineering industries and 11.7 per cent in chemical industries followed by other sectors. During the post-liberalization period, the number of collaborations in the area of electrical and electronics industries is in the range of 14-17 per cent of the total FC approvals, mechanical engineering industries in the range 8-11 per cent, and chemical industries in the range 14-16 per cent. An interesting feature to be observed is that while 'Other sectors' category remained only 22per cent during the decade 1981-90, it has increased to almost 45 per cent during the post -liberalization period. This category covers a host of areas, such as consultancy, management and financial services, tourism, garments, food items, leather items, granite, mineral water, aqua-culture and sea food. Service sector is attracting more collaborations now than before.

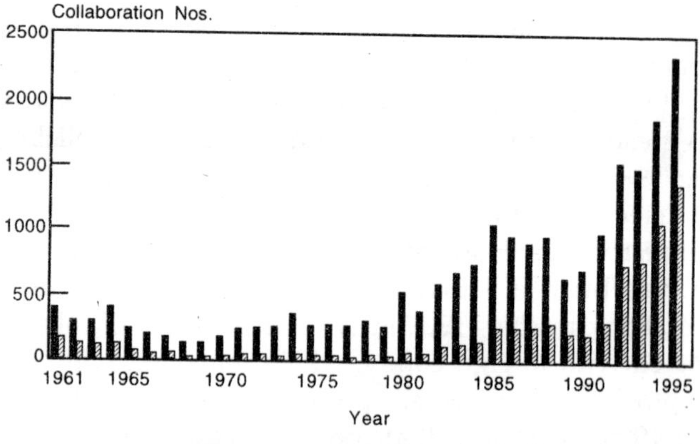

Fig. 1 : Foreign Collaboration Approvals (1961 - 1995)

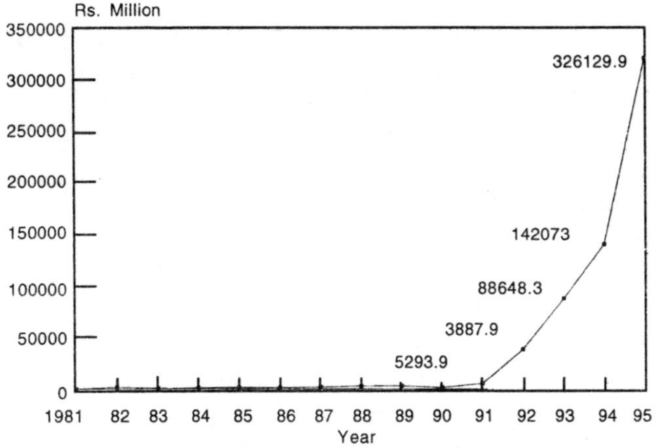

Fig. 2 : Foreign Investment Approvals
(1981 - 95)

Another noteworthy feature in the post-liberalization period is that more and more foreign technicians and consultants are involved in the process of technology transfer. Some of the multinational companies have shown interest in setting up 100 per cent owned R&D companies in the country in selected areas. The number of export-oriented companies also has shown an increasing trend. It may be further observed that some of the existing companies with foreign equity participation have increased the level of such participation.

Initiatives Towards Making Indian Industry Technologically Competitive to Meet the Challenge

The new policies of the government, since the middle of 1991, are mainly aimed at attracting greater foreign investments, improving the quality of infrastructure and products, increasing competitiveness in industry and boosting our exports in the global markets. These call for a higher degree of technical, technological and professional inputs and greater investments for strengthening domestic technological capabilities and greater commitment for R&D by industry.

Support Measures and Incentives

The Department of Scientific and Industrial Research (DSIR) has played an important role towards enhancing domestic technological capabilities. Technology status reports, technology norms reports, technology profiles of several countries, consultancy capabilities and in-house R&D capabilities reports numbering over 200 were brought out for the first time in the country. The Department has also nurtured the growth of in-house R&D system in the country and has facilitated provision of several fiscal and other incentives for industry and others towards enhancing domestic technological capabilities. DSIR has also started direct funding of R&D projects undertaken by the industry jointly with national R&D institutions. The annual allocations are to the tune of Rs. 100 million and over 100 projects are already funded on co-financing basis. National information centres aimed towards providing information services for technological advancements have been set up by DSIR in various areas, such as chemicals, drugs and pharmaceuticals, textiles, leather, food, etc. National Research Development Corporation (NRDC), a public sector enterprise under DSIR, has adopted a new approach towards development, commercialisation and export of domestic technologies.

The Council of Scientific and Industrial Research (CSIR) has reoriented its priorities and programmes towards undertaking innovative research and maximizing support to the industrial base in the country. Linking CSIR research to market needs, mobilizing and optimizing resource base and investing in high quality science that will be the harbinger of future technologies have been identified on the road map of new CSIR. CSIR has also entered into MOUs with industry associations and financial institutions for forging strategic alliances.

The Department of Science and Technology (DST) has evolved a number of new initiatives to support futuristic R&D for technological advancements in frontier areas, such as advanced materials, robotics and others. Technology Information, Forecasting and Assessment Council (TIFAC) has established institutional mechanisms and information networks for promotion and development of technologies. TIFAC has also established and has designed a computerised interactive on-line nationally accessible technology information system (TIFACLINE). It has also brought out a number of technology status reports in major sectors of industry and provided assistance for projects aimed at encouraging commercialisable home-grown technologies. The Department of Biotechnology has supported a number of programmes in industry relating to biotechnology. It has also supported the creation of a vast infrastructure for biotech research and human resource development. The Department of Electronics has also started direct funding of R&D projects in industry.

The Income Tax Act has, from time to time, provided fiscal incentives for encouraging investments in R&D by way of tax concessions, accelerated depreciation allowance for plant and machinery set up based on indigenous technologies and weighted tax deduction for sponsored research. The Drug (Prices) Control Order in force also envisages encouragement to development of domestic technologies for manufacture of bulk drugs from the basic stage by providing exemption from price control for a period of 5 years. These incentives are administered through DSIR.

Government-R&D-Industry Cooperation

Effective interactions and stronger linkages among various agencies, such as industry, universities and IITs, R&D institutions, financial institutions, consultancy organisations and government need to be nurtured and strengthened for maximising the output with optimum inputs. There is considerable opportunity for enhancing the existing channel for communication as well as evolving new mechanisms. There are good examples of such linkages in the past and experience from this needs to be diffused further to provide a monolithic base of S&T infrastructure to meet the challenging needs of the industry.

Increased Investments in R&D

Larger investments in R&D are important to improve the R&D infrastructure and make faster technological advancement. Investments in R&D at present are low and should be increased substantially if the industry has to become technologically sound and globally competitive. Presently, the share of industry in the national R&D expenditure is 26 per cent. This should increase to at least 50 per cent by the turn of the century. This can be achieved through voluntary measures by the industry on their own and through some promotional measures such as increased tax concessions or through suitable legislations if need be.

Human Resource Development

The quality and skills of S&T personnel need to be improved on a sustained basis to keep pace with the latest developments. There should be greater opportunities for scientists and technologists to participate in national and international S&T fora. Also, greater motivation and encouragement need to be provided to scientists and technologists. This can be achieved by offering better career prospects and national recognition coupled with accountability.

The skills of the technical support staff in industry should be upgraded in line with the technological requirements. Towards this, the ITIs and other technical training centres need to revise their syllabi, equip themselves with necessary modern facilities, etc. The faculty also need to be exposed to technological developments and modern teaching methods to enable them to improve the quality of training. This equally applies to the large number of polytechnics and engineering colleges.

Information Systems

Several organisations and departments have created S&T information networks such as National Information System for Science and Technology (NISSAT) , National Informatics Centre (NIC) , Technology Information, Forecasting and Assessment Council (TIFAC) , BISNET of Federation of Indian Chambers of Commerce and Industry, Information Centre on Patents under the Department of Industrial Development, etc. For the industry to have prompt access to the databases and latest technical and technological information, it is important to have a centralised network connecting all the databases in the country. Towards this, TIFACLINE has been set up by the Department of Science and Technology. This needs greater support to play a major role towards meeting the information needs of the industry.

Technological Support to Small Scale Industries

The existing institutional mechanisms, such as Test and Development Centres, Prototype Testing and Development Centres, and Small Industries Service Institutes, need to be modernised by way of imparting advanced training to the manpower as well as upgrading the facilities to assist the small scale industries in their technological upgradation. Creating common R&D facilities, common test centres and common library facilities for use by the small scale industries will also facilitate their technological upgradation. Consortium approach between small scale industries and national laboratories or technical universities should be encouraged. These support facilities should be easily accessible at affordable cost to the small scale industries.

Management of Technology Development and R&D

Proper planning and efficient management of technology development and R&D are vital for the success of any enterprise in the competitive environment. The following points deserve utmost consideration for making the industry technologically competitive:

* Prioritisation of technology development programmes
* Strengthen and support engineering and product design capabilities
* Greater emphasis on environmental concerns and clean technologies.
* Emphasis on energy efficient technologies
* Emphasis on quality and ISO: 9000 series certifications.
* Provide rightful place to R&D in corporate planning.
* Commitment to competitive technology development.

Conclusions

The changed policies of the government have provided an opportunity and a challenge to the Indian industry to compete in both domestic and global markets. The R&D scenario in Indian industry appears to be promising and the industry can take up the challenges. However, investments in R&D by industry should increase substantially and R&D should become an integral part of corporate planning. While the government will provide catalytic support to R&D and competitive technology development, a major role in the challenge of globalisation has to be played by the Indian industry.

5. The Crux of Deepening Reform of Chinese Scientific and Technical System: Strengthening the Links of Technical Innovation

Chen Wenhua

Abstract

Looking back at the ongoing reforms of Chinese scientific system, this article seeks to highlight the reasons for reforming the scientific system. Points out that the purpose of reform is to liberate and develop the scientific and technological productive forces, and develop socialist economy at a fast pace. The core objective of the reform is to promote the linking of science, technology and economy in an organic way to suit the needs of socialist market economy. It is pointed out that technical innovation is the medium of science, technology and economy. The characteristics of such a Technical Innovation Medium have been described. On this premise, the crux to deepen the scientific and technological system is to strengthen the new links of technical innovation to meet international competition.

Introduction

For more than 40 years, scientific undertakings in the new China have made great achievements of international significance. These have strongly promoted development of the country's society and economy. They have also markedly enhanced its international standing. The rigid system of science and technology, however, did not link technology with economy. This problem has been a matter of concern for China for a long time. After the passing of "Decision on Scientific and Technological Reform by the Communist Party of China" on 13 March 1985, China has been carrying out all-round reform of scientific and technological system.

Reform of the S&T System

For over 10 years, the guiding principles in China have been that "science and technology is the first productive force"; and "economic construction must depend on science and technology and the work of science and technology must serve the economic construction." As a result, the country has made considerable achievement in repect of adjustment of structure, talent distribution, promotion of the organic linkages of science, technology and economy, establishment of the new operating mechanism and so on, through systematic scientific and technological reforms. Brief details of the reforms made in the science and technology system are given below.

S&T Structure and Personnel Distribution

By implementing the policy of "Secure one part and give free rein to the rest," we gradually reduced the number of scientific research institutions supported through state finance, and promoted the mobility of scientific personnel. "Secure one part" means to establish a group of crack, high level research workers for carrying out fundamental research, applied research in long-term national interest, hi-tech research and research in main areas of social sciences of public interest. These researchers are mainly supported by the State. The institutions also follow the policy of "opening, mobility, competition and co-operation" to strengthen the vigour of research work. "Give free rein to the rest" means bringing market mechanism into full play under the State policy guidance. Based on the principle of self-support, most of the research institutes adopt the policy of engaging at least 80 per cent of technical power to serve the cause of economic construction, or merge with large or middle size enterprise group; or carry out the policy of "one institute, two systems" to change themselves into scientific enterprises, or establish new enterprise groups of various forms.

Technical service enterprises provide consultation and information and form the Third Industrial Field. They make available various forms of profit-giving services. Through structural adjustment, for promoting exploitation of scientific and technological research, the Center of Enterprise and Socialized

Service has been established. It helps in smooth development of technical enterprises and contributes to economic construction. For example, Shanghai Technology Physics Research Center of Chinese Academy of Sciences began to work under the principle "one institute, two systems" in 1986 and established "Shanghai De Fu Light and Electricity Technology Company." It turns out, more than 30 postgraduates every year and at the same time some important research achievements have won national awards. Each year, the institute gets municipal awards for about 10 achivements and applies for patents for about 10 innovations. As a company, it has made contracts with 12 enterprises. Its sales amount to 213 million yuan per year and makes profits of about 10 million yuan. Gross income is over 50 million yuan. The research institute and the company mutually support each other and thus develop in an organic way. For creating modern institutes system, they thoroughly exploit new ideas, organizational structure and operational mechanism.

Plans and Programmes

Since mid 1980's, China has formulated and implemented a series of scientific and technical plans for furthering the cause of both technology and economy. These include the "Spark Plan," "Prairie Fire Plan," "Bumper Harvest Plan," and "Technology Helping Poverty Plan" aimed at promoting agriculture and country's economy. Over 100 technical areas have been covered under the Spark Plan. Grain production is increasing year by year. (In 1994 the total grain production in China reached over 4,500 billion kilograms, which was 709 billion kilograms more than in 1985). Town and township enterprises are developing vigorously. "Scientific and Technological Key-Task Plan" was formulated and implemented to guide the personnel in institutes and universities to do research in significant scientific and technical fields of overall importance to the State economy. In the period of the Seventh Five Year Plan (1986 - 1990), 10462 enterprises earned profits, 1339 new experimental production lines were established, 872 industrial pilot plant bases and 2513 county experimental bases were established. The direct economic profits amounted to 406.8 billion yuan.

China also formulated and implemented the "hi-tech development plans" (863 plan) and "The Torch Plan" which helped the country to achieve progress in hi-tech areas. The State has set 287 state torch enterprises and established 52 state hi-tech industrial areas. In 1992, industrial output of these areas was valued at 359.9 billion yuan, total technical industrial-trade income was 346.2 billion yuan, the interest tax was 54.4 billion yuan and earned 53.6 billion yuan in foreign exchange.

Achievements

Since 1985, in China, over 10 thousand significant technical achievements have been getting provincial and departmental recognition every year and their number is increasing year by year. In 1993, they numbered 33,000, which is three times that in 1985 (i.e. 10476). Technical finance is increasing year by year. Its source and organization have become more diverse. In 1992, the income of technology-exploiting type research institutes, (state finance supported enterprises) was only 22 per cent of the total. Capital was 84.4 per cent more than that at the beginning of reform (in 1985). In 1994, in universities (enterprises-supported) entrusted funds were 36.2 per cent of the total research funds, and were 74 per cent more than in 1991.

Strengthening Technical Innovation System

The purpose of reforming scientific and technological system is to liberate and develop scientific and technological productive forces, and promote the development of economy and society. Its objective is to meet the needs of socialist market economy and to establish a new system capable of contributing to scientific and technological progress.

Under the planned economic system, there exists a barrier or a gully among science, technology and economy. Their linkage is realized indirectly by the government and government actions(see Fig. 1). In fact, the aim of reform of scientific and technological system is "to remove the barrier and build a bridge," "to fill up the gully," and promote the organic linkages of science, technology and economy, and gradually integrate science and technology with economy.

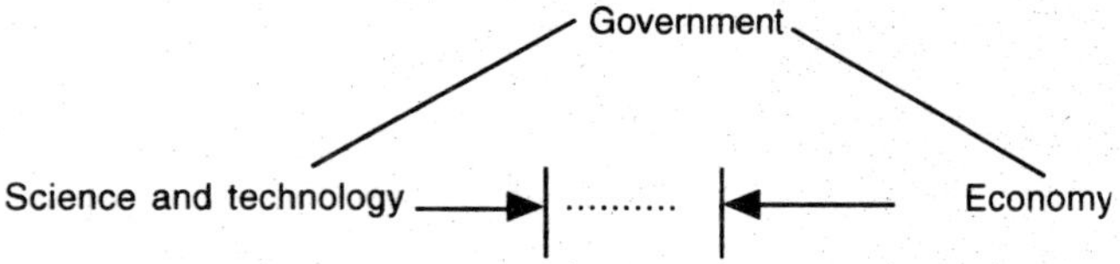

Fig. 1

Technical innovation (Technological innovation) is a medium of science, technology and economy. The concept of technical innovation was put forward by Austrian economists J.A. Schumpeter in the early days of this century. He thought that technical innovation is a process of re-organizing productive forces through the grasp of market opportunity by the entrepreneur. Afterwards, economists made a thorough study of the theory on technical innovation and regarded it as a resource of economic growth which can be exploited is an unlimited manner. We think that technical innovation is an integral process which transforms scientific and technological achievements into products and realizes their commercial value for the first time. Its function is to set up organic linkage between research exploitation and economic development. To be exact, technical innovation is an indispensable medium between technical invention and technical proliferation (expansion and application). Therefore, an essential characteristic of technical innovation is its role as a medium (Table 1). The table shows the important role and key usage of technical innovation in the development of society and economy. The achievements arising from technical innovation enter constantly into the fields of society and economy, and then spread to the whole society and promote the constant development of society and economy.

Technical innovation is an indispensable main medium in scientific and technological activity. In general, scientific and technological activity is divided into four stages - scientific research, technical research, technical innovation and technical proliferation (expansion and application)(Fig. 2).

Table 1

The Medium Characteristics of Technical Innovation

type / Content / items	Science discovery	Technical investment	Technical innovation	Technical dispersal
essential characteristics	exploration creativity	oriented exploration creativity	creativity first success in market	practice; success in wide market
people engaged in	scientific researchers	scientific researchers	enterprisers, scientific researchers	enterprisers workers
aim engaged	science achievement	technology achievement	gain commercial profit first time	gain widely social economical effect
relationship between reason & result	reason or result	result, reason	reason, result	result, or reason
work characteristics	work in lab. highest degree of freedom	work in lab. higher degree of freedom	experiment in medium-test base, certain degree of freedom	work in factory enterprise or field, low degree of freedom
value view	science value	Sci & Tech value	technology and commercial value	commercial value
valued standard	haven't social average labour	investment hasn't social average labour	negotiation in market	decided by social average labour
field	science	technology	technology. economic	society, economic
affected factors	science law	sci & tech. law	tech. economic law	social economic law

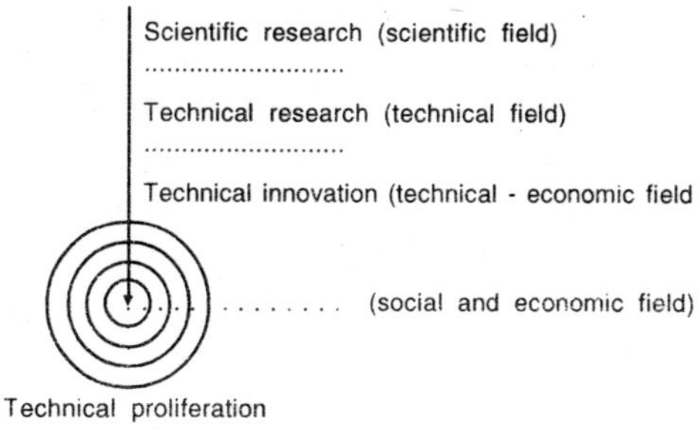

Fig. 2

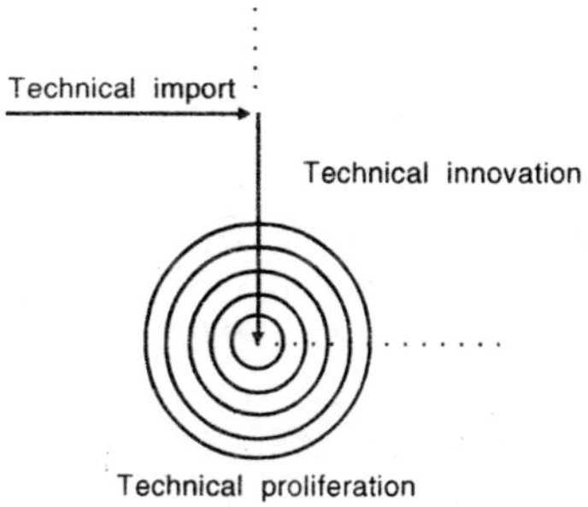

Fig. 3

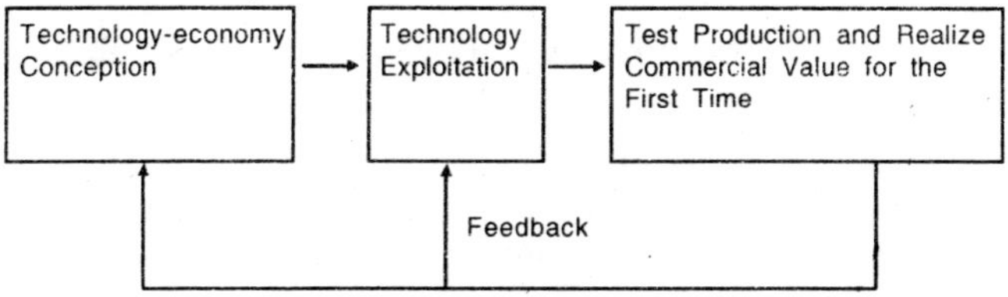

Fig. 4 : Process of technological innovation

In reality, not all countries start from scratch in developing scientific and technological activities. They usually start with technology import and then carry out "the second technical innovation"— absorption, digestion and innovation(Fig. 3).

Therefore, regardless of vertical or horizontal pattern, technical innovation is an indispensable bridge in developing scientific and technological activities. Technical innovation itself is a process which combines science, technology and economy. In general, this process is made up of three stages: technological and economic conception, technical exploitation and trial production and acquiring commercial profits for the first time. Fig. 4 shows that technical innovation is an active process which joins technology and economy. As a result, enterprise is the main body for technical innovation.

Discussion and Conclusions

Strengthening technical innovation is the key of reforming Chinese scientific and technological system. Since 1985, every year, China has had over 10,000 scientific and technological achievements which gained provincial and departmental recognition. (These were 33,000 in 1993, and 26,000 in 1994). But only 30 per cent of them resulted in production. The key factor responsible for this is the lack of links of technical innovation. This calls for adoption of following measures urgently:

Firstly, it is necessary to change the direction of expenditure made. In the mid-1980's, the proportion between research, technological exploitation and production in China was 1:0.1:100. In developed countries, however, it was 1:10:100. Seeing from the investment structure, in 1987, the break up of scientific research funds for research institutes, enterprises and colleges was 54.4 per cent, 29.6 per cent, 15.9 per cent respectively. In contrast, in the developed countries, the enterprises invested more than the research institutes, and the research institutes more than the colleges. For example, the R & D expenditure made by enterprises in America, Japan, Britain and West Germany was more than 60 per cent and only 10 - 15 per cent of the total R&D funds were spent by research institutes and colleges. Funds for research institutes in Britain and France are a bit more- 25 per cent of the total.

Secondly, the pattern of distribution of scientific research personnel should be changed. The distribution of scientific research personnel in China is very unreasonable. Very few scientific research personnel are there in the field of agriculture only 6.6 per 10 thousand, and on the average there is only 1 in 160 town and township enterprises. In 1986, there were 373.8 thousand independent industrial enterprises in China; 69.8 thousand of them were under the ownership of people; and 7558 of them were large or medium sized enterprises. There are only 3674 R & D institutes which account for 48.6 per cent of large or medium sized enterprises. On the average, only 41.2 engineers per institute are engaged in R & D work. There are 4836 R & D institutes under ownership of people, accounting for 6.93 per cent of the total. Therefore, most of the researchers are concentrated in research institutes established by the government and colleges. To change this situation, we should adjust research structure and their ownership, besides developing the horizontal cooperation among research institutes, colleges and enterprises, and encouraging researchers to work in enterprises. The main contention is to reduce the number of state owned research institutes. Most of the research institutes should become research-leading enterprises, some institutes may belong to enterprises, and the rest become the independent institutes to tackle problems of society.

Thirdly, the middle-level links between research and production should be strengthened and some institutes with technical innovation as a relatively independent activity should be established owning sole responsibility for profits or losses.

References

1. CHEN Wenhua: *Science, Technology and Development Metrical Research.* Changsha; Central South University of Technology Publishing House, 1992.
2. CHEN Wenhua: Science and Technology is the First Productive Force. *Leading Learning,* 1995, 1.
3. CHEN Wenhua, WAN Zhongman: *The Phenomenon of 40° N.* The second STS Symposium Proceedings, 1994 (November).

6. Progress of Science and Technology Reforms in China

Fang Xin

Abstract

The two motives of science and technology (S&T) progress are human beings' creative pursuit of secrets of nature and meeting practical demands of economic and social development. These motives should be activated to the maximum by various mechanisms and managerial methods. In this context, the author discusses the various organizational structure and systems reforms being carried out in China over the last decade. These reforms are directed towards institutional autonomy, market orientation and increased openness for exchange of personnel and information. However, a reasonable balance is being emphasized for basic research, applied research and market orientation. The problems of brain drain, aging, lack of successor talent, shortage of funds, etc. have surfaced threatening long term development of S&T in China. These problems need to be solved urgently. The present strategy is to emphasize on the promotion and development of high technologies in certain specific areas and remodel the traditional industries with modern S&T.

I. Introduction

Our society in a centennial turn is undergoing enormous transformation. Along with this, the contemporary progress of science and technology (S&T) is also unprecedented in depth, scope and rapidity. The interaction between national economic construction and S&T has also become unprecedentedly close. Under such a background, how to develop the national S&T, and what strategies to choose, becomes a research focus for S&T community in China.

As is known, and also emphasized by us, the progress of S&T has its own logic and physical principles, which should be obeyed in formulating S&T policies. Two motives for the progress of S&T are human beings' creative pursuit of secrets of nature, and meeting practical demands of economic and social development. The two motives should be activated to the maximum by various mechanisms and managerial methods, and directed to the progress of S&T. Further, the scope, rapidity and stress of S&T progress in a country are decisively related to the national conditions and national strength. The basic national conditions in China are as follows:

(1) An over-populated nation: The population of China has almost doubled in the recent 40 years, to nearly 1.2 billion—more than one-fifth of the total population of the world. Moreover, 80 per cent of the people are farmers and the "family planning" attempts in the countryside are still not proving very effective. As a matter of fact, the population is still prone to rapid increase. The clothing, food, housing, traffic, education and medical care of such a huge population pose severe challenges to the development of the economy and S&T.

(2) A relative shortage of resources: Although China has a territory of 9.6 million sq km, a rather large part of the territory is composed of desert, extremely cold or drought areas, which are unsuitable for agriculture and animal husbandry. The per capita cultivated land is less than 1.5 mu (less than 0.1 hectare). China's total water power resources exceed those in any other country in the world, but the per capita resources are only one-fourth of the world average. 13 per cent of the territory is covered by forests, but if this ratio is divided by 1.2 billion population, it becomes only one-sixth of the world average. The same is true of other resources like coal, titanium, tin and rare-earth metals, which have total reserves richer than those in many other countries, but the per capita reserves are below the world average. Besides, there is shortage of some resources and minerals like petroleum, natural gas, iron and copper.

(3) Harsh environmental conditions: Excessive but less-educated population leads to excessive and plundering

exploitation of the resources, such as land, minerals, and forest wealth. Hence, the ecological environment is getting damaged. Many enterprises which are poor in production technologies cause serious pollution. All these impose immense pressure on continuous development of the economy.

(4) Traditional industries urgently need modernization: Although China has set up a complete and independent industrial system which acts as the technological basis for industrial modernization, but economic development has been at the cost of excessive resource consumption. Many industries are backward in technologies and management, with low economic efficiency and inferior-quality products. In fact, China is in the group of countries featured by utmost resource consumption per unit of output value, social productivity being rather poor. Therefore, China is faced with the difficult tasks of remoulding traditional industries and effectively improving productivity by means of modern science and technology.

As a result of the above-mentioned basic national conditions, socio-economic development of China is expected to follow the following pattern: establishing a resource-saving national economic system, actually directing the economic construction to rely on S&T progress and improving the vocational quality of the people so as to guarantee continuous, stable and healthy development of the national economy.

The System of Science and Technology: Organizational Structure and System Reform.

The system of S&T is usually studied in respect of two aspects: (i) the organizational system consisting of the managerial system and the executive set - up, and (ii) the operational mechanism, i.e. administrative system.

Compared with the S&T managerial system of other major countries of the world, the S&T managerial system prevalent in China is a fairly centralized one, which is principally controlled by uniform policies. Anyway, because of the longitudinal and latitudinal partition among the central and local departments, which resulted from the administrative system and financial system, China's S&T system is a complicated network- a multi-layer, multi-channel, and multi polar network involving both central and local departments, both managerial and executive set-ups, as well as various disciplinary fields and vocational lines. The government, as the core of the network, formulates S&T strategies and provides S&T resources to the research institutions affiliated to it executing S&T activities. The State Council, as the supreme decision-making and managerial department of the country's S&T undertakings, takes the responsibility for formulating the guiding principles for the S&T progress. The State Science and Technology Commission, the State Planning Commission, the State Commission of Science, Technology and Industry for National Defense, the State Education Commission, the Chinese Academy of Sciences (CAS) and some executive departments of the State Council act as S&T managerial organizations at the level of the central government, while those at local level include the commissions and bureaus affiliated to local governments (provinces, prefectures, counties). All these compose a up-to-down multi-level managerial system. However, for historical reasons, the systems of social sciences in China, both research and management, are still separate from that of natural sciences.

In natural science and technology, the institutions in charge of scientific research and technology development are various R&D institutions affiliated to government departments, universities and colleges, large and medium scale enterprises, and those sponsored by collectives and individuals (Table 1).

Among these institutions, those affiliated to the CAS and the State Council, although not many in number, are superior to others in matters of funds, personnel resources, and S&T capabilities, acting as the backbone R&D institutions. In recent years, with the ongoing reforms of the economy and the S&T system, the technology development institutions sponsored by collectives and individuals have undergone a rapid growth in number, but have seldom made any significant achievements till now.

These various institutions have different research orientations. Basic research is mainly concentrated in CAS and some important universities and colleges. Basic research on some specific subjects

Table 1: R&D Institutions by Sectors

Numbers

	1990	1993
Govt. Institutes	4961	4996
CAS	123	123
Higher Education	806	814
Institutes belonging to Enterprises	8116	10477
Private Institutes	8523	55000
Total	22529	71410

is also carried out by institutions in medical sciences, agriculture and industry, but these institutions mainly concentrate on R&D related to general and key technologies which are closely related to their specific discipline and specific applications. Enterprise - sponsored institutions mainly serve for the technological progress of their own enterprises, while the collective and individual sponsored ones are devoted to market - oriented projects which yield quick benefits and are of short-term duration as well as easy to develop.

In the S&T organizational structure also, as a result of the reforms, some new components based on established institutions have been evolved. Since 1984, about 155 national laboratories have been set up based on CAS and universities with support from the State Planning Commission (SPC). Meanwhile, CAS and the State Education Commission (SEC) have also set up their own open laboratories. These laboratories have not only consolidated China's research achievements in some disciplines, but have also pioneered research in some newly emerging disciplines. They also form a base for training and collecting talented individuals and academic experts belonging to the new generation. By the end of 1993, 67 R&D centers for engineering technology had also been established with the support of the state for the purpose of improving R&D for engineering technologies and strengthening the linkages between R&D institutions and enterprises, thus promoting the transfer of research results to the industry. In addition, 100 enterprises set up their Technology Development Centers (TDC's) during 1993 and 1994. Another new component in the S&T organizational structure is the so-called centers for the mobility of post - doctoral researchers. In order to attract and cultivate senior S&T talents, the State has established 299 such centers in 96 universities and 69 research institutions. These new organizational methods have had great impact on the national S&T activities and have stimulatted the emergence of new S&T mechanisms.

Besides the many R&D institutions, there are some 45,000 service organs in China responsible for handling S&T information and documents, conducting reconnaissance surveys, meteorological observations, earthquake monitoring, mapping, environmental protection, patent, management of technological market, and so on.

The above set-up forms a huge infrastructure for science and technology in China. The main issue with the Chinese government is how to carry out effective management of this huge S&T network and give full play to the potentials of S&T talents and extensively and rapidly apply the S&T achievements to production. In this direction, the decade-long reforms of the S&T system could be summed up as follows:

(1) Applying different methods of fund allocation and management to different R&D institutions, gradually reducing the allocation for operational expenses and increasing investment for specific projects. A foundation system is being adopted for basic research, and projects are supported on the basis of competition and evaluation by experts from the same discipline. Various programmes and special funds have been created to support institutions doing research on environmental resources, ecology, basic data collection, and some general key technologies. The institutions working mainly on technological development are encouraged to enter into the market and collaborate with enterprises. Thus, these institutions obtain funds for their development through technology transfer to the production system. Meanwhile, the State provides necessary but limited support to such institutions.

(2) Expanding the autonomy of research institutions for independent decision - making, and reducing

administrative interference from the government, and introducing a market mechanism for competition.

(3) Encouraging the R&D set-ups to open to the entire country or even to the other countries. Encouraging the circulation of talents and strengthening latitudinal coalition between R&D institutions, universities and enterprises and mutual exchange of personnel and information among themselves.

(4) While emphasizing to work for the national economy, continuing to attach importance to basic research and maintaining a reasonable proportion between basic research, applied research and market-oriented development.

The on-going reform of economic and S&T system in China is concerned with the reform on the problems which had accumulated over a long time in history. This includes value concepts, living methods and working manners which people had taken for granted. This is an extremely difficult task. The main problem at present is how far should the R&D institutions go in the direction of market orientation. If R&D set-ups are compelled to change their track and financial resources are reduced without carrying simultaneous reforms in their surroundings, it may, probably, cause quick draining of their internal accumulated resources. This may affect the long-term development of science and technology. On the other hand, over-emphasis on market orientation, may lead to self-degeneration of the R&D institutions. This may also result in reduced coalition between S&T and the economy. Fortunately, the State has come to realize these problems and is making appropriate adjustments in the policies.

Input and Output of S&T Activities

Input and output of S&T activities in China with respect to the S&T personnel, funds, achievements and papers are discussed in this section in order to reflect the scope and strength of S&T activities in China.

S&T Personnel

S&T personnel play a decisive role in the development of S&T undertakings. In 1993, a total of 2,426,300 professionals were engaged in S&T activities in China (exclusive of those from service departments and technicians). Among these, 1,484,300 were scientists and engineers, accounting for 61.2 per cent of the total. In the same year, there were a total of 917,000 R&D personnel in the country, among whom 598,000 were scientists and engineers, accounting for 65.2 per cent of the total. Among the R&D personnel, 344,000 were from specific R&D institutions, accounting for 37.7 per cent, while enterprises and universities accounted for 27.5 per cent and 26.3 per cent of the total respectively (Table 2).

The vitality of S&T undertakings lies in an incessant supply of new talents from universities and colleges, which could be regarded as cradles for scientists and engineers of the future. In 1993, there were a total of 2,536,000 university undergraduate students, among whom 571,000 graduated from universities in the same year. Again in 1993, there were a total of 107,000 graduate students, among whom 18,000 were for Ph D degree, and 28,000 graduated in the year. Since 1990, those majoring in sciences and engineering have been accounting for about 40 per cent of all the undergraduate students. Graduate students, account for 63.7 per cent.

Table 2: S&T personnel in China by sector (1993)

(Thousand)

	R&D Personnel (A)	Scientists & Engineers (B)	B/A (%)
Total	916.8	598.2	65.2
R&D Institutes	344.1	248.1	72.1
Among CAS	50.6	36.2	71.6
Enterprises	252.1	84.9	33.7
Higher Education	240.7	224.9	93.4
Others	79.9	40.3	50.4

Table 3 : National R&D Expenditure

(billion yuan)

	1987	1988	1989	1990	1991	1992	1993
R&D expenditure	7.40	8.95	11.23	12.54	14.23	16.90	19.60
Real growth (%)	NA	8.17	15.11	4.89	7.39	12.92	0.66
GERD/GDP	0.65	0.64	0.70	0.71	0.70	0.70	0.62

Although the number of S&T personnel, undergraduate students, and graduate students in China is large, their per capita ratio is still lower than in some developing countries, let alone developed countries. The reform of S&T system carried out in recent years has provided opportunities for personal development of the S&T personnel, enabling them to find positions suitable for both - their actual individual status and the social demand, where they can give full play to their abilities. This, in the meantime, has had some negative effects. The relatively poor working and living conditions of S&T personnel and the influence of the social environment under the reform, have caused a serious drain on S&T talents. A rather large number of them, especially younger ones, have gone abroad or into business. This has caused a great damage to the community of S&T personnel in China. As at present, problems like drain of talents, aging, and lack of successor talents have tended to threaten the long-term development of science and technology in China and need to be solved urgently. The concerned authorities have realized the enormity of such problems and a series of measures have been adopted by the State governments and CAS to attract excellent talents, especially the young ones, to join the S&T undertakings, by creating preferential conditions for giving free rein to their abilities and wisdom.

Funds for S&T Activities

Fund is an important index reflecting the promotion of S&T activities, and is an important sustaining factor for the development of S&T undertakings. S&T funds, especially funds for R&D, not only reflect the S&T strength of a country, but also indicate how much support is given to S&T undertakings by the government and the society.

In 1993, the total expenditure on R&D was 19.6 billion yuan (Table 3), an increase of 2.7 billion over the previous year, but only 0.66 per cent higher than in 1992, if calculated at fixed price. This rate of increase was the lowest since 1988. Since 1989, R&D fund has been accounting for about 0.7 percent of the GNP, but it decreased to 0.62 per cent in 1993. This change, on the one hand shows the rapid economic growth and the sharp growth of GNP in China, but on the other hand reveals insufficiency of investments for R&D, which is much less than in the developed countries and lower than even in India.

In China R&D funds are invested mainly by the government, with a small part coming from enterprises. It can be seen from Table 4 that R&D institutions affiliated to the government constitute the predominant part of the R&D funding in China. The R&D facilities in enterprises are rather weak, and universities having considerable potential for R&D needs to be exploited. This government - dominated pattern of R&D activities is, of course, closely related to the present stage of economic development, but is to a great extent decided by the S&T system of China. With the intensification of the reform process, we believe the situation will change positively.

Among the total R&D funds, the funds for basic research, applied research and technological development have been coordinated at a stable ratio since 1990, with the funds for basic research accounting for nearly 7 per cent, lower than the average international standard. Generally speaking, the insufficiency of R&D funds has severely hindered further development of science and technology. It is expected that by the end of this century, funds for R&D will be increased to account for 1.5 per cent of the GNP, with the basic research funds accounting for 10 to 12 per cent of the total R&D funds.

Output of S&T Activities

The number of S&T papers published and the frequency of their citation are indicators of the output and level of S&T activities, and relative strength of S & T in comparison to other countries.

According to the statistics compiled by three major international indexing systems, i.e. SCI, ISTP and EI - 20,178 papers by Chinese authors were covered in the indexes in 1993, an increase of 9.2 per cent over the number in 1992. For two consecutive years, China ranked 12th in the world in the number of S&T papers published. As for the status of being cited, 32,503 papers were included in SCI from 1987 to 1992. In 1993, 7059 papers were cited 12,896 times, the rate of citation being 0.22, which is lower than the average world rate.

Table 4. National R&D Expenditure by sectors

(billion yuan)

	R&D expenditure	(%)
R&D Institutes	9.79	49.9
Enterprises	4.45	22.7
Higher Education	3.47	17.7
Others	1.89	9.7
Total	19.60	100.0

Patent is another important index of S&T output. In the past two years, the patent regulations and laws were further improved and remarkable progress was made in patent undertakings. In 1993, the Patent Office of China accepted 77,000 patent applications, and ratified 62,000 of them, representing an increase of 34.2 per cent and 97.4 per cent respectively over 1992. Among the applications put in during the year, inventions constituted 25.4 per cent, with utilities 61.5 per cent and designs 13.1 per cent.

Since the mid-1900s, science and technology have been developing rapidly and entering into economy at an unprecedented scale. Science and technology have become an internal variable of economic development. Historical facts have demonstrated that science and technology are responsible for and capable of contributing to economic growth. And this contribution is of multiplying type. In China, scientific research is providing expertise, theories, methods, ideas and talents for economic development, and a large number of research results are getting transferred into production systems. In promoting the progress of industrial technologies, the new products, new techniques and new designs provided by research are the sources for high technology products and high technology industries. Furthermore, in the exploitation and utilization of resources, and in environmental protection, scientific research provides basic principles, basic data and theoretical information. Even in fields related to people's food and health care like agriculture and medical sciences, scientific research provides the basis for cultivation of new agriculture breeds, development of new farming technologies, effective prevention of plant diseases, and control over insect pests. For development and production of new medicines and new medical technologies, theoretical knowledge arising out of research is of great importance. The developments mentioned above are difficult to quantify. Calculations made on the basis of Solow model have shown that since 1990, the progress of S&T has contributed to the extent of 26.7 per cent to economic growth in China. This percentage is far lower than that in developed countries. But comparing it with the past and especially when considering it under the current state of China's economic development, we can say that science and technology are increasingly contributing to economic growth.

Strategies and Policies for the Progress of Science and Technology

As a large developing country, China's science and technology have to achieve substantial progress in the following directions. First of all, S&T undertakings should aim at economic construction and market demand. Modern science and technology, especially electronic-based information and automation technologies, should be exploited to remodel the traditional industries, and modern agricultural and biological technologies should be exploited to remodel agriculture. Secondly, high technologies and their commercialization should be selectively promoted. Thirdly, in the important fields featured by the adjustment of the relationship between human beings and nature, especially population control, environmental protection, and comprehensive exploitation of energy and resources, significant results must be obtained. Lastly, massive thrust is needed in both basic and applied research.

In the 90's, governed by the national strategic objective of social and economic development, the S&T undertakings in China sought to work under the strategy of selectively developing high technologies and commercializing them, giving stress on industrial technologies and equipment of large-scale production; steadily strengthening basic research and making every effort to increase awareness about science and technology in the entire nation.

Basic research is of great significance for economic construction and improvement of the general outlook of the entire nation. If a large country like China does not improve the scientific and cultural level of the people, and depends solely on imported technologies, then it cannot hope to reach advanced international level, let alone make innovations. Therefore, it is necessary for the country to do its own basic research commensurate with its national strength as a developing country. The demands of national economic construction and social development must be considered when selecting subjects for basic research. Otherwise, steady progress of basic research cannot be assured.

Based on such an understanding and taking into consideration the trends of development of modern science and technology, we put forward the following suggestions in respect of long and medium-term basic research to be conducted. The quality of S & T researchers should be improved significantly, to lay the foundation of sciences and develop talents for advancement of the country. Basic research in some priority fields must keep pace with international scientific progress, and innovations should be made in some front-line areas and some well-developed fields. In the meanwhile, applied basic research should be paid special attention so as to enhance the scientific capability for solving key issues concerning national economy, social development and national defence. Basic research and applied basic research should be open to the entire country even to countries all over the world. Talented S & T personnel should have mobility among different set-ups. Cooperation between CAS and universities must be promoted to nurture talent and increase vitality. The State supports and guides basic research in the entire country through the National Natural Science Foundation of China (NSFC), and the Climbing Programme.

Development of high technologies is a window reflecting the strength of a nation. High technologies and their commercialization have become one of the most important aspects of competitiveness of contemporary S & T in the world. The development of high technologies in China must be based on the guideline of "emphasis on selective targets" by giving special support for projects and areas holding promise of significant breakthroughs and extensive applications. In the historically advantageous fields, breakthroughs and innovations should be made matching with the advanced level in the world. In the meantime, efforts should be made to foster the development and commercialization of high technology. International cooperation and market development of high technologies should be promoted by taking advantage of international resources to accelerate commercialization, industrialization and internalization of high technologies. In the 90's emphasis on high technology development should be focussed on:

(a) Concentrating efforts on development of microelectronics, technologies relating to computer software manufacture, robot technologies, technologies for computer manufacture.
(b) Special attention should be paid to the cultivation of new breeds of animals and plants, new biological products, medicines and new bacteria, and comprehensive utilization of recyclable resources,
(c) Conducting R&D on new materials, such as composite materials, structural and functional materials, amorphous materials, super-conducting materials and opto-electronic materials,
(d) Strengthening research and development work on astrodynamic and propelling technologies, measuring and controlling technologies, manned astronautic technologies to maintain China's appropriate position in this field.

The State supports such efforts mainly through the 863 Programme, the Torch Programme, etc.

Under the conditions prevalent in China, the population growth should be controlled and the quality of the population improved. Natural resources should be exploited rationally and the ecological environment should be effectively protected. The sciences and technologies concerning social development must be emphasized, and R&D on population, medicines and sanitation, social services, public infrastructure, environmental and ecological protection must be strengthened so as to create favourable natural and social surroundings for the people in their effort to explore a way of steady development with Chinese characteristics. After the last International Convention on Environment and Development, Chinese government has formulated a programme named "Agenda for the 21st Century China" with the purpose of fostering R&D in areas of environment, ecology and resources.

To remodel the traditional industries with modern sciences and technologies is an issue urgently needed to be solved for economic development of China. This is an essential responsibility to be shouldered by S&T personnel. Development of industrial S&T must be focussed on improvement of economic efficiency, with modern sciences and technologies and modern management systems, especially electronic information technology, technologically remoulding various industrial departments, improving the technological standard of mechanical equipment, saving energy, reducing the consumption of raw materials, improving quality, developing different varieties of products, improving productivity and enhancing international competitiveness of products, as well as optimizing industrial structure and product structure. Facilities for agricultural research must be strengthened substantially. Emphasis needs to be laid on the extension of advanced agricultural technologies, with proper adjustment to the industrial and employment structure. For this purpose, the State has launched the 'Programme for Tackling Key S&T Issues', the 'Programme for the Coalition of Industries, Universities and Research Institutes,' and the Spark Programme.

To ensure implementation of the above strategies and realization of objectives, the State has formulated a series of policies and measures, such as reforming the management of S&T funds, giving energetic support in respect of key projects in some priority fields; exploiting personnel resources, cultivating and taking advantage of new talents in science and technology, giving full rein to the abilities of the S&T personnel, guiding and promoting the S & T progress in enterprises and in the countryside through economic measures like taxes, price adjustment, credits and loans, and depreciation, creating stable, democratic and free academic surroundings for progress of S&T, actively fostering international cooperation and exchanges.

References

1. STATE SCIENCE AND TECHNOLGY COMMISSION: *China Science and Technology Indicators*-1992. Beijing; Science Press, 1993. p7-21.
2. STATE SCIENCE AND TECHNOLOGY COMMISSION: *China Science and Technology Indicators*-1994. Beijing; Science Press, 1995. p6-41.
3. STATE SCIENCE AND TECHNOLOGY COMMISSION: *Guide to China's Science and Technology Policy.* Beijing; S&T Documents Press, 1986. p298-304.
4. STATE SCIENCE AND TECHNOLOGY COMMISSION: *Guide to China's Science and Technology Policy.* Beijing; S&T Documents Press, 1990. pl-16.
5. FANG Xin: *Analysis and Evaluation on the Management System of Science and Technology of China.* (Report Submitted to ADB, 1992)
6. ZHOU Chengkui et al; *Study on the Reform of China's Science and Technology System.* Machinery Industry Press, 1992. pl-43.

7. Strengthening University - Industry Linkages: Need for Technological Competitiveness in India and China

V.P. Kharbanda

Abstract

With globalization and international competition, university - industry symbiosis has gained importance, particularly in India and China. Both these countries have established a strong research base in terms of institutional infrastructure and manpower, which are comparable with the best research institutions in the world. However, these countries lack the capacity to translate innovations in terms of economic gains. On the other hand, R&D capabilities of the firms have also failed to fully exploit the indigenous research base. Industry mainly depends on import of technologies for its survival. In this scenario, the author surveys the efforts made by these countries to overcome the problem of linking the knowledge base with industry and proposes some measures to face the onslaught of multinational giants in this globalized world.

Introduction

University - industry symbiosis, over the years, has proved to be of immense significance for achieving international technological competitiveness in the developed countries.[1] Emergence of new high technology knowledge intensive areas, especially micro-electronics, computer and information processing, biotechnology, and new materials, etc., has further accentuated their interdependence. Industry harvests the potentials of fundamental research pursued in universities, whereas universities

[1]PAPADAKIS Maria: The Delicate Task of Linking Industrial R&D to National Laboratories. *Technovation 1995,* 15(9) p. 569 - 583.

benefit from the industry in terms of, (a) selection of research programmes of vital significance to the society at large; (b) selection of teaching material/courses in accordance with the needs of industry; and (c) financial resources required to pursue research programmes. Following globalization of trade during the 1990's, this symbiosis has gained further importance in view of: (a) international technological competition; (b) increased cost of research; and (c) decreasing time lag in research and application. In this scenario, the national governments have to provide an appropriate environment and conditions under which academic institutions and the industrial sector can fully realize their potentials and capabilities[2]. Since late 1980's and beginning of 1990's, both China and India have been pursuing their science and technology policy reforms for strengthening this symbiotic relationship. In India, while the need for strengthening such linkage was felt almost 35 years ago (Thacker Committee, 1961)[3], little has been achieved[4]. Similarly, Chinese institutions of higher education had been severely affected by the political events in the recent past, which made them to work in isolation from the industry[5]. The present policies in China, particularly after

---

[2]KHARBANDA V.P.: Analysis of Technological Change at the firm level- A case study of an Indian firm. *Journal of Scientific and Industrial Research* 1993, *52*(9) p.585-594.

[3]*Report of the Review Committee on Postgraduate Education and Research in Engineering and Technology*. (Ministry of Education and Culture, Govt. of India, 1980). p.11-12.

[4]Recent studies carried out at National Institute of Science Technology and Development Studies(NISTADS), New Delhi; and National Institute of Educational Planning and Administration, New Delhi; have revealed that, over the years, such linkages have, in fact weakened. See, for instance: Pruthi S, JAIN Ashok: Image of Indian Science and Scientific Community. Journal of Higher Education 1996, *19*(1) p.9-30; AHMED Rais: Scientific research in Indian Universities and IIT's. New Delhi; NIEPA, 1989; University - Industry linkages. *University News* 1991,March 11, p.16.

[5]KHARBANDA V. P., QURESHI M. A.: *Science, Technology and Economic Development in China*. New Delhi; Navrang Publishers, 1987. 206p.

the initiation of 'Reforms in the Science and Technology System' since 1985, are oriented towards rectifying such a situation.

How can these linkages can be strengthened? What are the different forms and methods by which these can be made more effective? What lessons can be drawn from the experiences of India and China? What are the prospects for regional cooperation in terms of globalization? In the present day context, these questions call for a detailed review of the nature of these linkages. This may help in obtaining some insight for strengthening this most sought after symbiotic relationship.

The present study is divided into three parts. Part I throws light on the research base in these countries; Part II reviews the existing linkages and problems; and Part III deals with policy options to improve the linkages and conclusions.

Research Base

Over the years, a strong research base has been built up in both India and China. This includes building up of a huge institutional infrastructure, training of the human resource to fulfil the needs of R&D, and availability of financial resources on a constant basis to execute R&D. A broad picture of the research base as it has evolved in the two countries over the years, and how university research in particular was emphasized is depicted below.

India

Since independence in 1947, India has developed a vast infrastructure for Research and Development (R&D) in basic research; for application of science and technology development; and for training S&T manpower through a planned approach. As a result of major efforts through various plans, the number of universities has increased from 27 in 1950-51 to about 211 in 1994-95, including 10 institutions of national importance. There are over 8,210 colleges with a total enrolment of around 5,000,000 of which 37.7 per cent belong to science, engineering and technology[6]. These institutions are primarily involved in teaching, fundamental research and training of manpower. However,

[6]Annual Report 1994-95. *Ministry of Human Resource Development,* New Delhi; Department of Education 1995 (Part I), 303p.

scientific research in universities has evolved rather slowly since independence. Although Kothari Commission[7] in mid-sixties drew attention to inadequate efforts provided for scientific research in universities, it was only after the introduction of the New Education Policy(1986) that efforts have been directed towards improvement of higher education and research. The objective has been to promote culture of research in universities and enable the faculty to keep abreast of latest developments in their subject areas. It called for cooperation, collaboration and networking relationship between institutions at various levels and with the user system[8]. An assessment of the inputs to R&D and related S&T activities in the educational institutions has been made at NISTADS, which gives an indication of the enhanced expertise for scientific research in universities. The financial inputs to R&D more than doubled from Rs 3.78 billion in the 6th Five Year Plan (FYP) to Rs 7.93 billion in the 7th FYP[9]. Funding of extramural projects to academic institutions has increased considerably over the recent years (Table 1). University Grants Commission (UGC) and various other government departments like Department of Biotechnology (DBT) and Department of Science and Technology (DST) through its Science and Engineering Research Council (SERC) have taken up the major responsibility for funding scientific research projects/programmes in universities and R&D institutions (Table 2). During 1994-95 the total number of research projects being financed by the various central agencies was 1557 at a total approved cost of Rs. 1313 million as compared to only 607 million during 1985-86. Of this, the academic sector received 65 per cent and the national laboratories received 27 per cent of the extramural funding. Of these projects, 101 were by DBT, 458 by DST, 182 by CSIR, 211

---

[7]*Education and National Development* (Report of the Education Commission (1964-66), Ministry of Education, New Delhi, 1966, p 420-421.

[8]*National Policy on Education-1986.* New Delhi; Ministry of Human Resource Development, 1986. p.18.

[9]AHUJA J. K.: R&D Statistics - Source-wise funding of R&D undertaken by Educational Institutions in India. *Current Literature on Science of Science,* 1988, *17* (5) p.35.

by ICAR, and 163 by UGC[10].

Table 1: Distribution of Extramural R&D Projects by type of Institutions

(*Rs.million*)

Agency	1990-91		1994-95	
	No. of Projects	Cost	No. of Projects	Cost
Universities/ Colleges	567	253.38	891	562.98
Deemed Universities	64	79.83	109	132.34
Institutes of National Importance	187	131.46	205	162.06
National Labs	202	394.95	243	349.72
Others	205	135.11	109	106.28
Total	1225	994.73	1557	1313.38

Source: Department of Science and Techonology (DST): *Research and Development Statistics* - 1992-1993; and 1994-95. New Delhi; DST, 1996.

[10]*Research & Development Statistics*, 1994-95. New Delhi; Department of Science and Technology, 1996. p.43.

Table 2: Organizations under the Central Government which provide major funds to Educational Institutions in India for R&D

Per cent

Name of the organization	1980-85	1985-90
1. ICAR	30.5	26.2
2. Dept. of Education (UGC &Tech. Edu. Div.)	21.3	22.7
3. Ministry of Science & Technology		
(a) DST	13.6	11.1
(b) DBT	-	5.0
(c) CSIR	7.1	6.2
4. DNES	11.0	10.7
5. DRDO	5.0	4.4
6. ICMR	3.5	5.9
7. DOEn	2.3	2.3
8. DOE	3.0	1.8
9. DAE	1.2	0.8
10. DOS	0.8	0.6
11. DOSt	0.05	0.6
12. DOM	0.1	0.6
13. DOC	0.1	0.4
14. DOP&NG	0.1	0.2
15. DOD	0.05	0.2
16. Others	0.3	0.3
Total	100	100

Source: AHUJA J. K.: R&D Statistics - Source wise funding of R&D undertaken by Educational Institutions in India. *Current Literature on Science of Science* 1988, 17(5) p35.

Notes: Indian Council of Agricultural Research (ICAR), Department of Science and Technology (DST) , Department of Biotechnology (DBT), Council of Scientific and Industrial Research (CSIR), Department of Non-Conventional Energy Sources (DNES), Defence Research Development Organization (DRDO), Indian Council of Medical Research (ICMR) , Department of Environment (DOEn), Department of Electronics (DOE), Department of Atomic Energy (DAE), Department of Space (DOS), Department of Steel (DOSt), Department of Mines (DOM), Department of Coal (DOC), Department of Petroleum and Natural Gas (DOP&NG), Department of Ocean Development (DOD).

UGC, established in 1956 under an Act of Parliament, has the prime responsibility of regulating the standards of higher education sector in the country. The scientific research capability in universities has been strengthened by opening a number of Centres of Advanced Studies (CASs) in botany, zoology, chemistry, astronomy, etc. At present, there are 41 CASs in 13 universities. These CASs pursue research work of excellence to realize international standards. As per recommendations made in the New Education Policy (NEP) 1986, UGC, since 1987, has been assisting universities in developing education and research capabilities in both basic and applied research areas. For example, projects in the areas of super-conductivity, atmospheric sciences, biotechnology, environment, energy, electronics and futurology are being implemented as independent programmes in collaboration with DBT, DNES, and DOE. Emerging area programmes were initiated in 1993 in the areas of electronics, computer applications, remote sensing, environment and energy. Co-operative research facilities have been established by UGC in high priority areas through inter-university centres, e.g. in Nuclear Science, Astronomy, Astrophysics, MSD Radar, and Crystal Growth. To strengthen university research, some special efforts have been initiated by UGC which include programmes on "Strengthening of Infrastructure in Science and Technology" (COSIST) and Special Assistance Programme (SAP) under which about 200 S&T departments in universities have been covered. These programmes are basically oriented for co-ordinating research in frontier areas and training scientists in universities and colleges in research methodology, etc.

About 38,000 postgraduate and 4000 doctorate degrees are awarded every year by the university system in natural sciences, including pure sciences, engineering & technology, medicine, agriculture and veterinary sciences. Analysis of data on PhD degrees awarded during 1993-94 reveals that this constitutes 49.7 per cent of the total doctorate degrees awarded in all fields. These have a total faculty of about 25,000, including professors, readers, and lecturers[11]. Over the years, higher education sector has generated a mammoth force of S&T manpower which totalled 6.3 million in 1996. However, of this only 114,403 personnel are engaged in R&D activities, 77,113 (67.4 per cent) being employed in the institutional sector and the rest 37,290 (32.6 per cent) in the industrial sector.

Concerning expenditure on R&D, the educational system receives about Rs. 590 million, which is only a meagre 1.16 per cent of the total expenditure on R&D in the country in 1992-93. This decreased to 1.13 per cent in 1994-95. Besides, the percentage share of expenditure on university and other higher educational institutions in total education expenditure has also, over the years, declined from 14 per cent in 1983-84 to 12 per cent in 1994-95, although in absolute terms it increased from Rs. 7.77 billion to Rs. 32.54 billion during the same period[12].

Apart from this, scientific research, mainly of applied nature is being conducted in 531 R&D institutions funded by the central government[13]. During 1996, there were also a number of in-house R&D units set up by the industry, primarily involved in industrial research, which totalled 1230 in the private and public/ Joint sector, besides 470 Scientific and Industrial Research Organizations (SIR0s).

The total national expenditure on R & D has increased from a meagre Rs. 46.8 million in 1950-51 to Rs. 68.21 billion in 1994-95. Though in absolute terms the total R&D expenditure has shown an increasing trend, R&D expenditure as per cent of GNP has also declined from 0.967 in 1988-89 to 0.81 per cent in 1994-95, much below the expenditure of 2-3 per cent of GNP incurred by most of the industrialized countries[14]. Field-wise, during the year 1994-95, the per cent share of basic research has increased from 14.50 per cent in 1990-91 to 18.1 per cent in 1994-95 (Table 3).

---

[11]Ibid. p.28.

[12]Ibid. p.25.

[13]For details see *Directory of R&D institutions,* 1994. New Delhi; Department of Science and Technology, 1994. 181p.

[14]*Research & Development Statistics,* 1994-95, Op cit. p.51-52.

Resource-wise, the Central Government continues to shoulder the major share of national expenditure on R&D. During 1994-95, the share of the government was 73.50 per cent(Table 4). The share of industrial sector in the total R & D expenditure is still very low; it was only 26.5 per cent in 1994-95 amounting to Rs. 18.08 billion. This is in contrast to what one observes in developed countries, where the share of industrial sector in the total R&D expenditure is about 70 per cent. Further, industrial expenditure on R&D as percentage of sales turnover of the industries continued to be as low as 0.6 per cent in 1994-95 again far below the figure of 7-10 per cent in the developed countries[15].

Table 3: Share of National S&T Expenditure by type of work in India

(per cent)

Type	1990-91	1994-95
Basic Research	14.50	18.10
Applied Research	39.00	38.80
Experimental Development	34.70	31.20
Other Activities	11.80	11.90

Source: Department of Science and Technology (DST): *Research and Development Statistics* - 1990-1991; and 1994-95. New Delhi; DST, 1996.

Table 4: Share of Resource wise National R&D Expenditure in India

(per cent)

Agency	1990-91	1994-95
Central Government	68.90	64.90
State Government	7.90	8.60
Industrial Sector	23.20	26.50
Public Sector	10.60	10.1
Private Sector	12.60	16.40

Source: Department of Science and Techonology (DST): *Research and Development Statistics* - 1990-1991; and 1994-95. New Delhi; DST, 1996.

[15]Ibid. p.vii.

China

After liberation in 1949, the Chinese government, at the initiation of the First Five Year Plan (FYP) (1953-57), made a nation - wide reorganization of the higher education system under the scheme, Readjustment of Colleges and Departments, (Yuanxi tiaozheng), essentially on regional basis. After this reorganization, most provinces had one university and their own colleges of engineering, agriculture, medicine and teacher training. During this period, the Chinese system of higher education was heavily influenced by the Soviet system. The Chinese Academy of Sciences (CAS) was formally established on 1st November 1949, on the basis of research institutes of the former Academia Sinica and the National Beiping Research Institutes. CAS quickly developed from a complex of 14 research institutes with 224 researchers in 1949 to a network of 47 research institutes with 2977 researchers (including 448 senior ones) in 1955. The Chinese Academies of Medical, Agricultural and Geological Sciences were established in 1956, 1957 and 1959 respectively, drawing a number of leading scientists from universities. The de-emphasis on university research relative to the centralized academies closely followed the Soviet pattern and inevitably gave rise to complaints and even protests from university professors. Their voices were strongly echoed in a number of national conferences on higher education convened in the fifties. In September 1953, a National Conference on Comprehensive University Education proposed that universities should also be the institutions for research. Following this conference, strong impetus was given to the development of research in universities[16].

However, the next two decades witnessed havoc with educational institutions, except in the brief period of 1960-1964, primarily to meet political objectives. In 1958, the country entered the Great Leap Forward (GLF) phase, which was characterized by mass movements in every field from academic research to industrial production. The number of institutions of higher

---

[16]HUANG Shiqi: *University Research in China.* Paper presented at the International Seminar on Current Policies for Higher Education Reforms, Beijing, 21-25 June 1988. p.7.

education increased swiftly from 229 in 1957 to 1289 in 1960, with intake of students increasing from 1,05,581 in 1957 to 3,23,161 in 1960. Obviously, such high rate of growth was too hectic to be sustainable[17]. During the Reorientation Period (1960-64), the State Science and Technology Commission (SSTC) drew up a programme for scientific and technological development (1963-72) which listed 374 key R&D programmes, including 41 basic research programmes. By the end of August 1964, 184 institutions of higher education were involved in about 70 per cent of the research programmes.

The Cultural Revolution (CR) (1965-1976) seriously disrupted the functioning of and inflicted heavy losses on institutions of higher education[18]. After the downfall of the Gang of Four on 29 July, 1977, Deng Xiaoping indicated that the Ministry of Education (MOE) must encourage a number of key universities which should be both centres of education and centres of research. Since then, higher educational institutions have been given autonomy in respect of enrolment, establishment of organization, appointments and removals, fund utilization, professional title appraisals, salary distribution and international cooperation. The government has shifted from direct administrative control of institutes of higher education to macro-management through legislation and policy guidance[19].

In 1993, there were 1065 universities or colleges, of which 34 were under State Education Commission, 670 under Provincial authorities and 361 under ministries/departments and commissions. In these institutions, during 1993, there were a total of 2.536 million undergraduates, of which 263,000 graduated with 2.6 per cent in natural sciences, 22.3 per cent in engineering sciences, 5.7 per cent in medicine, 2.3 per cent in agronomics and 0.6 per cent in forestry. During the same year, some 29,000

---

[17]Ibid. p.9

[18]For details of disruption of higher education and research during CR see KHARBANDA V. P., & QURESHI M. A.: (1987) Op cit.

[19]CUI Lili: New target for educational reform. *Beijing Review* 1993, 36 (22) p. 13-18.

received postgraduation. There were 588 colleges and research units to confer Master's degree and 248 for doctorate degrees.

China's R&D base consists of about 72,167 R&D units which include research institutes in CAS(123), units attached to various government departments (4996) , institutions of higher education and their affiliated R&D units(2616), technological development units of medium and large industrial enterprises (9432), and technological units run by collectives and individuals(55,000). Basic research is mainly carried out by CAS and key universities. During the reforms since 1985, about 155 national laboratories, jointly managed by CAS institutes and the universities, and some open laboratories under CAS have been set up to carry out research in some newly emerging areas and train outstanding S&T personnel. Research units in medicine, agriculture and industry conduct some mission-oriented basic research, but their R&D work mainly concerns subjects with potential applications. R&D units attached to industrial enterprises mainly serve the technological development of their enterprises, while R&D units run by the collectives and individuals focus on short-term projects that can give quick profits. However, most of the research units are very small in terms of R&D funding as well as R&D personnel. In the country, about 65 per cent of R&D manpower and 68 per cent of R&D funding is concentrated in just 11 per cent research units under the State, attached to CAS and higher education research institutions. Some 70 per cent of the research units have less than 20 personnel. Most of these are collective and private institutions. Although these have shown rapid growth during the last 10 years, their contribution to R&D is insignificant. R&D efforts by the industrial enterprises are also not very significant. Research units under the enterprises spend only 22.7 per cent of the total R&D funds and engage only 27.5 per cent of the total R&D manpower[20]. Thus although there are not many institutes belonging to CAS and the State Council,

[20]FANG Xin: *Programme of Science and Technology in China*. Paper presented at the Workshop on Science and Technology Development Strategies in India and China, organized jointly by NISTADS, and Central- South University of Technology, Changsha, China from 25-26 September, 1995, at New Delhi. 13p.

but they are the major actors of R&D and enjoy major share of funds and personnel resources and act as backbone R&D institutions.

National Natural Science Foundation of China (NSFC) founded in 1986 has the prime responsibility to direct, coordinate and finance basic research in the main and part of applied research; and to identify and train talented persons. It has six science departments and several functional agencies, which cover all disciplines of natural sciences, including mathematical and physical sciences, chemical sciences, life sciences, material sciences, engineering, information and management sciences. NSFC receives and processes R&D project applications from the whole country. Regional foundations have also been set up by the State departments and local governments. Till now, 28 science foundations have been set up in 26 provinces.

NSFC has contributed a lot towards stabilizing high level science and technology workers for the country. A series of policies and measures have been drawn up to support the young scientists' research. Some special funds have been set up to encourage and assist young scientists in their work, such as the Fund for Young Scientists, the Special Fund for Middle-aged and Young Talents, and the Science Fund for Outstanding Young Scientists. In 1986, only 1.3 per cent scientists below 35 headed the NSFC supported research programmes. In 1994, this increased to 26.9 per cent.

In 1986, the Foundation received a fund of 80 million yuan ($9.5 million), which reached 500 million in 1995. The fund to support regional foundations has reached 200 million yuan ($25 million). It also receives funds from various societies and scientists from both home and abroad.

Apart from this, beginning from 1985, about 928 Mobile Post-Doctoral Research Stations (MPRS) had been established by 1995 in 212 universities and 69 CAS research institutes with a view to providing young doctorate recipients with opportunities of research, increasing their mobility and helping them to find the most suitable places to settle down for permanent employment[21].

[21]*China Science and Technology Newsletter*, 1995 (63) Dec. 15, p.4.

In the country, in 1996, there were a total of 19.13 million S&T workers; among these, 1.48 million were scientists and engineers. Of these 9,17,000 were R&D personnel 65.0 per cent in State-owned R&D units and institutions of higher education and 26.2 per cent in industrial enterprises.

The level of funding has gradually grown from 20 m RMB yuan in 1963 to 44 m RMB yuan in 1966, but it became negligible during the period of CR(1966-1976). In the aftermath of CR, the funding was restored to 40m in 1979[22]. After this R&D funds have increased steadily from 7.40 billion yuan in 1987 to 22.2 b yuan (US $ 2.7b) in 1994[23] . However category-wise, expenditure on basic research has declined from 12.47 per cent in 1986 to 6.7 per cent in 1994. Like in India, government continues to be the major source(70 per cent) for expenditure on R&D. As percentage of GDP, there was a decease from 0.70 per cent in 1989 to 0.50 per cent in 1994, which indicates that increase in R&D inputs was falling behind the rate of development of the Chinese economy and its rapid growth in GDP. However, it is hoped that by the end of this century, the expenditure on R&D as per cent of GDP may increase to 1.5 per cent and the proportion of funding to basic research may be of the order of 10-20 per cent of the total R&D expenditure by the State.

Having built up an infrastructure for scientific research and development in the years immediately after independence in India and liberation in China, both have attempted gradual reorientation of their policies and programmes to forge closer linkages between this infrastructure and the pressing tasks of socio-economic development. What has been the outcome of these attempts? A comparative picture of the broad indicators of the research base developed in India and China shows that although India and China had similar approaches of initially concentrating on the invention end (knowledge creation) to establish S&T capabilities as an essential infrastructure, over the years they moved closer to the diffusion or production end; the resulting capabilities in India and China show interesting differences.

---

[22]HUANG Shiqi(1988) Op cit. p.17.

[23]*China Science and Technology Newsletter*, 1995 (45) May 30, p.8.

In other words, both India and China launched their science and technology activities from a policy regime closer to the inventive end of the technological capability spectrum and then, over the year, have attempted reorientation so as to cover the entire spectrum, from invention to diffusion or production, i.e. they have moved towards establishing better linkages between academia and the industry. In this respect, China has performed better than India[24].

Existing University-Industry Linkages

India

In India, in 1961, Thacker Committee stressed the need for the establishment of various forms and mechanisms of cooperative relationship between industry and universities in India. These included sponsorship of research projects by industry in universities and other research institutions, arrangement to share technical data and information, provision of consultancy services from faculty in universities on specific problems of industry, making available facilities of research laboratories to industry, introduction of courses to train specialists of utility to industry and offering of endowment fund or special chair of professorship by industry. Kothari Commission (1964-66)[25] and Bhide (1969)[26] also suggested some useful methods promoting of cooperation among the universities, R&D organizations and industry. The efforts indicated restructuring of coordination among some of the existing national level committees like discipline- oriented expert committees of UGC and National Council of Science Education; research committees of Council of Scientific and Industrial Research (CSIR) , Indian Council of Medical Research (ICMR), Indian Council of Agricultural Research (ICAR), etc. The Review Committee on Post-Graduate Education and Research

---

[24]JAIN Ashok, & KHARBANDA V.P.: Science and Technology - Issues for the Twenty- first Century. *Indian Journal Social Science*, 1993, 6(3) p.275.

[25]Education and National Development (1970), Op cit. p.369-421.

[26]BHIDE V. G.: Basic and Applied Research in a Developing Economy. *Indian J. Public Administration* 1969, 15(3) p.338-343.

in Engineering and Technology (1971), and Satish Chandra Committee on UGC (1980) and the National Policy on Education-1986 (Programme of Action - 1992), all these Committees strongly re-emphasised the need for close university - Industry collaborations.

In this direction, a number of steps have been taken by the Ministry of Human Resource Development, All India Council for Technical Education (AICTE), UGC and DST to promote academia-industry linkages. These include:

1. *Industry - Varsity Cells:* A scheme to establish Industry-University cells has been implemented by AICTE, to coordinate all the activities pertaining to industries and universities.

In this direction, a number of agreements have matured between universities and industrial enterprises. For examples, Roorkee University has signed an MOU with the Confederation of Indian Industries for promoting partnership with industry[27]. Birla Institute of Technology and Science (BITS) (Pilani) has signed an MOU with Ashok Leyland for upgradation of shopfloor personnel. Thus, education at the work centre is being promoted to ensure human resource development without loss of manhours. Constant interaction between engineers and faculty members would also keep abreast of the latest developments in the industry[28]. The Electronics Corporation of India Ltd (ECIL) and Jawaharlal Nehru Technological University have signed an MOU to establish Advanced Photonics Technological Centre at Hyderabad. The main thrust of this centre will be on developing practically implementable devices in the area of Fibre optics technology and communications, Opto Electronics, Fibre optics sensors, Electro-optics and Non-invasive bio-photonic diagnosis[29].

---

[27]RAMACHANDRAN M.: Proceedings and Recommendations of Workshop on Industry-University Partnership. Organised by the Department of Higher Education with the cooperation of AICTE. Yojna Bhavan, Lucknow, March 22, 1995. *J. Higher Education* 1995, 18(3) p.496-534.

[28]BITS, Ashok Leyland sign MOU. *University News* 1996 (Jan 8) p.22.

[29]Advanced Photonics Technological Centre. *University News* 1996 (Oct. 21) p.17.

A large number of Industrial Liaison and Training Cells are also being established in engineering colleges and universities. For example, the Foundation for Innovation and Technology Transfer (FITT) has been established at Indian Institute of Technology (IIT), Delhi. Similarly, other IITs and universities like BITS (Pilani), and Indian Institute of Science(IISc), Bangalore have established industrial consultancy and extension centres to facilitate utilization of domestic R&D. IIT, Kharagpur has entered into a joint venture with the Bhopal-based 'Optel Communications, for the establishment of an Optical Fibre Research and Development Centre. The 'Optel-IIT R&D Centre' inaugurated in 1996 at Kharagpur, would undertake project development of technology for manufacturing erbium-doped fibres[30]. Asea Brown Boveri Ltd(ABB) in India is reported to have instituted a Research Chair in Electrical Engineering at the Indian Institute of Science(IISc), Bangalore[31]. The ABB Chair Professor or any of the IISc faculty members may participate in mutually beneficial areas of work. These may be in the form of short-term course, seminars, technical units, HRD programmes, or co-ordination of projects. In this direction, AICTE has also framed some guidelines for the institution of National Chairs in Engineering and Technology in institutions of national importance. The expenditure on the AICTE National Chair will be met by the designated institutions from their funds raised from industry, philanthropists and others[32]. Punjab University, Patiala plans to set up a centre to meet the needs of industry dealing in biotechnology, computers, pharmaceuticals and business management[33]. The university will provide short - term courses, consultancy, expert services and human resource development in the concerned fields. Anna University has

---

[30]IIT, Kharagpur-Optel tie up. *University News*, 1996 (January 15) p.19.

[31]ABB Research Chair. *University News*, 1996 (September 9) p.27.

[32]National Chairs in Engg & Technology. *University News*, 1996 (April 1) p.23.

[33]Centre for University - Industry dialogue. *University News*, 1996 (September 16) p. 17.

proposed to seek help from industries in the areas of leather, chemicals, ceramics, and textiles sectors which are crucial players in foreign trade[34].

2. *Science and Technology Entrepreneurship Parks(STEPs):* The scheme was initiated by National Science and Technology Entrepreneurship Development Board(NSTEDB) under DST in 1984. The objectives of STEPs are:

(a) To forge a close linkage between universities, academic and R&D institutions on the one hand and industry on the other;
(b) To promote entrepreneurship among S&T persons;
(c) To provide R&D support to small scale industry mostly through interaction with research institutions.

The scheme is being jointly promoted by DST, IDBI, IFCI and ICICI, concerned state governments and the host institutions. So far, 12 STEPs have been established in various parts of the country and they are in various stages of development. The STEPs have been successful in promoting nearly 400 enterprises, many of which have grown significantly. These have been established mainly in the areas of electronics, computers, mechanical engineering, machine tools, automation and controls, chemicals, material sciences, biotechnology and environment. A few enterprises are engaged in manufacturing hi-tech products having either import substitution or export potential value and meeting requirements of agencies such as Indian Railways, Bharat Heavy Electricals Ltd., Indian Telephone Industries etc. So far STEPs have been instrumental in development and commercialization of more than 200 products/processes[35]. However, implementation of this programme seems to have slowed down after initial success. While 11 of these were established during 1984-89, only one was established (in 1994) after that. Only nine of them are fully functional. Only two STEPS have performed exceptionally well, accounting for 110 technologies out of 152 technologies

---

[34]University-Industry Interaction. *University News,* 1996 (July 22) p.22.

[35]Department of Science and Technology-Annual Report(1995-96). New Delhi; DST, 1996.

developed so far by all of them. Further, only 64 of them have been commercialized so far[36].

3. *Establishment of Consortia*: Establishment of consortia is another important channel to promote multilateral linkages and success rate of the projects. A consortium comprising Indira Gandhi Centre for Atomic Research (IGCAR), Kalpakkam, Madras University and the industries in the southern region is proposed to be set up in Madras for promoting multilateral cooperation among the three constituents. Initially, Madras University and IGCAR have decided to form a structure for the proposed consortium to be called the Universities, Industries and Department of Atomic Energy Consortium (UNIDAC). Its activities will include establishment of science and technology window, a cyclotron as a multi-disciplinary research facility and an Environmental Science and Technology Laboratory. Setting up of the consortium would involve a capital outlay of Rs. 200 million including that for land, building and research facilities. It will have a governing council as a policy making body with members from DAE and the University[37].

4. *Home Grown Technologies*: TIFAC, an autonomous body under DST, has started a Programme of Home Grown Technologies (HGT) under which it monitors technologies available with the laboratories and matches these with the interests of industries willing to venture into the development, design, and commercialization efforts. It covers a wide range of sectors like steel, sugar, biotechnology, agrofood processing, environment, chemicals, energy, etc. Under the HGT programme, identification of the projects is based on the industry requirements, user needs and the expertise available (individual and institution) . Only part funding is given by TIFAC, even this fund is generally to be returned. So far, about 12 projects have been funded and another 20 projects have been cleared. Some of the funded projects are showing good results. Confidence level on commercialisation

---

[36]*Science and Technology Entrepreneurship Parks (STEPs)-Status Report,* 1996. New Delhi; DST, 1996. 64p.

[37]Varsity-IGCAR-Industry Consortium. University News, 1996 (April 15) p.17.

has been reported to have gone up considerably. For one project on high energy magnets, an industry is about to enter into large scale production. HGT projects have forged intensive collaboration among institutions and the industry. For example, the project on Synthetic Substitutes of Kerosene in Pigment Printing is being executed through an integrated approach. The working group consists of an industry: Gujarat Narmada Fertilizer Corporation (GNFC) and R&D Centre, and Ahmedabad Textiles Industry's Research Association (ATIRA), Ahmedabad. Development of the synthetic compound has been carried out at GNFC. Formulation of the compound is tested and checked at ATIRA with simultaneous feedback from other institutes like IIT-Kharagpur and the user industries[38]. Further, in cooperation with AICTE, it has launched a programme to tune courses and curriculum to the needs of industries[39].

5. *Programme Aimed at Technological Self Reliance(PATSER)*: Initiated by the Department of Scientific & Industrial Research (DSIR), this programme aims at supporting industry for technology absorption, development and demonstration. Partial financial support is offered for research, development, design, engineering (RDDE) projects undertaken jointly by industry and national R&D organizations and academic institutions. About 70 projects have been supported so far under this programme. Some of the projects have involvement of IISc, Bangalore and the IITs. These projects cover products and processes in various industries, such as metallurgy, electrical, electronics, instrumentation, mechanical engineering, etc. The total project costs are currently running at Rs 260 million, of which the contribution of DSIR is Rs 60-70 million[40].

---

[38]RAJAN Y.S.: *Problems of Absorption of Technologies in India.* Paper presented at the Joint Workshop on Science and Technology Development Strategies in India and China, organized by NISTADS, and Central- South University of Technology, Changsha, China, 25-26 September, 1995, New Delhi.

[39]RAJAN Y.S.: Technology Forecast and Higher Education. *J. Higher Education*, 1995, 18(3) p.463.

[40]*Programme Aimed at Technological Self - Reliance* (PATSER) Guidelines for Project Support. New Delhi; DSIR, 1994. 18p.

6. *Technology Development Fund:* To accelerate the development and application of indigenous technology to production process, a provision has been made to credit the 5 per cent cess on payments of royalty on imported technologies which was hitherto collected under the Research and Development Cess Act 1986, into a new Fund for Technology Development and Application. This fund is placed at the disposal of DST to help the indigenously developed technologies reach the stage of commercial application. In this concern, a 'Technology Development Board' under DST has been constituted. The main objectives are: (1) To provide equity capital to industrial concerns and other agencies attempting commercial application of indigenous technologies or adapting imported technologies for wider domestic applications; (2) to provide financial assistance to research and development institutions engaged in developing indigenous or adaptation of imported technologies for commercial applications[41].

Besides the above, some scattered efforts are also being made for example a practice school programme is offered by BITS (Pilani) in which a student spends seven and a half months of his five year degree course working in an organization[42]. A Science and Technology Park has been established at University of Poona. Tamilnadu is planning to create a Science City Authority- a platform that will enable business, research institutions and the academia to interact with one another. The main function of the authority will be to tailor the education system in the State to the needs of industry and business. Suggestions from the industry shall be studied regarding the change in curriculum and study materials and introduction of new courses[43].

Government Incentives

The government is trying to lure the corporate sector to increase investment in R&D to produce state of the art products through:

[41]*Research and Development in Industry* (1996) p.31.

[42]University—Industry linkages. *University News*, 1991, March 11, p.16.

[43]TN to set up Science City Authority. *Business Standard*, New Delhi, 1996 (23 August) p.4.

Five year tax holidays, three year excise duty holidays and zero custom duty facilities on specified goods imported for use in R&D projects partially financed by any government department or agency[44] . In this direction, a scheme has been launched to promote 'Commercial R&D companies' which will take industrial research as a business itself. These companies can avail a five year tax holiday under Section 80(i) (A) of the Income Tax Act. The government will provide full exemption from excise duty for three years on goods manufactured based on indigenously developed know-how and patented in any country of European Union, USA or Japan.

Problems

However, the problems remain. The link of national research institutions and educational institutions with industry has been extremely weak[45]. Little has been achieved, although the need for such linkages was felt almost 25 years ago. Active participation of universities in industriəl research has not materialized on a large scale, although some universities have established effective linkages with industry. Recent studies[4] have highlighted a number of problems concerning the absence of linkages between the academia and the industry in India. Isolation of universities is borne out by the fact that while UGC, CSIR and other government departments like DST, DBT etc. are the major supporters of university research (about 60 per cent), industry has a meagre share of only 2-6 per cent. The studies point out that although 70 per cent of the research in universities is applicable, most of the theses produced year after year mainly stand on the library shelves. This is mainly because those who research in universities are not well aware of industrial problems and those who face problems are not conversant with the possible methods of tackling them. The university sector of research is the single largest sector of research in the country with 20, 000

---

[44]Budget lures Cos. into R&D with concessions. *Times of India* 1996(lst August) p. 13.

[45]*National Policy on Education-1986*. Programme of Action 1992. New Delhi; Ministry of Human Resource Development, Department of Education, 1993. p.66.

scholars, and more than 10, 000 supervisors (60 per cent of them are PhDs). These figures compare very well with 1,14,403 R&D personnel primarily involved in R&D in the country with about 12,500 PhDs among them. Yet, this huge sector is allowed to vegetate. Further, the available technologies in universities, IIT's, and engineering colleges are not put to fuller utilization. In most cases, this is left to chance.

Major schemes like STEPS, HGT and PATSER, although initiated in mid/late eighties, have been pursued half-heartedly. It is disheartening to note that after an encouraging start of the scheme under STEPs in 1984, 11 STEPs were established during 1985-1989. And only one STEP was registered until 1994, which still needs the approval of the financial institutions. Further, at present only nine of them are fully functional. Only two STEPS have performed exceptionally well, accounting for 110 technologies out of 152 technologies developed so far, by all of them. Out of 152 technologies developed so far, only 64 have been commercialized. A systematic evaluation of the impact of these schemes as well as individual efforts made by the universities on the industries is urgently needed to mitigate the problems coming in the way of promotion symbiotic relationships between universities and the industry.

The other major cause is the lack of confidence of the industry in research institutions regarding their capabilities, as most of the technologies developed by them are either half-baked or of low quality as compared to international standards[46]. After independence, academic research in universities did not get the much-needed attention. University research was clouded by other priorities in the form of mission-oriented research and development of nuclear and industrial research institutes to solve immediate problems of development[47]. That university research

[46]KHARBANDA V.P., & JAIN Ashok: Indigenization and Technological Change at the Firm Level-A case of Black and White TV Picture Tube. *Technovation* 1997, 17(8) P. 439-456.

[47]RAINA Dhruv, & JAIN Ashok: The advent of Big Science and the Evolution of the University and Scientific Research System in India. In John Krige, Dominique Pestre (Eds): *Science in the Twentieth Century.* Harward Publishers, Forthcoming, 1997.

is being neglected is evident from the fact that there has been a declining trend in the output of MScs and PhDs during the nineties[48]. An inherent constraint of weak linkages between universities and industry is the narrow base of in-house R&D units in industrial enterprises and viable research groups in academic institutions. Studies conducted at NISTADS reveal continued existence of such weak linkages[49]. Due to unwillingness on the part of the government and lack of confidence of the industry in the academic institutions, there is often a time lag in the transfer of knowhow from laboratory to industry with the result that investments for further R&D on product development are low and the interest of the scientists is killed[50].

CHINA

In China, up to 1985, close academia-industry interactions were almost non-existent. The S&T reforms initiated in 1985 were directed to intensify such a relationship. During the reform period, a number of mechanisms have been evolved to effectively link academia with the industry. There has been significant progress in linking research institutes with production which has facilitated efficient technology transfer through various measures. Over the years, the following approaches have been adopted:

a. Contract Research

In 1985 alone, 7469 technology transfer contracts were signed by academic institutions with industrial or other recipients. Classified by category of recipients, large and medium sized state-owned enterprises accounted for 47.9 per cent, small state-owned enterprises for 29 per cent, collective enterprises for 7.4 per cent,

---

[48]BHARGAVA S. C., JAIN Ashok, & SHARMA Lalita: Declining trend in MSc and PhD Manpower-Modelling and Forecasting. *Current Science* 1994, 66 (6) p.398-400.

[49]*A Study on Linkages of Research Institutions with Academic and Industrial Organizations.* New Delhi; NISTADS, 1989. 101p. (NISTADS-REP-72 (TD) 89).

[50]KHARBANDA V.P., & JAIN Ashok: 1997. Op cit.

village and township enterprises for 9.6 per cent and other entities for 6.1 per cent[51]. Statistics reveal that about 80, 000 S&T research projects are being conducted in China's more than 1000 institutes of higher learning. An investigation of 17 colleges and universities in Beijing, Shanghai, Jiangsu and Hubei shows that during 1991-92, more than 5000 inventions by college students have been put to use. These are reported to have been received well by the industrial circles[52].

b. Teaching- Research-Production(T-R-P)

For university-industry cooperation, a number of *T-R-P Associations* have been formed. These associations help in facilitating transfer of technology for new products and processes developed by the academic institutions to small and medium size or village and township enterprises as they lack R&D capability. University cooperation with large and medium size enterprises is growing steadily in order to: (a) update their technologies; and (b) to assimilate the newly introduced foreign technologies. A number of universities in China are shifting to manufacturing to finance their research and educational programmes. For example, Beijing University has established a number of companies, e.g. The Bei Da New Technology Company; The Bei Da Fang Zheng Co.; and Bei Da Weimin Bioengineering CO.[53]. According to reports, per capita R&D expenditure by these university-run enterprises was about US$ 850 in 1995[54] which is much higher than the average per capita expenditure on R&D in China.

The success of these college-run enterprises set up by the universities is evident from the fact that some of them have been listed among the country's top 100 high tech enterprises. These are Beijing University's Founder Group, Xi'an Jiaotong

---

[51]HUANG Shiqi (1988) p.18.

[52]Scientific Inventions by College Students. *Beijing Review* 1993, 36 (6) p. 21-23

[53]CUI Lili: The Metamorphosis of Beijing University. *Beijing Review* 1993, 36 (13) p.19-22.

[54]*China Science and Technology Newsletter* 1995 (63) Dec. 15, p.3.

University's Kaiyuan Group, Qinghua University's Ziguang Group, and Fudan University's Fuhua Group[55].

c. School Based R&D Centers

Several School based R&D Centres have been set up jointly by the enterprises and the academic institutions. For example, the Ministry of Petroleum and the General Petrochemical Corporation have set up seven such Centres in close cooperation with nine academic institutions. Salient features of these Centres are as follows: (1) Liberal investment from industries combined with the existing facilities of academic institutions is very cost-effective in organizing R&D activities; (2). It is easier to have a well conceived research programme; and (3). It is possible to realize better integration of basic research, development, design and production[56]. Some 67 R&D Centres for Engineering Technology were established by 1993 by the State for improving R&D for engineering technologies and strengthening the connection between academic institutions and the enterprises and promoting transfer of research results to enterprises.

d. High Technology Development Zones(HTDZ)

These HTDZs (similar to S&T Parks in developed countries) are being promoted through Torch Programme(TP) initiated by SSTC in August 1988, with the objective to develop new and high-tech products in close cooperation with universities and CAS research institutes[57]. TP helps in commercialization of high tech and new technologies resulting from '863 Programme' and 'National Key Technologies Programme' . The Beijing New Technology Industrial Development Zone founded in 1988, was the first of its kind in China. So far, it boasts of more than 2000 new and high tech enterprises employing 50,000 people.

---

[55]*China Daily* 1996 (Oct. 30) p.4.

[56]HUANG Shiqi (1988) p.21.

[57]These HTDZs are essentially Science and Technology Parks (STPs) and is an industrial complex, close to the place of learning like universities, colleges or polytechnics or research laboratory having formal or informal links. It is designed to encourage formation of knowledge-

contd..

In 1992, it realized trade income of 16 b yuan and industrial output of 5 b yuan.

By 1992, 52 such national level zones had been established; these house about 12,900 high/new tech enterprises. About 1000 of them are affiliated with the educational sector. At present, China has 120 HTDZs at different levels. These zones have been developed with different specialities. For example, the 'Shenyang Zone' features integration of machinery and electronics, while the 'Nanning Zone' focuses on the development of agro-biotechnology. 'Daqing Zone' established in 1992, places high priority on promotion of development of oil and petrochemical equipment and refined chemical technology. This zone enjoys cooperation from Qing Hua University in Beijing, to develop low temperature nuclear heating systems to cope with severe winters. It has also linked itself with Micro-organism Research Institute of CAS in respect of a key project which relates to direct pouring of micro-organisms into the strata for raising oil recovery by 30 per cent .

The first Torch belt was set up in 1991 in the Yangtze river belt with the approval of SSTC. In 1992, it undertook 120 Torch Programme projects. Projects listed in the Torch Programme are entitled to secure funds from the State budget or preferential loans from banks for start ups. Funding by the State increased from 750 m yuan in 1990 to 1.6 b yuan in 1992. It is estimated that through the implementation of the TP, the total industrial

---

contd.

based industries in a high quality and competitive environment. It has a management function for transfer of technology and business skills to enterprises on site; aims to reduce the time gap between scientific invention and its commercial application. The first STP was established in Stanford University, USA around 1950. However, the STP movement picked up only in late 1980's. At present, there are 200 parks in the USA, 52 in UK, 86 in Germany, 27 Technopolis in Japan, and 120 High tech zones in China. In UK, the 52 parks have come together to form a Science Park Association. (NAIK B. M.: Technology Park and its relevance to India. University News 1994, (November 7) p. 65.

output value may increase from 15b yuan in 1992 to 250 b yuan by the end of the century. Out of the total of 12,000 projects, one third are export-oriented[58]. The major technologies being developed in these zones relate to new materials, biotechnology, electronics, information, integration of machinery and electronics, efficient new energy sources, energy and environmental protection technology. Some high technologies, such as those in the fields of fibre optic communications and electronic information have grown into major industries in Wuhan, Hubei and Beijing HTDZs[59].

e. High/New Technologies Pioneer Service Centres(PSCs)

These PSCs (also known as incubators) have been established in high tech zones to transfer scientific research results to production. The first such Centre was founded in Wuhan in Central China in 1987. Since then, more than 70 incubators have been established in various provinces and cities. These Centres provide scientific workers a place to turn scientific results into technical applications. These, thus act as incubators for research institutes and enterprises. Firms and research institutes need only to pay low rents to develop their products. These products are exempted from tax for two years. It is claimed that 'Pioneer Service Centres' in various provinces and cities have become an important part of the Torch Programme to the extent that the Shanghai Scientific and Technical Innovation Centre has set up a 'Pioneer village, where all institutions of higher education and research are entitled to enjoy preferential policies granted by the State. Services such as industrial and commercial registration for enterprises, housing lease and facilities for loan are available[60].

---

[58]WEI liming: Torch Programme - Making rapid progress - *Beijing Review*, 1993, 36 (10) p. 18-23

[59]*Beijing Review* 1994, 37 (18) p.7.

[60]WEI liming (1993) : op cit p.18-23.

f. Joint Exhibitions

Another important form of linkage is holding of Joint Exhibitions by the academia and the industry. One such exhibition was co-sponsored by the Central Committee of the Chinese Communist Youth League, the All China Student Federation and the 'Si da' Technology Development Centre - a non-governmental corporation in Beijing. The 470 items on display were selected from 5000 inventions submitted by 600 Chinese colleges to a national contest for student inventions in various fields like machinery, electronics, chemicals, computer sciences and telecommunications. Nearly 1000 industrial units entered into negotiations for technology transfers. Of the prize entries, eight were put up for auction. For example, the ultrafine magnetic production technology invented by Gu Hongchen and two other students from East China Chemical Engineering College earned the top bid of more than 2 m yuan and was bought by Shanghai Printing Ink Factory. According to experts, bio-adhesives developed by students at Jilin University are at a par with advanced international standards. Organizers of the exhibition felt that institutions should track markets and economic development and adjust curriculum accordingly[61].

To promote such relationships, China has established its first University - Industry Cooperation Committee to accelerate the translation of scientific research results into commercial products. In this direction, Law on Transformation of S&T Achievements is also proposed to be enacted[62].

The above-mentioned efforts do indicate that conditions are being created to have an efficient link between knowledge generating institutions and the industry in China. However, several problems still exist.

Problems

The first problem is the lack of coordination between educational training and industrial demands. When making plans to accept students, schools

---

[61]Scientific Inventions by College Students (1993) : op cit p.21-23.

[62]University, Enterprises Establish Committee. *China Science and Technology Newsletter*, 1995(Oct.15) p.4.

consider only funds and faculty resources instead of social demand. Further, the educational system remains backward in many ways. Specialties that are not associated with social development are given undue weightage. For example, colleges in Beijing now admit about 200 students per year to study Chinese medicine. However, the actual annual demand is only about 50[63].

The second problem is that of mobility. According to the present regulations concerning personnel, academic institutions cannot freely employ additional personnel for R&D without permission of the planning and personnel department. Lack of mobility of R&D personnel is still a great handicap. Jobs are still assigned by the State instead of encouraging graduates to find jobs by themselves[64].

The third problem is that of extremely low investments by industry in R&D. While the present reforms have made scientific research institutions to invest in applied R&D, insufficient enterprise reforms to make them invest in R&D inhibit efficient transfer of technologies. Further, the interface of technology transfer is still tilted towards R&D institutions, which is one of the reasons for insufficient transfer of technology to enterprises. This calls for more R&D investments by the industry, which would make them participate at an early stage of the perception of projects. One sided reforms of the scientific institutions have already led them to pursue short-term R&D activities. According to a survey, the projects have already become smaller and scattered. This may result in continuous degeneration of the comprehensive capabilities of technology supply and competitiveness[65]. To cite one example, in the past decade, China has put billions of yuan into '863 Programme' and till date has achieved about 1,398 research achievements in biotechnology, space, information, lasers, automation, energy and materials. But the problem remains of how to transfer the results into successful products; a major part

---

[63]ZHOU Bian: College Graduates Facing Market Challenge. *Beijing Review*, 1993, 36 (12) p. 27.

[64]ZHOU Bian: Ibid. p.28.

[65]JIA Xinmin: *Study on the Interface of Technology Transfer in Transition Economy*. Beijing; Chinese Academy of Sciences, 1995. 73p.

of research remains in the hands of researchers. This is mainly due to the fact that the scientists and technicians are unfamiliar with the work of transfer of results into goods[66]. Weak linkage between research institutions and the industry is also obvious from the fact that negative balance of hitech imports and exports has been continuously increasing over the years. It reached 11.23 billion US$ in 1993 as compared to only 4.47 billion US$ in 1987 (Table 5) . This shows that not enough efforts are being put in for absorption, adaptation and further innovation on the imported technologies. Statistics indicate that only 20 per cent of high technologies developed in China find applications in industry, and according to Ma Hong a famous economist in China, it is still lower[67]. Linkage with knowledge base is an essential requirement to manufacture 'knowledge intensive products' and to narrow down the technological gaps and reduce negative balance in technology trade.

Comparison

An overview of the efforts made in China and India in creating formal mechanisms to strengthen linkages between the academia and the industry indicate the following:

- Although Chinese efforts in establishing such linkages began only in mid-eighties, government support to promote such linkages has been more diverse in the form of High Technology Development Zones, R&D centres, technology incubators, university enterprises and promotion of scientist entrepreneurs. Though Indian efforts in this direction began very early in sixties, government support and implementation have not been forthcoming and whatever has been achieved has been mostly based on individual initiatives of research institutions and the industrial firms. Even after mid-eighties, the progress has been very slow.

---

[66]*China Daily,* 1996 (April 18) p. 4.

[67]XIAO Lu: Technology can Revive State Firms. *China Daily Bussiness Weekly,* 1996 (July 28 - Aug. 3) pl.

- Development of High Tech Zones (known as S&T Parks in western term) has been much faster as compared to India, although both the countries started almost at the same time (during mid eighties). India's STEP scheme (similar to S&T parks) has had very sluggish and discouraging pace of development as compared to its Chinese counterpart.

Table 5: Import and Export of Hi-Tech Products (1987-1993)

million US $

	1987	1988	1989	1990	1991	1992	1993
Total	6309	9418	8701	9653	12316	14708	20585
Exports	918	1286	1848	2686	2877	3996	4676
Imports	5391	8132	6853	6967	9439	10712	15909
Balance	–4474	–6846	–5005	–4281	–6562	–6716	–11233
Ratio in total goods, I&E (%)	7.6	9.2	7.8	8.4	9.1	8.9	10.5

Source : *China Science and Technology Newsletter*, 1995 Dec. 30 (65) p4.

- Both the countries still face acute problems of successful transfer of technologies from academic research institutions to industry. There are four reasons for this:

(a) Technologies developed by academic institutions do not match the interest of the enterprise, as the latter is not involved at an early stage of initiation of the project;

(b) Technologies developed by the research institutions are not tested for commercial applications because of lack of funds;

(c) Insufficient funding by the enterprises either for In-house R&D or to the academic research institutions to execute the projects. With the present level of R&D funding by the industry in both the countries, as compared to that in advanced countries, these cannot compete effectively in the international market.

(d) Funds devoted to R&D are too scattered in both the countries, with the result that critical investments for R&D fail to mature into commercial applications.

Policy Options for Improvement and Conclusions

A number of studies have revealed that the performance of enterprises in S&T parks is far superior to that of similar "Off Park" enterprises. Employment generation is fast, besides increased industrial output and exports[68]. Further, it provides a high class infrastructure and environment in the vicinity of educational institutes, where qualified scientists and engineers grow professionally. It is not enough now to provide engineers and technologists, but it is equally essential to offer them an environment where they can grow academically and professionally and not migrate. It is fully realized that without building up such capabilities for technological innovation, it is next to impossible to survive in the present-day globalized competitive world. This capability can be built up only if one exploits the knowledge base to the fullest extent and consequently builds up proper linkages between the academia (the generator of knowledge), the industry (the potential users), and the financial institutions to provide the necessary venture capital. Further, these linkages can survive only in a conducive environment with suitable government policies, rules and regulations. While both India and China have built up sufficient infrastructure for generation of knowledge base, they still lack sufficient linkages to generate technological innovations to compete in the international market. Although China has done comparatively well in building up linkages with industry, Indian efforts still need to be focussed. This calls for intensive government actions to promote industry to invest in R&D through contract, sponsored, or joint projects/programmes, building up of consortia and establish effective linkages with academic institutions. It is well recognized that the governments alone cannot bear the total burden of financing higher education as well as research. This is what is exemplified by the advanced countries where industrial R&D expenditure is 70-75 per cent of the total expenditure on R&D; the government spends the rest 20-25 per cent. In India

---

[68]NAIK B. M.: Technology Park and its relevance to India. *University News* 1994, (November 7). p. 65.

and China, it is just the reverse. Further, the multinational corporations in advanced countries spend about 4-5 billion US$ per annum on R & D, which is more than the total expenditure made by India and China. In such a scenario, it is very difficult for Indian and Chinese firms to compete in the international market unless some concrete and selective steps are taken. Therefore, the developing country firms, in order to compete, apart from mobilizing additional resources, have to pool resources and select high priority areas of their competitive advantage through networking. While China has already taken steps to allow universities to go in for contract research, establishing joint companies, pooling of resources with other research institutions, High Technology Development Zones, and S&T incubators, India still has to take some concrete steps. There have been some efforts by DST, AICTE, UGC, IIT's, etc. in this direction but these need to be exploited on a large scale. The Programme of Action (POA) related to the National Policy on Education, 1992, while acknowledging the crucial role of higher education in training manpower for national development, emphasises that it has become necessary for institutions of higher education to consider measures for raising internal resources to revitalize[69].

With the new phase of liberalization, globalization and privatization, industries in both the countries have to face a stiff competition with large multinational giants. It is well recognized that only knowledge-based industries can be successful in the present scenario of global competition. It is thus necessary that the research projects undertaken by R&D institutions must meet the societal demand apart from keeping abreast of new knowledge generated. Research institutions, industry, financial institutions and the government must work together in consonance with one another to be really effective in the present scenario. Several steps need to be taken in this direction.

---

[69]Annual Report 1994-95. Ministry of Human Resource Development, New Delhi; Department of Education 1995 (Part I), p.245; The Punnayya Committee Report, UGC, New Delhi,1993; Swaminathan Committee Report, AICTE, New Delhi, 1994.

(1) At present, the industrial funding for research in both the countries is extremely low as compared to advanced countries. Till recently, Indian industries have been investing only 0.7 per cent of their sales turnover on R&D programmes as compared to 7-10 per cent in advanced countries. In Japan 80 per cent of the budget for R&D of each university is made available by the industry. With the squeeze in the government financial support, the industries must come forward to extend financial help to research programmes of the universities and promote teaching/training/ exchange programmes to solve their problems at an early stage.

(2) There is a need for close industry—institute partnership through strategic alliances. To a large extent, Chinese institutes have been able to forge linkages with industry through various national programmes like Torch Programme, Specialized R&D Centres, Pioneer Research Centres(incubators) and joint exhibitions. However, Indian efforts need more attention. This should be aimed at facilitating industrial exploitation of infrastructural facilities and human resources available in universities/institutes/ national laboratories, matching up with industrial needs, curriculum, joint R&D, information sharing/data bases, joint seminars, exhibitions and training of teachers and industrial personnel.

(3) The government must continue to invest in strategic areas of basic research to keep abreast of new knowledge generated and its possible future applications.

(4) There is an urgent need to establish National Innovation Systems by creating regional Techno-Economic Networks to meet the above objectives.

PART IV : ISSUES OF TECHNOLOGY TRANSFER AND INTELLECTUAL PROPERTY RIGHTS

8. Problems in Absorption and Adaptation of Technologies in India

Y.S Rajan, Sunita Wadhwa

Abstract

Technology transfer process is important for industrial growth, and involves a number of critical steps in technical aspects as well as in management. The paper gives case studies on technology development and transfer efforts in different sectors in India, bringing up some issues and problems in technology transfer. In the country, a number of useful and meaningful technologies have been developed at lab-scale which can be brought to commercial reality. A few agencies are making efforts in this direction, through funding programmes, such as DSIR's PATSER, NRDC's schemes, TIFAC's Home Grown Technologies, etc. to develop linkages between R&D groups and industrial production groups. The experience and studies reveal that the degree of technology dynamism of the firm is the determining factor for its technology transfer and absorption capability.

Introduction

There are a number of successful cases of technology development, absorption and adaptation in India. These have made an important contribution to industrial and economic growth. The overall process of technology transfer involves a number of critical steps in technical aspect as well as in management. In this paper, we have primarily addressed the problems faced, reasons for failure and methodologies that were or could be adopted to tackle them. Success in the development of technology is judged as its application strength, i.e. it should succeed in the market place or with end users. In India, about 80 per cent of science and technology-related development work is being undertaken by government-funded S&T departments, agencies and institutions. The contribution by the private and public sector industries is not yet significant. The main source of technology in Indian industry has been through import of technologies. Technology development for industries to some extent takes place in the in-house R&D units of the industries, non-profit Scientific and Industrial Research Organisations (SIRO's), cooperatives and research institutions. This paper addresses a few examples of technology transfer to help derive a few broad conclusions as to the nature of the processes involved.

The Innovation Chain

As shown below, technology development and transfer is a step - wise operation starting from basic research to application by regular use of the technology, though it is not necessary that all steps are to be followed in all cases. The steps need not be linear either.

Linkages in Process Technology Development and Transfer

Research.. →Finalise use
(Operation or Business)

O .... O .... O .... O .... O

A B C D E

←.............................Technology..............................→
Development

Research

Regular
use

An effective linkage between these steps is necessary for successful development and diffusion of a technology. As indicated in the diagram above, the major part of technology development takes place around steps B and D. The bulk of funding by 'grants giving' S&T departments and universities addresses mainly the step between A to B. There are only a few agencies which fund the portions between B and D, e.g. DSIR's PATSER scheme, NRDC, TIFAC's Home Grown Technology activities and a few financial institutes. These aim at promoting linkages among centres/agencies/institutes involved in generating research ideas to go through various stages of technology development and commercialisation. There are also cases where a few industries develop in-house some critical capability from B to D in selected areas, often after import of the main technology. Very often, they operate between D and E as problem solving or material substitution for imported equipment or process.

Technology Transfer: A Few Case Studies

This paper gives some case studies on technology development in India and the associated technology transfer efforts. One case study is related to a large sector in India where industrial strengths as well as R&D capabilities in the institutions (including the portions of B to D) are relatively strong, i.e. in the chemical sector. The second case study is of automobile sector in which industrial production capability is relatively strong; however, there are no major R&D institutions for technology development, as in the case of the chemical sector, and, therefore, all activities are mostly within the industry after acquisition of technology mainly through imports. The third case is a specific case of technology import and further adaptation by a public sector electronics industry which has a good tradition in indigenous technology development. The fourth case is a consideration of a whole sector of computer industry at macro-level, especially to look at the impact of policies (of import, taxes, foreign investment, etc.) on the internal process of technological development within firms. These case studies are drawn from secondary and/or tertiary sources. On the whole, the four cases cover a broad range of activities.

The first study referred to is that of the chemical industry in India[1]. The chemical industry in India has a strong R&D base and also a set of good industry-oriented R&D institutions. In this industry, scaling up of the lab-scale development is very critical. It is very difficult to draw a sharp line between technology development and transfer. There are only a few cases in India where development of indigenous technology merges totally into

---

[1]DORAISWAMY L.K.: *Technology Development and Transfer in India: Agony and Ecstasy. Indian Chemical Industry- Emerging Scenario.* Indian Chemical Manufacturers Association (ICMA) 1992.

its transfer and eventually commercialisation, going through the whole cycle indigenously. Therefore, an essential requirement for technology transfer is emphasis on the scale of operation in which the initial developer proves the processes, and this requires good engineering and innovative use of new design strategies. Fig. 1 depicts two categories of development and transfer, one adopted in chemical process development and other in mechanical/electrical/electronic devices equipment transfer. In the chemical process, there is scale-up effect, but in the other, strategies of mass production of a prototype become the controlling feature. Technology transfer becomes particularly difficult in systems subject to scale-up influences.

Fig. 2 shows the conceptual feature of the possible modes of technology transfer in India for the chemical sector. The model brings out three modes of laboratory-industry transfer of technology. The transfer from a laboratory to an industrial R&D (Mode-1) is the most effective mode, for it is transfer between two like cultures. Transfer between the laboratory and the pilot plant (Mode -2) is reasonably effective, but there are barriers arising out of the mismatch between what the pilot plant engineers expect and what the R&D scientists are prepared to give. The third method of transfer (Mode—3) as indicated in the model, is from an external laboratory (after validation in its own pilot plant) to the commercial section of the industry directly through a firm of project engineers; such a method can be effective only in cases where there is a flourishing project engineering culture in the industry. Many multinational companies transfer their technologies by this mode. Dr. Doraiswamy who was an eminent chemical technologist and also former Director of National Chemical laboratory, Pune from whose paper we have taken these examples, draws several conclusions from the experience of transfer of technologies in the above mode with industries. He emphasises the like-mindedness of working cultures as an important factor. He has also cited examples in addition to process development, where equipment developer and user (as in the case of catalyst) play an important role. He concludes that it is essential to have them associated in the early phase itself.

The second case is about the Indian automobile industry which is an example of a poorly developed R&D capability within the country but has a strong production capability. We draw from a study conducted on the "Foreign Collaboration and Industrial R&D in Developing Countries: Case of Indian Automobile Ancillary Industry"[2]. Investigation of factors related to Foreign Collaboration Agreement (FCA) reveals that though both foreign equity and the number of foreign collaboration agreements have some positive impact on the firm's Development, Adaptation and Absorption (DAA) capabilities, the impact is rather weak and limited. A number of factors restrict DAA, such as unfavourable clauses in FCA, profitability of the venture, level of wages and salaries, experience in the related production line, etc. These are described in the paper referred to.

There is another interesting feature that has come out of this study: When the comparative analysis of product development/ adaptation by FCA and non-FCA units was done, it was found that inspite of weaker R&D base, the non-FCA units in relative terms spend more on R&D and develop/adapt larger number of products than the FCA units.

The third case study presented is one in which know-how for adaptation, improvement and application was obtained from abroad by a public sector electronics company to develop and produce the X-ray Baggage Inspection System. Problems were faced during the transfer of technology, and in the manufacturing process while adapting for indigenous sources of components/ parts/accessories, and further development of certain subsystems with different new technologies which would have made the equipment competitive in the commercial market. But it was not possible because the technology for important sub-systems was refused by the original technology suppliers and there were several restrictive clauses in allowing modifications. This could be due to lack of vision on the part of the top management. Also, the customers' need and satisfaction were overlooked, and

[2]CHUNGAN P. K. : Foreign Collaboration and Industrial R&D in Developing Countries: Case of Indian Automobile Ancillary Industry. *Economic and Political Weekly* 1995, (August 26).

no attention was paid to making efforts to diversify the technology into related areas like industrial radiology and medical radiology. This case study indicates that technology transfer for a particular equipment to be sold commercially should not be viewed as a mere one-shot process; rather a strategic action plan for diversified uses and continuous injection of new technologies is required for long term commercial survival, especially in the fast-growing area of electronics.

The fourth is the case of Computer Industry in India which has grown through varying phases since 1960's and has a large turnover now, though it is still small compared to international standards. Indian Computer Industry has come a long way since its total reliance on imports from multinationals in 1960's. Now, India is heading towards development of industrial production, exports and technological capabilities. However, the growth has not always been in the direction of building internal technological strengths. There were brief periods in which industrial R&D in a few areas was strong and a few unique products were produced. However, the pressure from global market demands of users, and the need for quick financial returns are generating a different pattern of growth of industries; their emphasis is no longer on building technological niches because of expenses and uncertainty about returns. These are described in a recent paper rather exhaustively[3]. The following table reproduced from the paper provides broad features of the findings. The possible relationship of the patterns with the policies and markets are discussed in the paper.

[3]HEEKS Richards: Import Liberalisation and Development of Indian Computer Industry. *Economic and Political Weekly* 1995 (August 26).

Model within The Indian Computer Industry

Period	Dominant Model for Trade	Dominant Model for Consumption/ Production	Dominant Model for ownership
1960s	Imports	Consumption	Multinationals
1970s	Import substitution	Production	Public sector
1979-83	Import substitution	Prod. & Cons.	Private sector
1984-87	Imports	Consumption	Private sector
1987-91	Import & import substitution & some exports	Consumption & Production	Private sector & multinationals
1990's	Import & some exports	Consumption & some production	Multinational & private sector

Home Grown Technologies (Programme for intensive R&D institutions-industry linkages).

The above examples give some clues about the issues and problems in technology transfer. It may also be borne in mind that a lot needs to be done in the country in translating research strengths to commercial reality and these leading to enhanced competitive capability of the industries. Now, therefore, there is an increasing realisation that a number of meaningful technologies developed at the lab-scale need to be pursued to commercialise them. There are a few programmes in existence, such as DSIR's PATSER scheme and NRDC's efforts to develop and strengthen linkages between R&D groups and industrial production groups. Government policies are also oriented to make institutions to work with industries. We describe here a few efforts made by Technology Information, Forecasting and Assessment Council (TIFAC). TIFAC, an autonomous body under the Department of Science and Technology, has a primary function to do technology forecasting and assessment studies. In the process, it has excellent networked strengths with industries, institutions and government bodies. There is a demand on TIFAC to follow up with agencies to take actions on the technologies forecast. As a part of this effort, TIFAC has started a programme to support Home Grown Technologies (HGT), which supports joint technology development efforts between institutions and industries. Under this programme, technologies available within the laboratories are identified and matched with the interest of industries willing to venture into development and design commercialisation efforts. Under the HGT programme, identification of projects is based on industry requirements, user needs and the expertise available (individual and institution). Only part funding is done by TIFAC; even this fund is generally to be returned; this is a part of the condition to create a commercialisable approach. HGT projects largely cover some of B to C to D portions on the technology cycle indicated earlier. It covers the crucial phase where technological risks are perceived to be more as an incentive to the industry to work with the institution and also contribute part of the funds to the joint effort. So far, about 12 projects have been funded and another 20 projects have been cleared; various levels of detailed MoU's or agreements are being worked out to start the projects. Some of the funded projects are showing good results.

Confidence level on commercialisation has gone up very high. For one project on high energy magnets, an industry is about to enter into large scale production. An analysis of the large number of proposals received, those which are funded, and have common features, if any, etc. was done and discussed by the group which screens and gives recommendations on funding. It is found that it is not easy to adopt a single model. The approach varies from area to area. A lot depends on the informed judgement of the groups which obtain proposals, and judge them through the entire chain of innovation. An example of a project is given below:

As an example, HGT project on the Synthetic Substitutes of Kerosene in Pigment Printing can be elaborated. The project has been taken, because residual kerosene pollutes the environment, and there is a desire to move towards adopting environment-friendly processes and products. For execution of the project, an integrated approach has been adopted: the working group consists of an industry: Gujarat Narmada Fertilizer Corporation (GNFC) and R&D Centre; and Ahmedabad Textiles Industry's Research Association (ATIRA), Ahmedabad. Development of the synthetic compound has been carried out at GNFC. Formulation of the compound is tested and checked at ATIRA with simultaneous feedback from other institutes like IIT-Kharagpur and the user industries. The project has been in progress for over a year and has already achieved 75 per cent success. Efforts for achieving further improvement are still on. Completion of the other projects and undertaking of new ones will give better insights into such efforts to orient R&D for commercial ends. Government has recently supported a number of technology projects in a mission mode which encourage joint participation of industries and institutions in technology development.

Conclusion

Mastery of the technology transfer process is going to be the key factor for competitive advantages whether we talk of indigenous or imported technologies. Sometimes it is not easy to effect technology transfer even when it is within the same organisation or company as R&D groups and production groups have different goals. Mastery of the transfer process has not been easy in the past, though there are success stories. Technology development, particularly absorption and adaptation by industry, depend on and are controlled by a number of factors, such as human resources strength, policies, market, administration, etc. Often it is difficult to find out a universal formula even within a sector. The role of alert individuals and groups who are ready to take risk and look through the innovation chain is going to be increasingly important. Ability to live with technology dynamism[4] within the firm and institution is also going to be a crucial factor in determining their transfer and absorption capability.

---

[4]RAJAN Y.S. and WADHWA Sunita: *Technology Dynamism and Human Resource Development*. Paper presented in the International Seminar on Human Resource needs for Change in R&D Institutes held on 15-17 February, 1995 organised by WAITRO in New Delhi.

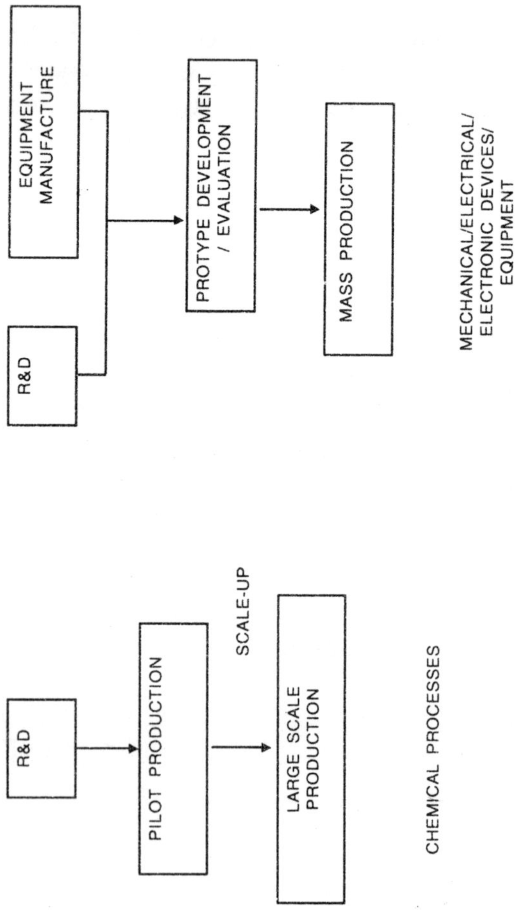

Fig. 1 : Two Categories of Development and Transfer

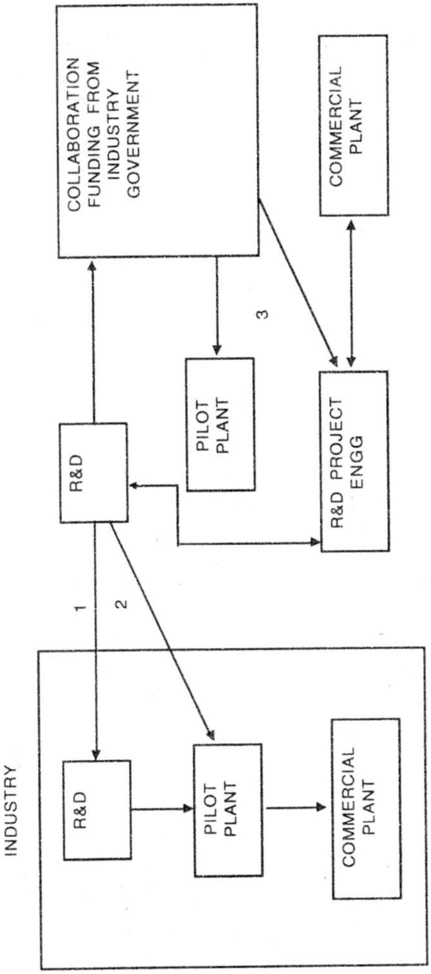

Fig. 2 : Modes of Technology Transfer in India

9. Survey and Analysis of Technology Transfer in the Chinese Transitional Economy

Jia Xinmin

Abstract

In early 1995, a survey and analysis of technology transfer in transitional economy of China was carried out with the support of Unesco. Presented here, in brief, are the findings of this survey. The author presents a model of mechanism of selecting and shifting the position of 'interface of technology transfer' between the research institutions (the supplier) and the industry (the consumer). It is concluded that while the ongoing reforms have laid the foundation for China to finally enter the market economy, the reforms have been one-sided. While the S&T system has witnessed widespread reforms, the requisite reforms of the enterprises are lagging behind, which has forced the interface of technology transfer, to shift downwards where early participation of the enterprise in the R&D project is negligible. This has resulted in widespread wastage of institutional resources. Concludes that if reforms of the enterprises do not coincide with institutional reforms, the current problem will continue, thereby threatening the development of China.

Introduction

Technology Transfer (TT) has been one of the most prominent areas for policy studies in many countries during the past decade. In fact, it has been considered worthwhile to study it as an independent topic, as also as a composite part of technology innovation theory. With this change, more and more literature and works on interface of TT have been published which have provided the basis for the present study.

This study was carried out with financial support from Unesco with the purpose of describing and evaluating China's TT interface under the conditions of a transitional economy. A number of research studies carried out in the past have been of value to our study on TT. For example, when designing the questionnaire for the study, we used the questionnaire designed by Prof. R. Nelson as reference at the recommendation of Prof. T. Turpin of Wollongong University, Australia. Our primary task was to find answers for the questions related to the impact of China's institutional reform on TT and to the current status and pattern of TT activities in China. An obvious fact is that because of the ongoing reforms in China, aimed at shifting its planned economic system to a market - oriented one, China's TT activities including the mechanisms involved, the selection of operational pattern adopted and the related cause - effect relations are much more complicated than those in foreign countries. A study of the existing literature reveals that they are the products of a significant historic revolution. A relatively simple and defined institutional system can hardly determine the forms and orientation of TT in China. This includes not only the co-existence of two kinds of economic systems, but also the institutional conflicts and frictions that accompanied the revolution. To make our study meet the requirement of users and to depict the complicated nature of the actual events, we carried out exploration and innovation on a theoretical and conceptual framework, and applied the results of transaction theory of institutional economics to the research on technology innovation and transfer. Based on this, we designed the model depicting the TT interface. This paper summarises the results of a detailed survey and analysis of technology transfer in the transitional economy of China.[1]

Theory and Method

In this study, we have introduced the transaction cost theory of institutional economics into the analysis of technology transfer

[1]*Study on Interface of Technology Transfer in Transition Economy.* Beijing; Institute of Policy and Management, Chinese Academy of Sciences, 1995, 72p.

mechanism[2,3]. On this basis, we have constructed a model of the mechanism of selecting and shifting the position of the interface of technology transfer.

The reason for our introducing the transaction cost theory is that technology transfer not only means the process of transformation of the technology itself from knowledge to capital form, but also the substitutional phenomenon of a process-propelling subject brought about by specialized work division accompanying with this process.

The process of capitalization of knowledge is composed of the seven stages: basic research/applied research/experimental development/pilot experiment/pilot production/standardized design/large-scale production, which have different functions but are linked up with one another from the first to the last. Under the traditional planned economy, such specialized work division and function linking are looked after by the Central Government of China through its directive plan. However, with the implementation of market-oriented reforms and the property rights protection and transfer, the government is no longer the sole owner of general property rights, including technology rights[4]. All matters concerning technology transfer wholly undertaken by the government have already lost objective foundation and basis. Whether technology transfer will take place or not and at what capital cost and in what ways, have become matters of concern to the rights owners. Without observing the technology transfer phenomenon from the angle of the transaction cost theory, it will be difficult for us to obtain an understanding of and mastery over the nature of the phenomenon of technology transfer taking place under the new environment.

When we use transaction cost theory analysis for mastering the technology transfer phenomenon. the aim in constructing

---

[2]COASE R.H.: The Nature of the Firm, *Economica* 1937, 4 (Nov.) p.386-405.

[3]WILLIAMSON Oliver E.: *Markets and Hierarchies: Analysis and Antitrust Implication*, 1975, p26-30.

[4]SHENG Hong: *Transitional Economics of China*. Shanghai; Sanlian Shudian and Shanghai Renmin Press, 1995.

the model of selecting and shifting the position of the interface of technology transfer' becomes application of the general adaptability principle to technology transfer.

The position of the interface of technology transfer means the time when the transfer of technology rights takes place between the parties interested in technology transaction, the stage at which the process of knowledge capitalization lies, or the level attained by it. We consider that the more close the stage at which it lies nearer to large-scale production, the more close the position of the interface of technology transfer gets to the lower reaches (backward or lower position). On the contrary, we consider that farther it is from the large-scale production stage, and is more close to the laboratory stage, i.e. the more close it gets to basic research, the more close the position of the interface of technology transfer gets to the upper reaches (advanced or higher position).

We take the position of the interface of technology transfer as a characteristic parameter for appraisal of technology transfer under transition economy. The reasons can be as follows:

First, technology transfer is an important factor having an independent influence under the conditions of transitional economy. Under such a condition, price is not the sole influencing factor for possible technology transfer. It also depends on the extent to which this technology been developed, namely, when entering the process of transaction, it depends on the stage at which this technology lies in the process of knowledge capitalization (basic research - large-scale production), and after this technology has been purchased by technology user, it depends on the extent to which the technology can be directly used in the large-scale production process. If it cannot be directly applied to production, and/or applied for the purpose of getting profits for the enterprise, then even if this technology is priced very low, the transaction cannot be continued. The reason is, that under this circumstance, not only would the price sought by the seller not be the total cost involved in completing innovation of the technology, but after paying the cost of purchasing the technology at the price demanded by the seller, the manufacturer also must continue to invest on completing the remaining work of knowledge capitalization left by the seller, i.e. further maturing of technology. Furthermore, even if the total is the cost paid for purchasing technology and the expense incurred on subsequent development, it might also not be the total content of technology cost for the innovation of this technology, because it will also depend on whether the buyer already possessed the general technology development ability needed for undertaking subsequent development work. If the buyer has this ability, then, roughly speaking, the aforesaid total expense on technology purchase charge and its subsequent development may be equal to the cost of technology innovation. But if the buyer does not possess this ability, then its cost composition will also include the expenditure to be incurred on training for acquiring general technology development capability. And the scale of this input is usually far larger than the cost paid for obtaining a specific technology right. As for this input, the manufacturer obviously cannot expect to obtain full return from the income earned from exploitation of this specific technology. It follows that if an enterprise (the technology buyer) lacks the pressure of long-term and sustainable demand for the innovation of the technology, then the motive of its innovation is aimed only at a specific technology, namely, occasional and only once. In that case technology transfer to be made on the upper reaches (advanced) position hoped by the seller obviously will never take place. And the only way to do the transaction would be for the seller to make a compromise on the problem of the position of the interface, and to undertake the liability which originally might have been undertaken by the buyer, catering to the capability of the buyer, and pushing the position of the interface to lower reaches. If both parties involved in the transaction stick to their original intentions, then this technology transfer has to be abandoned. Obviously, the position of the interface is a unique phenomenon of technology transfer under transition economy. Without paying due attention to it, we cannot obtain an understanding of the mechanism of technology transfer under this environment[5,6].

---

[5]FANG Xin: *Analysis and Evaluation on the Management System of Science and Technology of China*, 1992, (Mimeo).

[6]ZHOU Chengkui et al: *Study on Some Issues of S&T Reform*. Machinery Press, 1990.

However, selection of the position of interface of the technology transfer is not an arbitrary decision. The decisive factor for selection of the position of the interface of technology transfer is the key to understand the mechanism of technology transfer under transition economy. Of course, general marginal analysis may offer information of value for solving this problem, because the so-called closing up intermediate function fault essentially is an investment decision. Therefore, the respective inputs of the seller and the purchaser will decide, who will undertake this obligation, and to what scale, so as to decide the position of the interface of technology transfer. But the question is, the results obtained from this analysis always do not tally with actual observations. Sometimes there are large deviations. The reason for this is that in a general frame of marginal analysis, an important factor is always neglected, namely, the opportunity cost in the investment shared by the seller and the purchaser. As we know, the factors influencing actual investment decision include not only the ratio of cost to income of a specific selection itself, but also include whether the ratio of cost to income of this selection is larger than the ratio of cost to income of other investment opportunities. When the answer is positive, decision will be taken to make the investment. When no alternative opportunities except this one are available, it even surpasses the level of marginal scale to make investment, until the gross cost and the gross income cancel out each other. If the answer is negative, it will give up such a selection, to seek the selection of higher income at the same cost, or to seek the investment opportunity with still lower cost at the same income. When the parties interested in technology transaction are confronted with certain problems related to technology transfer function, some investments have to be made to overcome these problems following the above principle. For example, when for its existence and development the supplier (university or R&D institution) has stable and favorable resources, and does not have to depend on income from technology transaction, even if its technology is far from the level acceptable to the user, it is difficult for it to justify the motive of further expanding the scope of its obligation to the lower reaches to satisfy the requirement of the user. The user has to confront strong and long-lasting pressure of technology competition, and has to innovate technology sustainably and effectively. There is no other way out for it, except to make investment of its own to build up its own research and development capabilities (namely, the capabilities of forming the cognition, response, and development involving capability of the technology opportunity of the upper reaches). Otherwise, this technology transfer has to be terminated. Obviously, the position of the interface of technology transfer is decided by the opportunity cost of technology transfer met by the supplier and the purchaser under respective environments. The party with lower opportunity cost of technology transfer has a stronger motive and demand to rectify the functional fault in the process of knowledge capitalization, and propel shifting of the position of the interface of technology transfer. A party with higher opportunity cost lacks the above-mentioned motive. If the latter is the supplier party, then the position of the interface has the characteristics of downward deviation. If it is on the contrary, it will be reversed.

The meaning of the realistic policies concerning the position of the interface of technology transfer is: the technology transfers made on different positions of interfaces are not equivalent. Therefore, it performs the function of appraisal of the quality of the mechanism of technology transfer. The reason that people usually neglect the position of interface is: that they feel, that so long as a technology has completed the whole process of capitalization, no matter on what stage the transfer takes place, the social benefits would always be positive. The question of profit and loss is a matter of concern for the parties interested in the transaction. However, this view is wrong.

Usually, the reason for the interruption of technology transfer is breaking of the link between the above-mentioned functional chain, besides the technology itself. People may think that technology transfers at different positions might be equivalent, but it is not so, because the intermediate links (pilot experiment, pilot production, etc.) in the process of technology transfer are different from the typical state of the upper reaches and lower reaches in the process. At the typical position for upper reaches (in the stage of research and development), although the demand for resources is largely deviated from the resources endowment of the technology user (technology buyer), yet it highly tallies with the resources endowment of the technology supplier (technology seller). Similarly, at typical lower reaches position (large-scale production), although the demand for resources has comparatively great deviation from the resource endowment of the technology supplier, yet it highly tallies with the enterprises. But at the middle position, although its demand for resources is not very obvious as the cases mentioned above with regard to the deviation of the resources endowment of either the supplier or the purchaser, yet it is inconsistent with the endowment of either of the two parties. This means that without increased investment, there will be lack of an entity with ability to undertake intermediate function only by relying on its resource endowment.

This is, of course, reasonable. However, so long as we further deepen the gradation of the analysis, it will not be difficult to find out that the problem is not so simple. The key point is: if different investors offer funds to construct intermediate links for technology transfer, and link up the channel of transfer, then the nature of the capital used by them, and the characteristics of the resources formed, are very different. The intermediate links are constructed with funds invested by the supplier, namely, when the position of the interface appears in the lower reaches. The resources endowment of the supplier (university or R&D institution) is the fundamental research and development capability, and the gap in the resources needed for the intermediate functions is the knowledge of technology and goods of capital nature (factory building, equipment and raw and processed materials for pilot experiment, pilot products, etc.). To close up this gap, except diverting public welfare inputs held by the supplier for the society to support research and development, education undertaking, etc., the supplier cannot put to use any other existing stock. Under such circumstances, although technology transfer can be realized by shifting the position of the interface towards the lower reaches, yet it can be done at the price of consuming the comprehensive capability of technology supply of the State, namely, consuming basic resources of the society. Decided by the nature of the fund used by it, social income is negative. Then, let us look at the characteristics of the resources formed. As we know, as the entity of intermediate function, every pilot experiment has a very strong property right value. It has very low general adaptability with the new technology continuously turned out by the R&D institutions with the aim to make innovation. This means that each time a new technology is turned out, it will be accompanied by large scale capital renewal, with strong dependence on the increased investment. No doubt, this will devour and erode the quota of Solow's remainder and greatly reduce the net income which the society should have obtained from technology progress[7]. But under the opposite circumstances, namely, when the links of the intermediate function are constructed by the enterprise (the user), then the position of the interface is pushed upward to the upper reaches; what the enterprise has put in is its own resources. What it intends to construct is the R&D capability needed by the upper reaches position. Since production of R&D resources is a public function, these cannot be monopolised by the investor. His function is to turn private resources into resources meant for public welfare, and the public income is positive. On the other hand, since the capital goods concerned with intermediate links form a part of the capital of the enterprise, they can be mutually replaced through effective use of the remainder. Moreover, different from R&D institutions, technology innovations of enterprise have stability of orientation, accumulativeness, and continuity. The possibility of repetitive use of intermediate links, such as pilot experiment bases formed only once, is very large. All these have reduced dependence on the newly increased investment, favourable to the quota of Solow's remainder, thus enlarging the increasing net social income. Obviously, owing to the difference of the positions of the interfaces, and the great divergence of social income of technology transfer, asset rental of exclusiveness will by no means

[7]SOLOW R. M.:Technical Change and the Aggregate Production Function. *Review of Economics and Statistics* 1957, 39 (August) p312-320.

be completely divided between the interested parties. This shows the significance of appraisal tool of the position of the interface of technology transfer for the mechanism of technology transfer.

To sum up, the following five points are obvious:

1. Technology transfers at different positions of the interfaces are not equivalent;
2. No matter whether the gross income is positive or negative, it will not be divided between the interested parties in the transactions formed by the supplier and the purchaser, having obvious external nature;
3. When the position of the interface assumes the tendency of the upper reaches, the social income is positive, otherwise it is negative;
4. Besides being influenced by the resource endowment of the parties interested in transactions and the external supporting structures, to a larger extent, the position of the interface of technology transfer is stipulated by opportunity cost of technology transfer under respective environment of both the parties.
5. Owing to the above-mentioned advantages of the interface of technology transfer of upper reaches orientation type, the kernel formed by the mechanism of high-quality technology transfer is whether the reform has largely raised the unreplaceable position of technology innovations in the enterprise, i.e. whether the reform has effectively reduced the opportunity cost of technology innovations in the enterprise. Therefore, by the cognition of the position of the interface of technology transfer, we may not only be able to expand the analysis of the state of technology transfer to the level of system factor, but also in the meantime it enables us to appraise the developmental level of market mechanism and orientation of the reform.

Survey and Observations

The samples for the survey includes 150 universities, research institutions and 200 enterprises[8]. The data collected covers

[8]Data in the paper from reference 1.

interface of technology transfer, natural endowment of resource from both the parties, social infrastructure, share of transfer expenses and relevant opportunity cost of both parties. The results of the survey are given below:

1. The Position of Technology Transfer in Transition Economy

The interface where TT (technology transfer) occurred was classified into five types. The data provided in the questionnaire returned by enterprises and R&D institutes are summarised in Table 1 showing typical characteristics of downstream trends in technology transfer.

Table 1: Position of TT Interface

	Has been used in China	Has not been used in pilot plant	At R&D stage	At the initial stage of research
Enterprises	81.7	53.9	33.6	17.6
Univ. & R&D institutions	40.3	22.9	12.8	7.1

Table 2: Funding Source Constitution of Univ. and R&D Institutions

	Adminis-trative grant	Science funds	Project funds from the govern-ment	Project funds trusted by enterprises	Techni-cal services income	Others
%	18.6	5.2	22.1	23.1	16.0	15.0
Rank	3	6	2	1	4	5

Table 3 : Factors Strengthening Enterprises's Competitive Advantage

	State owned	Collec-tive	Joint venture	Share company	Total
Improve technology	3	1	1	1	2
Change product structure	2	2	3	2	3
Improve management	1	3	2	3	1
System reform	4	5	5	5	4
Diversify marketing	5	4	4	4	5
Others*	6	6	6	6	6

* *For "others", most answers explained it as "to pay more attention to manpower resource" cohesive force of enterprise" and "quality of senior managerial staff".*

This trend in China is closely related to the enterprises' resources features, especially the human resource features, which arise from historical factors. Under the planned mechanism, Chinese enterprises were not technologically self-sufficient, because technology upgrading was organized by national planning.

But this is not a unique reason. The highest frequency of technology transfer is at the position which "has been successfully employed by domestic enterprises". Strictly speaking it is not technology transfer but technology diffusion which has no or little risks of both technology and market and is obviously not linked to natural endowment of resources. In addition, for over 10 years of reform, the government has depended on a number of policies to optimize the function for developing technology in enterprises. Therefore, we have to carry out our study deeply to be able to explain this phenomenon.

2. Impact of Market-Oriented Reform on Demand and Supply of Technology

We used a series of criteria for investigating the impact of policy on technology suppliers and consumers. To be brief, we present two kinds of criteria here: for suppliers, investigating their structure of financial sources to evaluate the degree of their market dependence; and for consumers, investigating their positions during the course, while technology innovation affects their competitiveness. The results are discussed below:

(1) Impact of Reform on Universities and R&D Institutions

The purpose of China's S&T structural reform is to make S&T serve economic construction. The change in position of financial sources of universities and institutions can reflect the degree to which this purpose has been realised.

For the 150 universities and independent institutes, the general pattern of fund source is shown in Table 2.

By adding all the funds related to government sources, we get 45.9 per cent, i.e. less than a half of the total funds available to universities and independent institutes. In other word, more than 50 per cent of their funds have nothing to do with government agencies. We can say that the dependence of R&D activities upon government has been generally decreased, and market demand has become the main driving force for R&D activities.

(2) Dependence of Enterprises upon Technology

Table 3. shows that there have been no major changes in the importance of technology transfer in enterprises during the past 10 years of reform. Management factors are still the main reason limiting competitiveness of enterprises.

Cross - analysis giving the above results has no relationship with the size of the enterprise or whether it is a key enterprise or not, but has strong interrelationship with the features of the enterprise. Third ranking chosen by State Owned Enterprises (SOEs) to upgrade technology went to the first ranking in all other cases. The most noticeable distinction between SOEs and non-SOEs is related to ownership, which indicates that it is the root of the difference of management level. The management problem faced by SOEs is related to the more profound problem of management impetus and supervision cost which are concerned with ownership. This implies that before the reform in enterprise mechanism, which is focused on ownership, is completed, SOEs will have no solid foundation for a kinetic mechanism for stable technology progress.

Comparing the impact of reform in universities, institutions and enterprises, an obvious issue is the unbalanced result of reform. The lag in the reform of economic system, in particular the state-owned enterprises, is the key factor responsible for the unbalanced results. Under this condition, some reform can promote the demand for technology innovation in enterprises, but cannot establish the motive for enterprises to invest comprehensively in technology development.

These unbalanced results are likely to drive the interface downwards, which can bring about the reduction in R&D input. According to statistics for China, GERD/GNP dropped from 0.7 per cent in the early 1990s to 0.5 per cent in 1994. Meanwhile, because of market pressure, the phenomenon of short-term R&D activities was prevalent. According to our survey, it is a common view that projects became smaller and got scattered. Such a sustainable situation will also result in continuous degeneration of the comprehensive capabilities of technology supply and competitiveness.

In order to further verify the above facts, we analyzed the distribution of opportunity cost of both parties by investigating the interaction positions of technology suppliers and consumers. There are two main criteria: the pricing principles for TT transaction and methods adopted to solve the interest conflicts.

Data in Table 4 indicate that both parties have taken cost and return as the basis for fixing prices. The difference is that suppliers emphasize on cost and consumers emphasize on returns. Generally speaking, it is considered as the normal trend for profit made by subjects in their interactions. In practice, price determination is related to the methods used for payment. When prices are fixed by suppliers, lump-sum payment is preferred. On the contrary, payment in instalments is preferred by consumers according to the progress of technology development. This implies that technology suppliers shall share market risks with consumers. Because it is hard for suppliers to know whether consumers make profits or not, they cannot easily share profits or cannot cover their cost, even if innovation helps enterprises get returns. This shows the consumers' monopoly in the technology market. Data in Table 5 again show the consumers' superiority in interactions.

Table 4 : Principles of TT Fee Given by Supply and Demand

	Cost of R&D	Estimated profits	Payment power of enterprise	Regulat ions and policies	Generally accepted rules	Others
Supply	1	2	4	5	3	6
Demand	2	1	4	3	5	6

Table 5 : The Best Solution of Interest Conflicts by Supply and Demand

	Coordination between both parties	Application of law	Rely on each own upper authority	Wait for the other side to recess	Termination of cooperation
Supply	1	4	2	3	5
Demand	1	2	3	4	5

It can be seen from Table 5 that the most important method to solve the conflict of technology transfer is to coordinate profits between two sides. In case efforts at coordination fails, enterprises will attempt to utilize more active and stronger methods to solve the problem, while the R&D institutes behave mildly and passively. We learnt from our spot investigation that R&D institutes were not confident in their ways to settle dispute, except two-side consultations. One reason is that they did not think they had as many financial resources as enterprises to cover the expenses (visible and invisible) involved in contacting government agencies concerned and seeking legal protection. Another reason is that technology has its own limitations. So when there is dispute with one enterprise partner and there is no hope to settle it within a short period of time, they would rather change their partners and secure development costs as soon as possible.

Survey on Social Infrastructure of Technology Transfer

Social infrastructure determines raising element cost of technology transfer and effect opportunity cost of both parties from outside. We mainly investigated availability of two elements-information and finance - as well as getting cost and methods.

(1) Method and Channel to Get TT Information

From Table 6 we can see that market reform has largely reduced heavy dependence on government for information resources. The channels emphasized by both the supplier and the buyer are "special conferences" and "informal social relations". There remain big differences in the way orders for "publications", "professional conferences" and "enterprises consultants" are arranged. The differences in assessing publications probably arise from shortage of special publications which can systematically reflect technology demand of enterprises. There are unbalanced functions of consultants of institutions in enterprises who are responsible for exchange of information. This might be caused by unequal return from information supply for both parties.

Table 6 : Channels to Gain Technology Information by Supply and Demand

	Publications	Professional meetings	Informal Personnal Communication	Advisor to enterprises	Short-term exchange	Industrial associations	Medium organization	Advertisement exhibition
Supply	2	1	3	7	6	5	8	4
Demand	1	2	3	4	6	8	7	5

(2) Fund Sources, Raising Channels and Cost

Data in Table 7 show that reforms have increased the financial inputs and strengthened the motive of enterprises investment in innovations. We can see from the results that enterprises with self-owned funds took a leading role, while government was also a major fund supplier. Financing institutions played a secondary role in arranging external fund sources.

Again, different enterprises told different stories. Although self-owned fund was the first source for all enterprises, it is obvious that in respect of sources for external funds there are marked differences among different kinds of enterprises. The external funds of SOEs came mainly from government, which are usually non-reimbursable, while for collectives and joint-ventures, these funds mainly came from financing institutions, and the funds are reimbursable. As a result, there are marked differences in risk cost and transaction cost, and consequently in the total cost of TT. The SOEs enjoy an advantageous position.

Table 7 : Fund Sources for TT

Own fund of enterprise	Government special grant	Loans	Fund from supply & demand	Own fund of supply
1	2	3	4	5

In addition to fund source information, we studied the factors affecting the availability of loans. The results are given in Table 8.

Table 8 : Analysis of Factors which Influence Securing of Grants and Loans

	State owned	Collecti-ves	Joint ventures	Share company	Total
High profit in short-term	1	1	2	1	1
Good social benefit	5	5	3	3	4
Credit of the enterprise	2	2	1	2	2
Personal relationship with fin. inst.	4	3	5	4	5
Support of key person in the govt.	3	4	4	5	3
Scale of the enterprise	6	6	6	6	6

Appraisal of various indicators revealed that project benefits and reputation of enterprises were the most important parameters for credit fund evaluation. The influence of "governor's words" was significantly low. The indicators reflect achievements resulting from reform of commercialized financial structures. But the negative effect of reforms was increased pursuit of banks for short-term loans. This has an adverse effect on the capability of enterprises to innovate, especially since venture capital is not developed. The reforms can have a stronger affect on non state-owned enterprises, which rely mainly on credit funds. They can also influence opportunity cost of enterprise innovation by increasing financial transaction cost.

Another interesting fact is that non-SOEs did not care much about "governor's words". The reason for this ranking perhaps is that according to existing rules, state-owned banks, except in situations related to some national programmes, have no rights to offer loans to non-SOEs. So non-SOEs can get loans only from credit associations. In this case, a close relationship with non-state financing institutions is much more important for them.

To conclude, the chances to get fund for innovation are not same for SOEs and non-SOEs. The exclusivity of non-SOEs from government fund and state-owned bank funds has created the segmentation of capital market in China, which reduced efficient use of capital and increased the total innovation cost. It is worth noting that the victims of this unequal system of fund supply are non-SOEs who are usually regarded as the more efficient part of technology innovation. The outcome is more serious and disastrous than conventional capital market segmentation.

Conclusions

1. After over 10 years of reform, the dependence of Chinese technology transfer on government has come to an end. The motive for profit of both technology suppliers and consumers has been established. The present survey shows that the Chinese reform has rendered the economic planning system to be valueless and also expanded elements of market mechanism in transitional economy. This has laid the foundation for China to finally enter market economy.

2. The technology transfer currently in vogue in China is characterized by a downstream trend. Such trend if sustained, will lead to wastage of the institutional resources of China and seriously threaten or even disintegrate the national comprehensive strength and technology supply. This trend can become more acute with further opening of Chinese economy.
3. The factors responsible for creating the downstream trend of technology transfer in Chinese transitional economy might be relevant to the natural endowment of enterprise resource caused by planning economy. However, the more profound factors are the unequal progress of reforms in the S&T system and enterprises. The slower progress of enterprise system reform has restrained the demand of enterprises for technology innovation, leading to reduced technology development capacity for over 10 years of reform. Under these circumstances, the reason for maintaining technology transfer is increased pressure of the government on the reform of commercializing institutions. This makes the R&D institutions to divert R&D resources for filling the intermediate gaps in technology transfer. Therefore, under normal conditions, the motive of technology transfer in enterprises is weak, which hampers technology transfer. In China, the pressure imposed by the government on R&D institutions has built up the downstream trend interface of technology transfer.
4. As a special phenomenon of transitional economy, the above-mentioned consequences are unavoidable. But the situation can be changed with the reform of economic system and, particularly, the process of reform of the developed and well-established enterprises. Nevertheless, their requirements are harsh. If the lag in enterprise reform is not overcome, the current problems will continue and threaten the development of China seriously.

10. Current Status of Patent Protection in China and its Projected Future Development

Wen Xikai

Abstract

The present paper seeks to illustrate the status of patent protection in China, highlighting the present Chinese Patent System, related laws and regulations, enforcement of the Chinese Patent law and plans for the future. Points out that effective patent protection is ensured through both administrative enforcement and judicial mechanism at various administrative levels. With strict adherence to the reform policy, the Chinese authorities revised "their patent law in 1993 in tune with the international norms. On the basis of the experience gained so far, China is planning to further amend its patent law in the light of prevailing international practices.

Introduction

Ever since China adopted the policy of opening to the outside world at the end of 1970s, it has been keen to establish a patent system to protect IPR and to enforce the patent law. After more than 10 years of development, China, by 1996, had about 100,000 patent applications. This included applications for patents for inventions (28 per cent), for utility models (48 per cent), and for industrial designs (24 per cent). In general, 80 per cent of the applications were domestic applications (Table 1). But among applications for patents for inventions, about 44 per cent were by foreigners from about 80 countries and regions. This shows that a lot of foreign inventors and enterprises already know how to use the Chinese patent system to protect their rights. But there are many others who still do not know whether inventions could be protected in China, and how the patent law is enforced. The present article seeks to illustrate the status of patent protection in China. The aspects covered are: (1) The Chinese patent system; (2) Patent - related laws and regulations; (3) Patent law enforcement; and (4) Plan for the near future.

Table 1 : Patents Applied for and Granted in China

Year	*Total*		*Domestic*		*Foreign*	
	Applied	*Granted*	*Applied*	*Granted*	*Applied*	*Granted*
1985 (I April) to 1987 (31 Dec)	58958	9973	44754	9183	14204	790
1988	43011	11947	28582	11293	5429	654
1989	32905	17129	27367	15480	5538	1649
1990	41469	22588	36585	19304	4884	3284
1991	50040	24616	45395	21178	4645	3438
1992	67135	31475	61788	28311	5347	3164
1993	77276	62127	68153	56882	9123	5245
1994	77735	43297	67807	39777	9928	3520
1995	83045	45064	68880	41248	14165	3816
1996	102735	43780	82193	39721	20542	4059

Source: China Science and Technology Newsletter 1995 (59), Nov. 10.

1. The Chinese Patent System

The first Chinese patent law was adopted on 12, March 1984 at the Fourth Session of the Standing Committee of the Sixth National People's Congress. The law came into force on 1, April 1985. To meet the needs of the ongoing reforms and the open policy, the 27th Session of the Standing Committee of the Seventh National People's Congress examined and approved the bill submitted by the State Council on "Amendments to the Patent Law of the People's Republic of China". According to the amended Patent Law of China, which took effect on 1, January 1993, the term of protection has been extended to 20 years from the date of application; the patentable subject matter now includes inventions in all technical fields, including chemical substances and pharmaceuticals; the patentee enjoys the right of importation; and protection on patent right is intensified. With these amendments, the protection level provided for in the patent law is already up to the international standards. Facts have confirmed that in a little more than a dozen years' time since the end of the 1970s, and up to now, China has done tremendous amount of effective work to enhance patent protection. It achieved this in a little more than a dozen years; other developed countries took scores of years, or even a hundred years to do this. It established a relatively comprehensive legal system for the protection of patent rights, thereby attracting world-wide attention for its achievements not only in establishing the system but also in its enforcement.

2. Patent-Related Laws and Regulations and Patent-Related International Treaties to Which China has Acceded

During the past 10 years or so, apart from the important revisions done with regard to the patent law and its implementing regulations, the Chinese Patent Office (CPO) also formulated and issued a series of supporting rules and regulations, such as: "Guidelines for Patent Examination"; "Regulations for Handling Patent Affairs"; "Provisional Regulations for Investigation and Punishment of Patent-related Passing of Acts by Administrative Authorities for Patent Affairs", and "Rules for Implementing the Patent Cooperation Treaty", etc. Besides, the various localities and departments throughout the country all together issued and implemented about 300 patent-related regulations and rules. In July 1994, the State Council promulgated "Decision on Further Intensifying the Protection on Intellectual Property Rights", and established the Working Conference System of the State Council for Intellectual Property Protection. The General Administration of Chinese Customs also issued Broad Measures for the Protection of Intellectual Property Rights. In order to cope with these important measures of the State Council to intensify protection on intellectual property rights, the CPO has submitted a bill relating to the revision of Article 60 of the Patent Law to the State Council for examination and approval. With this revision, the administrative authorities shall be endowed with the right to seal, freeze, or detain articles involved in infringing activities while dealing with patent disputes. Regarding patent protection, China always lays stress on the strengthening of bilateral exchange and cooperation between China and other countries of the world in the field of intellectual property protection. Since its inception, China's patent system is geared to high international standards. China acceded to the World Intellectual Property Organization (WIPO)on 3 June 1980, and to the "Paris Convention for the Protection of Intellectual Property" on 19 March 1985. China was one of the signatory states of the Convention of Intellectual Property Protection with Regard to Integrated Circuits in 1989. On 1 Janauary 1994, China became member state of the "Patent Cooperation Treaty" (PCT) , and CPO became the Receiving Office as well as the International Searching Authority of the PCT. Then, on 1 July 1995, China became member of the "Budapest Treaty on the International Recognition of the Deposits of Microorganisms for the Purposes of Patent Procedures". China is also considering to accede to the "Strasbourg Treaty Concerning the International Patent Classification" and the "Locarno Agreement Establishing an International Classification for Industrial Designs". All these developments show that China has been constantly endeavouring to align herself with the international standards as well as international practices in the field of patent and intellectual property protection.

3. System of Enforcement of Patent Right

According to Article 60 of the Chinese Patent Law, any citizen, legal person or any other organization, when their patent right is infringed, may either institute legal proceedings before the People's Court through the judicial route, or ask the administrative authorities for patent affairs to deal with the matter through the administrative route.

(a) The Judicial Route

The various levels of People's Court in China are the backbone for the enforcement of patent rights. They attach great importance to the protection of patent rights, and have exerted unremitting efforts to faithfully implement the Patent Law. As early as February 1985, before the patent law came into force, the Supreme People's Court issued the "Circular on the Issues Relating to the Carrying out of Judicial Trial of Patent Cases". This Circular clearly defined the scope of the patent cases to be heard by the court, as well as issues relating to jurisdiction, litigation procedure, interpretation and determination of the technical facts, and so on. Thus, a patent litigation system which could be effectively operated was already in place. Later on, after the Patent law took effect on 1 April 1985, a series of judicial interpretation documents based on the judicial practice were issued, defining the territorial jurisdiction principle of patent infringement, concordance of procedures between patent infringement litigation and examination of request for patent invalidation, concordance between administrative mediation and litigation procedure in patent infringement disputes, calculation of the amount of compensation for patent infringement cases and so on. They greatly improved efficiency of patent litigation and ensured effective and impartial carrying out of the patent judicial proceedings. Besides, in order to facilitate the training of patent judges and strengthening the level of handling of patent cases, the Supreme People's Court stipulated that the management of patent cases be relatively centralized. Therefore, only 37 Intermediate People's Courts among a total of 337 Intermediate People's Courts all over China are designated to be the courts of first level for patent cases, and the 30 local Higher People's Courts are the courts of second level for patent cases. Starting from 1994, with a view to further intensifying protection for intellectual property rights, a number of higher and intermediate peoples courts have established Intellectual Property Right courts. Now, we already have a contingent of patent judges which is relatively stabilized and is basically qualified for judicial work. Up to the end of 1994, the people's courts had all together received 2516 patent cases, and closed 2100 of them. Among them, 65 per cent were patent infringement cases. Others were cases relating to disputes over right to a patent, or right to apply for a patent, whether a patent for invention should be granted or whether a patent for invention should be invalidated. Through the hearing of patent cases, the People's Court guarantees the legitimate rights of the patentee both at home and abroad.

(b) Administrative Route

In the light of the fact that China has a vast territory, large population and imbalanced economic development and with a view to strengthening the national patent work, it is stipulated in the patent law and its implementing regulations that local patent administrative authorities be set up in provinces, autonomous regions, municipalities directly under the central government, in some open cities as well as in the special economic zones; and that the responsibility for handling patent disputes be entrusted to them. Referring to "administrative enforcement of patent right", we mean that when anyone's patent right is infringed, the patentee or the interested parties may ask the administrative authorities to handle the matter. According to Chinese Patent Law, administrative enforcement of the patent right runs parallel to judicial enforcement of patent right, and it is optional for the patentee as to which one to choose. This is a feature of Chinese patent system. The administrative authorities for patent affairs handle patent disputes strictly in accordance with "Regulations on Handling of Patent Disputes by Administrative Authorities for Patent Affairs". They follow the same principles in the trial of patent cases as the courts, as far as the application of laws and regulations is concerned, and have adopted the summary procedure of the court in civil proceedings in their handling of the patent disputes. Where the parties concerned are not satisfied with the decisions of the patent authorities in the handling of patent disputes, they may appeal to the court within three months from receipt of the notice of the decision. This is to say that unless both parties are satisfied with the decisions of the patent authorities with regard to their dispute, the judicial route of settling the patent dispute is always open to the parties. By so doing, the consistency of handling patent disputes either through the administrative route or through the judicial route is guaranteed. All personnel handling patent disputes in the patent administrative authorities have a background of having graduated from an engineering or polytech university or college, and having been trained in patent laws and related laws and regulations. As far as ordinary patent cases are concerned, choosing the administrative route ensures quicker and less expensive settlement of disputes. According to a survey of 44 administrative authorities for patent affairs, up to the end of October 1995, altogether 3248 patent cases were received by these authorities, and 2847 of them had been disposed of accounting for 88 per cent of the total. According to incomplete statistics, among the cases received by the patent authorities, 60 per cent were patent infringement cases. The number of cases where the parties concerned were not satisfied with the decision of the administrative authorities, appellant to court for adjudication accounted for only 6.5 per cent of the total.

4. Effective Protection Close to International Level

According to Chinese Patent Law, legal responsibilities shall be borne by anyone who violates the law, including civil or criminal liability and exposure to administrative sanctions. The people's court and the patent authorities in China protect patent right earnestly in strict conformity with the patent law. Where an infringement case arises from unauthorized exploitation of a patent, the patent administrative authorities, while investigating and dealing with the matter, are empowered to order the infringer to stop all acts of infringement and compensate for the losses. Where anyone counterfeits a patented product or wrongly appropriates a patented technology, the patent administrative authorities will order him to cease all acts of counterfeiting, to provide the public with notification of his or her violation, and to pay a fine. In a case of serious violation, the criminal liability of the person directly responsible shall be investigated through application of relevant articles of the Criminal Law, and if found guilty, the person directly responsible shall be sentenced to a fixed term of imprisonment of not more than three years criminal detention or a fine. In a patent right infringement litigation, if the plaintiff thinks the facts of infringement are obvious, and his patent right is valid beyond any doubt, he can file an application with the people's court for property reservation or advanced execution so as to prevent further losses caused by continuous infringing acts on the part of the defendant. The people's court shall order the defendant to stop infringement activities, seal up and confiscate the infringing products or freeze a certain amount of the capital fund belonging to the defendant to ensure execution of the final judgement. In determining the scope of protection of a patent claim, the principle of appropriate explanation of the terms of the claim and the principle of "Doctrine of Equivalents" are followed by the people's court and the patent authorities. According to the latter, not only the technical features contained in the claims, corresponding to the technical features disclosed in the description and the appended drawings, are to be protected by the patent, those technical contents which are equal to those of the claim, should also be interpreted as belonging to the scope of protection of the patent. Where the people's courts or the patent authorities, after hearing and investigating, hold that the alleged infringement is true, they shall order the defendant to stop the infringing act and compensate for the losses on this basis. If the infringing act has done harm to the reputation of the plaintiffs patent right, the defendant should make a public apology and shall be ordered to eliminate the ill effects. In the hearing and investigation of foreign-related patent cases, the people's courts and the patent authorities stick to the national treatment principle. In case of discrepancy between Chinese law and the international treaty to which China is a party, the relative provisions in the international treaty shall prevail.

5. Measures China will Take to Further Perfect Her Patent System

Establishment of the Chinese patent system was an actual requirement for China's reform and opening up, to further perfect the patent system and to intensify the efficiency of patent protection, as well as to cope with the requirement of the country's dedicated goal to gradually improve the economic system of the socialist market. Therefore, in order to intensify patent protection, to punish severely the various patent infringing acts, the Supreme People's Court issued, in September 1994, "Circular on Further Strengthening the Judicial Protection of Intellectual Property Rights". China follows closely the new trends of international harmonization in patent protection. In respect of patentable subject matters, research on TRIPS of the GATT and protection of plant varieties are already on top of the agenda. China has taken keen interest in the development of information technology, and is always trying to update its automation project on patent work. To ensure better and more efficient patent protection, China is planning to amend its patent law again to further adjust and improve her examination and approval procedures of the patent office based on the country's own experiences and in the light of international practices. To conclude, China is willing to cooperate actively with all countries to further tap the potentiality of patent system, and to bring into full play its important role in enhancing the progress of science and technology and the development of national economy.

References

1. *Intellectual Property Protection in China.* Beijing; Information Office, State Council of the PRC. June, 1994.
2. WEN Xikai: *TRIPS and its influence over the Chinese Patent System. "Patent Law Research"* Beijing; Patent Document Publishing House, 1994. p.l-9.
3. *The Remarkable Achievements and the Basic Experience Gained in Implementing the Chinese Patent Law over the last ten years.* (Mimeograph) .
4. GAO Lulin: Speech at the Tenth Anniversary Meeting of the Entry into Enforcement of the Chinese Patent Law.
5. FEI Zhongyi: Procedures Before the Court Relating to IPR Cases in China. *China Legal Daily* 1986 (January 10).
6. *Annual Report of Chinese Patent Office(CPO):* 1989 - 1996 Statistics. Beijing; CPO, 1996.

11. Development and the Present Status of Intellectual Property System in India

N.R. Subbaram

Abstract

Traces the development of Intellectual Property Right System in India with the first legislation brought into force during 1856. Highlights the salient features of the present Patent Law, 1970 which came into force in 1972. Recognizing the importance of patent information, a Patent Information System was established in India in 1980. Discusses the measures being taken by the Indian government to revise its Patent Law, 1970 in the light of establishment of the World Trade Organization in January 1995.

Introduction

'Intellectual Property' is a class of property emanating primarily from the activities of the human mind. Historically, patents for inventions, industrial designs and trade marks for marketing a product were collectively known as 'Industrial Property'. These three forms of 'Industrial Property' together with Copyright in literary work form Intellectual Property. However, the term Intellectual Property is now used to include all property resulting from the exercise of the human intellect.

Just as the ownership of and transactions in other forms of property are governed by law, so is the case with Intellectual Property as well. Intellectual Property is, thus, protected and governed by appropriate legislation, i.e. the applicable law of the country.

Though many species of intellectual property are prevailing in many countries, in India currently the intellectual property system consists of patents, trade marks, designs and copyrights.

Development of Intellectual Property System in India

Patents

The existence of Intellectual Property system in India can be traced back to 1856. The first legislation on the subject was brought into force in the form of an 'Act on Protection of Inventions, based on the British Patent Law of 1852. As per this legislation, a system was introduced in India in the form of grant of certain privileges to an inventor of new manufacture for a period of 14 years. The Act was re-enacted with some modifications, as the Act XV of 1859, in which Patent Monopolies were called Exclusive Privileges. According to this legislation, an inventor of a new manufacture, by filing a specification disclosing the invention, could obtain the exclusive privilege of making, selling and using the invention in India and authorising others to do so for a term of 14 years from the date of filing of such a specification. In 1872, the 'Patents and Designs Protection Act' was passed, which was followed by the 'Protection of Inventions Act of 1883'. These Acts were consolidated as Inventions & Designs Act, 1888'. Subsequently, the Act of 1888 was replaced by the Indian Patents & Designs Act, 1911 (IPD).

The 'Indian Patents & Designs Act of 1911' established for the first time in India a system of Patent and Design protection under the management of a Controller of Patents and Designs. The 'IPD Act, 1911' was inforce when India became independent on 15, August 1947.

Based on the fact that the patent system has played a key role in the industrial development and economic advancement of many countries which are today recognised as industrially and economically advanced countries in the world, it was felt that the patent system that has been in vogue in India has not helped in achieving its fundamental objectives, namely, to stimulate and encourage innovative activities in the country and to encourage commercial utilization of the patented inventions to boost industrial development of the country and consequent economic advancement. Considering that India was on the threshold of major industrial development, it was felt that the provisions of the 'IPD Act, 1911' were not adequate and required modification.

At this stage, the subject was referred to Justice Rajagopala Ayyangar and he was requested to advise the government on the question of revision of the Patent Law. Justice Ayyangar submitted his report in 1959 with various recommendations for effecting radical changes in the Patent Law.

Based on the recommendations of Justice Rajagopala Ayyangar and also considering the changes proposed in the light of further examination by the Government of India, a comprehensive and consolidated Patent Bill was introduced in the Parliament in 1967.

The Parliament passed the Bill in its historical special sitting on 19 August 1970 after a long and gruelling debate. This Act (Act 39 of 1970) received assent of the President of India on 19 September 1970. The Patents Act, 1970, thus, came into effect from 20 April 1972 and is in force as at present.

The Patent Act, 1970 has many salient features; some of them are highlighted below:

1. Definition of patentable inventions
2. Definition of inventions which are not patentable.
3. Novelty of the invention based on worldwide knowledge
4. Exhaustive definition for drug/medicine
5. Requirement of providing information regarding the application being processed in different countries for the same or similar invention covered in the Indian application.
6. Only process protection available in the areas of pharmaceutical, food and chemical substances prepared by chemical processes (Section 5 of the Act)
7. Two types of term of patents (i) for drugs and food patents, the term being five years from the date of sealing or seven years from the date of the patent whichever period is shorter, and (ii) in all other categories of inventions 14 years from the date of the patent.
8. Provision of endorsing with the word Licences of Right in the case of inventions falling under Section 5 of the Act after the expiry of three years from the date of sealing.
9. No resident in India to file the application abroad without written permission of the Central Government in the event the application is to be filed without filing in India.

Alternatively, such an applicant has to file the application in India and he can file the corresponding application abroad after the expiry of six weeks from the date of Indian filing in case no secrecy direction on the application is issued by the Controller of Patents.

10. Grant of patent is subject to certain conditions.

Designs

The law for the protection of designs developed from the efforts to afford protection to textile or linens, cottons, calicoes and muslins. Over a period of time, this concept was extended to patterns and prints made by modelling, casting or embossment of an impression or ornament on any article of manufacture. Prior publication or use of designs before registration was not allowed and the concept of novelty was introduced as distinct from that of copyright in literary work and a separate law for copyright for industrial designs came to be established. The 'Patent and Designs Act' was passed in 1872 and was consolidated in 1888. Later, the Act was replaced by Indian Patent and Designs Act, 1911. After the Patents Act, 1970 came into force, modifications were incorporated to confine the 'IPD Act, 1911' to designs only. Thus, 'IPD Act, 1911' came to be known as the 'Designs Act, 1911' with effect from 20th April 1972 and is currently in force.

Trade Marks

Prior to India getting independence, more precisely on 11 March 1940, a legislation for the protection of trademarks was enacted in India for the first time. The Act came into force on 1 June 1942. Prior to the enactment of the Act, though there was no legislation governing registration of "Trade Marks" in India, such protection was possible by a Passing Off action under the Indian law of Torts based on the principles of the English Common Law relying on the guidelines of established British case laws. This Act of 1940 was later amended in 1958 and the revised 'Indian Trade and Merchandise Marks Act, 1958' was brought into force on 25 November 1959. This Act is currently in force.

Copyrights

The development of Copyright Law in India was closely associated with the British Copyright Law. The British Act of 1911 codified and consolidated the earlier Acts on different works. It also abolished the Common Law Copyright and introduced a term of life of the author plus fifty years. Further, it abolished the need for registration of copyright as a prerequisite for claiming right or enforcement of copyright.

In India, the first Copyright Act was passed in 1914, which was a copy of the British Copyright Act of 1911, with suitable modifications to make it applicable to British India. The Copyright Act, 1957, which is currently in force, replaced the above Act, which adopted many principles and provisions of the British Copyright Act of 1956. The Copyright Act, 1957 was passed on 4 June 1957 and came into force on 21 January 1958 with the intention to cope with the new problems in the law of copyright created by technological advances in the fields of communication, broadcasting, microfilming, photolithography, movies, cinemas and talkies.

A number of amendments to this Act were effected in 1983 mainly to avail of the benefits arising from the revision of the Berne Convention and the Universal Copyright Convention, to which India is an adherent. Further amendments were effected in 1984 with the main objective of discouraging and preventing the widespread prevalance of piracy in video films and records. Amendments made in the Act in 1992 extended the term of copyright protection from the life time of the author plus 50 years to the life time of the author plus 60 years.

The Copyright Act was further amended in 1994 mainly to define more clearly the terms computer and computer programme. The amendments mainly relate to:

(a) Making adequate provision for the special nature of computer programmes and for the protection of computer-generated works;
(b) Extending to authors, in respect of the original copies of their paintings, sculptures and drawings and original manuscripts of literary, dramatic or musical works resale share;

(c) Including within the scope of copyright in artistic works the right (subject to appropriate exceptions) to display the work in public;
(d) Promoting collective administration of rights through copyright societies in the interest of both the owners of rights and the general public;
(e) Dealing more effectively with infringement of copyright and related works;
(f) Further improving the functioning of the Copyright Board;
(g) Simplifying and improving the law relating to copyright and related rights, in the interest of the general public, in particular, of the users as well as the owners of such rights.

Information Contained in Patent Documents.

One of the important advantages of patent system is the variety of technical information it provides. The unique featires of the scientific and technical information contained in patents are enumerated below:

- They are uniform in structure
- They are available in a minutely classified manner
- They provide information not only about the concept of the invention but also its practical application.
- The information is not secret. It can be freely used for R&D and education without permission of the patentee

So far, approximately 30 million patent documents have been published world-wide; the annual increase is of the order of one million. Currently, many international agencies maintain the databases and help disseminate the information. The degree of use of information varies from user to user.

Establishment of Patent Information System in India

Recognising the importance of patent information, Government of India established a Patent Information System (PIS) in Nagpur in 1980. PIS functions under the administrative control of the Controller General of Patents, Designs & Trade Marks. Currently, PIS is being modernised under a project of UNDP and WIPO. PIS provides a variety of services connected with dissemination of information contained in patent documents. A glimpse of the picture of patents filed and granted in India is given in Table 1. The data since 1987 show that the number of applications for patents made every year is of the order of 3000 with the highest number of 3869 applications made during 1993-94. The applications made by Indians numbered around 1000 per annum. About 67per cent of the total applications for patents received in 1993-94 were from foreign nationals.

Table 1: Patents Applied and Granted in India

Year	*Total*		*Domestic*		*Foreign*	
	Applied	*Granted*	*Applied*	*Granted*	*Applied*	*Granted*
1987-88	3457	2104	930	588	2527	1516
1988-89	3598	3380	1077	795	2521	2585
1989-90	3661	1890	1040	519	2621	1371
1990-91	3764	1491	1180	379	2584	1112
1991-92	3552	1696	1293	551	2259	1125
1992-93	3467	1272	1228	251	2239	1021
1993-94	3869	1746	1226	442	2603	1304

Source: *Department of Science and Technology (DST): Research and Development Statistics- 1994-95. New Delhi; DST, 1996.*

Present Status of Intellectual Property System in India

Consequent to the GATT Agreement, the World Trade Organisation has been established with effect from 1 January 1995. Being a signatory to the establishment of the Agreement and GATT Final Act containing provisions of Trade Related Intellectual Property (TRIPS), India has to make certain modifications in the Patents Act, 1970 to honour the commitments made in the Agreement.

The provisions relating to TRIPS in the GATT Final Act cover seven categories. These are trade marks, copyrights, geographical indications, industrial designs, patents, integrated circuits and trade secrets. As explained earlier, in India, currently, laws relating only to trade marks, copyrights and industrial designs are in force. Only in the area of patents, the norms and standards envisaged in the GATT Final Act are significantly different from the existing Indian Patents Act, 1970. Therefore, India would have to amend the Patents Act, 1970 particularly with respect to allowance of product protection in all fields of science and technology, providing protection for micro-organisms and providing a uniform term of 20 years.

The GATT Final Act provides a pipeline period of five years for implementation of all provisions of the agreement. It also provides for an additional transition period of five years to those developing countries which do not presently provide for product patents in any field of technology, to extend product patent to those fields of technology. However, in respect of pharmaceutical and agricultural chemical products alone, the GATT Final Act requires that the member countries receive applications containing product claims in the above - mentioned areas from the date of the Act coming into force. If those products are granted patents and marketing approval in any other country, and the patent owner desires to introduce those products in the Indian market, he should be given exclusive marketing right for 5 years or till his pending patent application in India is approved or rejected, whichever is earlier.

To meet the requirements of granting Exclusive Marketing Rights for the applications for patents received in the areas of pharmaceutical and agricultural products, the Government of India had promulgated an ordinance on 31 December 1994. The bill containing the provisions of the ordinance was passed in both the Houses of Parliament in March 1999.

It is to be noted that the provisions of the Bill are confined to the grant of Exclusive Marketing Rights and matters related thereto and not for completely modifying the Patents Act, 1970 allowing protection for products in all categories of inventions and for the extension of the term of the patents. Government of India has also by the notification dated 3 January 1995 declared 72 countries as Convention Countries under Section 133 of the Patents Act, 1970. The countries included in the notification are those which are signatory to the GATT Final Act. These countries are in addition to those which had already been notified earlier, viz. United Kingdom, Australia, Sri Lanka, State of Ireland, Canada and New Zealand.

Protection for Plant Varieties

The GATT Final Act does not insist on the member countries to grant patents for plants and animals, but makes it obligatory for the countries to grant protection for micro-organisms and micro-biological processes and for the enactment of an effective system for protecting New Plant Varieties. The system for the protection of new plant varieties may be by way of patents, by *sui generis* system or a combination there of. India can decide on the form of protection for new plant varieties. India has opted for a *sui generis* system for the grant of protection for new plant varieties. A draft of the *sui generis* system has been drawn up by the Ministry of Agriculture, which is under consideration.

Trade Marks

With a view to consolidating the law relating to trade marks and providing better registration of trade marks for goods and preventing abuse of the use of fraudulent marks, the Government of India has drafted a bill, which has been introduced in the Parliament for consideration. The bill is currently being studied by a Select Committee of the Parliament. Some of the salient features of the proposed bill are:

(a) Providing protection for service marks in diverse areas such as banking, communication, financing, insurance, real estate, transport, energy, hotel industry, entertainment and advertising.

(b) Extension of the duration of the term of the trade marks and the subsequent renewals from 7 to 10 years

(c) Guidelines for rectification of the trade marks/service marks.

(d) Simplification of the procedure for registration by permitting registration in several classes of goods/services by means of single application.

(e) Confining the registration in one part of the Registrar of Trade Marks instead of the current practice of registration in either Part A or Part B of the Register.

12. Theoretical Approach for Choice of Appropriate Technologies for Rural Areas

M.A. Qureshi

Abstract

Three different approaches for choice of appropriate technology have been discussed. In the context of rural areas the study highlights the constraints faced in efforts to upgrade craft-based technologies through application of modern S&T. It is emphasised that a specific technology delivery system incorporating both the mechanism and process of technology transfer has to be established for rural areas. Two crucial issues are involved in evolving a technology delivery system. First, in view of the growing liberalisation of economy and increasing demand for environment-friendly technologies, the choice of sustainable technology emerges as an added dimension. Second, there is need to improve the social relations of different sections of rural society with S&T. This involves participation of technology receivers, R&D institutions, local authorities, financial institutions, and an interfacing agent. Based on this approach, results of an experiment on upgradation of traditional technology carried out successfully by NISTADS have been discussed. Details are given of a case study on rural leather workers.

Introduction

The increasing prominence of technology in contemporary society has been accompanied by increasing contention over the status and meaning of technology and its value for human beings. This contention over the role of technology has been exacerbated by the increasing rate of technological change, wherein technology itself has become a controversial issue. There are five major contentions against modern technology. (i) It has developed and got promoted in a utilitarian framework, which implies that knowledge is power and that it is being utilized to exploit the natural resources, wherever they are, for profit[1]. (ii) The rate and direction of technological development have been largely responding to market demands, resulting in environmental degradation (iii) The overdominance of technology in public affairs is leading to dehumanization of society (iv) The technology-practice evolved is heavily biased in favour of the social and economic conditions in the developed countries. Technology-practice is the ensemble of operations, activities, situations or phenomena which involve technology to a significant extent. (v) Technology has increased social inequality. Particularly in traditional societies, introduction of modern technology has completely destroyed their local knowledge systems and socio-economic structures. The critics of modern technology espouse a negative assessment of technology *per se*, which gave rise to the need for appropriate technology (AP) and ultimately led to emergence of the movement for appropriate technology.

Appropriate Technology

The definition of AP adopted in literature and taken to represent it, is as follows: Artefacts which have been tailored to function as relatively efficient means and to fit the psychological and biophysical context prevailing in a particular location and period. The thrusts of the appropriate technology movement in developing countries are different from those in developed countries. In developing countries, it has emerged as a response to problems concerning socio-economic development. Many of these problems are related to technological dependency, technology transfer, technological non-competitiveness, provision for meeting basic needs, employment generation, environmental degradation and social and economic inequality. In developed countries, bulk of the response to appropriate technology movement stemmed from the growth of a technological society. The popular view held was that modern industrial development

[1]UNESCO, *Meeting of Experts on Comparative Studies on Changes in Relations Between Science and Society*, Jamaica, 21-24 June 1983. New Delhi; National Institute of Science, Technology and Development Studies, 1986.

is becoming inimical to human health in both psychological and biophysical terms; emerging technological society is insidiously eroding people's control over their own affairs; and some decisive response to technology is required for the maintenance of human and environmental health[2]. Both the movements tend to be concerned about the direction of technology-practice and the nature of and control over technology.

Many related terms have come into widespread use to describe appropriate technologies; these include, besides the term appropriate technology itself, alternative technology, community technology, convivial tools, eco-technology, liberatory technology, light-capital technology, modest technology, participatory technology, progressive technology, radical technology, soft technology, utopian technology, vernacular technology, village level technology, and no-cost technology. Each of these terms reflects particular viewpoints and intentions and, therefore, there is considerable diversity among the meanings attached to the various forms of appropriate technology. In all these, a predominantly specific-characteristic approach is applied in the choice of technology. Each country, and within its territory, each region, has specific characteristics which need to be taken into account while making choice of technology. The specific-characteristic approach has specific and tangible contents[3]. A general-principles approach has the advantage of providing for flexibility in the choice of technology. It emphasizes the universal importance of examining the appropriateness of technology under each set of circumstances.

Appropriate technology has not so far become an integral part of the techno-economic structure of societal decision-making. It lacks political support and is yet to be imbedded in the cultural and ethical systems. The criticism of appropriate technology stems largely from its weaknesses which are mentioned in the literature as follows:

---

[2]WILLOUGHBY K. W.: *Technology Choice - A Critique of the Appropriate Technology Movement*. London; Intermediate Technology Publications, 1990.

[3]ECKAUS R. S.: *Appropriate Technology for Developing Countries*, Washington D.C.; National Academy of Sciences, 1977.

(i) By definition, appropriate technology is technically inefficient and by nature incapable of matching the supposedly superior productive capacity of the so-called high or modern technology;

(ii) Appropriate technology is not economically competitive; it is an anti-growth concept; it is based upon inadequate or spurious economic theory; the invisible hand of the market is important, where it requires public planning for promotion of appropriate technology, which is a relatively efficient mechanism;

(iii) Physical constraints make appropriate technology impractical, particularly for countries with poor access to sources of low-cost raw materials;

(iv) Appropriate technology has inherited anti-technology sentiments;

(v) Appropriate technology is really a social concept rather than a technological one;

(vi) Appropriate technology, from an intellectual perspective, is a mixture of incomplete and ambiguous ideas, combined with a collection of ambiguous symbols and poorly defined terms and hence it is not a coherent concept.

Choice of Appropriate Technology

Technology choice is a seed-bed for technological progress. Three conceptual approaches may be used for technology choice: Substitution Factor Approach, Epistemic Approach, and Social Relations of Technological Change Approach.

Substitution Approach

Recently, the possibility of substitution among factors, or more explicitly, the question of technological choice, has assumed an important role in the literature on economic development. Interest in the subject stems mainly from concern with providing employment to the growing labour force in the developing countries.

Search for appropriate technology has taken two directions, one theoretical and the other empirical. At the theoretical level, the focus has been on the criteria for selecting the appropriate technology. At the empirical level, the search has focused on whether choice of alternative techniques is possible. The underlying hypothesis is that greater the substitution possibilities, the greater will be the range of alternative technologies available on the shelf and, therefore, the greater will be the scope for investing in more labour - intensive processes that could generate employment. However, technology choice is more complex than is generally assumed. It is not only an issue involving decision-making units, with specified objectives of selecting a particular technique from the shelf of technologies subject to the state of knowledge and other economic-resource constraints, but also a decision about consumer preferences, quality of products to be produced, potential market, government policies and the entire socio-economic framework.

Most of the analytical models used in research on technology choice have been derived from economic theory of production. Traditionally, production technology is conceived as consisting of alternative methods of transforming inputs into production. Analysis of the nature of this process could, therefore, provide some insights into the technical constraints that limit the range of productive process of an individual firm or organization. Most of the models on choice of technology, therefore, hinge crucially on the assumptions about the variations between input and input-output ratio.

The neo-classical production function analysis with its smooth factor-substitution possibilities and multiple technological choices has been articulated by appropriate technology schools of thought to provide a viable model. The neo-classical analysis of technology choice is based on the mode of pure and perfect competition. The basic analytical tool is the production function, which expresses the relationship between the maximum quantity of output and the inputs required to produce it, and the relationships among the various inputs themselves[4].

Within neo-classical frame, a production function embodies an abstract technology. This technology determines at any time

[4]BAGACHWA M. S. D.: *Choice of Technology in Industry - The Economics of Grain-milling in Tanzania*. Ottawa; IDRC, 1991.

the input-output relationships and the relationships among inputs themselves. In particular, a technology has four attributes : (i) Efficiency attribute, which determines the output that results from a given set of inputs. (ii) Technologically determined economies of scale, which describe the extent to which a proportionate change in inputs generates a proportionate change in output. (iii) Capital intensity, which denotes the specific proportion of the quantity of capital to that of labour. (iv) Elasticity of substitution, which indicates the ease with which capital is substituted for labour.

Given the assumption of substitutability between labour and capital, the theory postulates that a given output may be produced by a variety of technologies, i.e., various combinations of capital and labour. In neo-classical world, there is thus a near-infinite array of technologies for producing a product. Because different technologies can use different combinations of capital and labour, the neo-classical model of choice of technology boils down to choosing technologies of differing capital and labour intensities.

A technology will be chosen if it maximizes the profit in the competitive setting given by the relative prices and suitability between the primary inputs. The neo-classical concept of technology is illustrated in Fig.1. For expositional purposes, only two inputs-Capital(K) and Labour(L)-are assumed. The smooth isoquant $I_o I_o$ reflects different efficient technologies that may be used to produce one unit of output. If pp′ is the relevant factor-price-line, profits are maximized at A, where the relative marginal productivity ratios are proportional to wage-rental ratios. The resulting equilibrium of factor proportion at A defines the optimum technology to be adopted.

The inappropriateness of a given technology is basically explained in terms of departure of factor price from that corresponding to this theoretical optimum. Thus, in Fig.1, if pp′ captures the correct price, only firm A is price efficient and hence uses an appropriate technology. Firm B, although technically efficient, is price inefficient and is said to employ inappropriate technology. However, there is considerable scope for manipulating capital-labour ratio and for inducing correct levels of appropriate technologies through specific public policy actions.

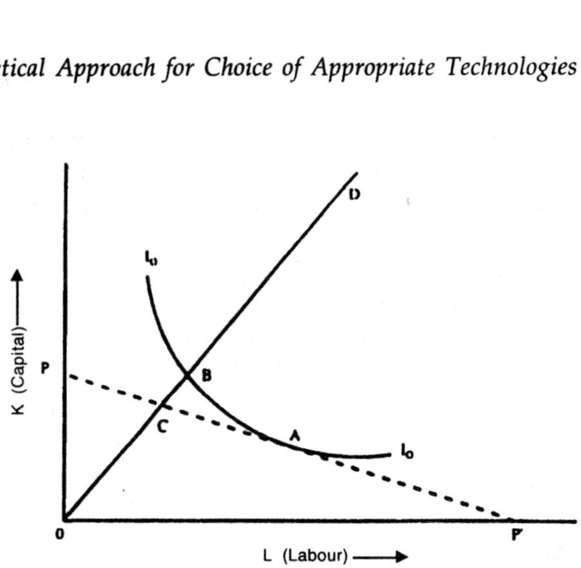

Source: Bagachwa, MS¢, choice of technology in industry - the economics of grain milling in Tanzania, Ottawa, IDRC, 1991

Fig. 1 : Neoclassical Choice of Technique

Epistemic Approach

The criticism of contemporary technological advance on ethical grounds and with a laudable desire to improve mankind's lot may be self-contradictory. But it does not do away with the question whether technology is good or bad. Technology must be justified by its purpose and the ways in which it is used. Defining a purpose suggests a choice and by the same token a set of values, and some decision-makers are able to get their choice accepted. The epistemic approach is to be applied to examine the utility of technology in a comprehensive social framework and is one of the approaches adopted for technology assessment. It can be illustrated by elaborating the concept of the possibility suggested by Henryk Skolimowski[5]. This concept emerges from the distinction between science and technology. The basic problems of truth in science are centred on the notion of

[5]SKOLIMOWSKI H.: Towards a Humanistic Technology. *Research Management*, 1971 (September).

reality as "described" in our theories. The basic problems of truth in technology are centred on the notion of "possible" in technical sense. In technology this notion becomes an epistemic category. If we markedly enlarge the scope of our possibilities, we can create new objects and thus render new correspondences valid.

Analysis of the scope of the "possible" leads to the distinguishing of five kinds of possibilities, the last three of which are specific for understanding the nature of technological knowledge.

P1 = The logical possibility- an object is logically possible if it does not contradict the laws of logic.

P2 = The physical possibility-an object is physically possible if it does not contradict the laws of natural sciences.

P3 = The technical possibility-an object is technically possible if there are technical means for its production.

P4 = The technological possibility-an object is technologically possible if it is technically feasible and can be mass produced, so that it is available to society at large.

P5 = The socio-technological possibility-an object is both technologically possible and socially desirable when it does not result in long-range, unintended, detrimental social and individual consequences.

Technological knowledge is represented by a system of statements which specify the means for achieving those ends; the correspondence is invalid if those are not accomplished. If facts, states of affairs or desired objects do not turn out to be what they were meant to be, and yet correspond to our blueprint, it means that the technological knowledge which serves as the basis for their production was deficient or inadequate.

Therefore, a valid objective of technology must be to satisfy the condition not only of purely technological possibilities but also of socio-technological possibilities, which is, to say, Valid Technological Knowledge (VTK), a function of five possibilities:

$$VTK = P_1 . P_2 . P_3 . P_4 . P_5$$

The necessary condition is that it must be technically feasible or structurally sound. A technological statement, a technological theory or a technological system, must be judged according to all five criteria for technological knowledge. A system of axioms which is concerned only with its internal consistency and not with empirical content, is a formal system and not a description of reality.

The epistemic model suggests rather than makes an explicit framework for evolving a methodology for technological choice. It represents a logical dissection of an object of technology so as to ascertain the intrinsic coherence of each facet of the "possible", which is then integrated in an overall notion of valid technological knowledge.

Social Relations of Technological Change

It has been noted more and more that all the social classes and strata in a society cannot use science and technology in the same way. Particularly in developing countries, there are wide variations in the capabilities of different classes and strata of society to derive benefits from science and technology. The results of scientific researches have not vitally affected the living conditions of the poor. Such maladaptive processes in the implementation of technological change could be explained by a complementary aspect of development, the behaviour of different classes and strata towards the use of science and technology. This phenomenon is defined as relations of classes and strata to technological change[6]. In respect of people in a class or stratum, these relations are measured by their:

(a) ability to have access to science and technology;
(b) objective interest in science and technology as the basis for development;
(c) capacity to generate scientific and technological results;
(d) ability to apply scientific and technological results; and
(e) motivation in the development of science and technology.

It is necessary to investigate these relations to understand the contradictions in development and the relative pace of progress in solving the socio-economic problems of people. This approach

[6]WAHL, D. & QURESHI M.A.: *Relations of Classes and Strata to Science in Developing Countries - The Case of India.* In Aqueil Ahmed and M.A.Qureshi (eds.) Social Relations of Scientific and Technical Change, Seminar Proceedings, 5-7 April 1983, Administrative Staff College at Hyderabad. New Delhi; UNESCO Regional Office, 1984.

is applicable only if the choice of technology, technology development and technology implementation are integrated. Assessment of social relations of classes and strata based on the relevant criteria provides a basis for evolving a strategy for developing appropriate technology for various sections of society. Based on social relations approach a technology delivery mechanism has to be set up. It envisages two simultaneous processes : (i) operationalization of networking of technology receivers, technology generators, financial institutions, governmental functionaries, marketing system and a technology agent, whose responsibility is to nurse the networking; and (ii) the process of continuous interaction between the technology receivers and the change agent, the latter has to articulate the status of technology receiver on the basis of the criteria mentioned above and mobilize the resources for designing the technology to suit the interests of members of the concerned social class. The role of technology change agent is so designed as to compensate for the communication deficiencies of all actors in the network. It ensures continuous upgradation of technology in an incremental manner and creates conditions for gaining mastery over appropriate technology. Live experiments being carried out at National Institute of Science, Technology and Development Studies (NISTADS) New Delhi, aimed at upgradation of traditional technology by blending it with modern technology are presented at the end in the form of a case study.

Integrated Framework

An overview of the literature on technology and efforts made to substitute the existing technology with one which suits the local conditions has raised a few questions which need clarification. Some of the more important questions pertaining to the concept of AP which require in-depth analysis are:

(i) Is the search aimed at a new way of life? Should this include a search for a new philosophy and value system? Or should it be limited to a stop-gap arrangement for the developing countries?

(ii) Should the scope be limited to rural problems using agro-based materials?

(iii) What should be the criteria of choice for such a technology and its assessment?

(iv) What are the possible means to develop, organize and utilize AP?

(v) Will a general-principle approach to AP be acceptable universally? Does it require international negotiations for evolving an international regime of technology?

The starting point for AP, therefore, should not be the modification or expansion of existing technologies to meet the existing or expanded demand. The response to these questions can be articulated through an integrated approach to appropriate technology. It must be viewed as involving three dimensions: a technical-empirical dimension, a socio-political dimension, and an ethical-personal dimension. The domain of each dimension is briefly mentioned below[7]:

(i) "Technical-empirical dimension" refers to the aspect of AP concerned with technologies as artifacts; the tangible components of technical design process; and the empirically derived knowledge which forms an objective part of technology-practice.

(ii) "Socio-political dimension" refers to the aspect of AP concerned with strategic action of classes of people; social institutions; organizations; and structures which form the main fabric of human corporate life.

(iii) "Ethical-personal dimension" refers to the aspect of AP concerned with normative factors; metaphysics; and matters related to the subjective experience of people or the inner experience of persons, particularly as it relates to their capacity for volition and to act as autonomous centres of power.

A framework which incorporates all three independent dimensions of technology-practice in appropriate technology is illustrated in Fig. 2.

---

[7]FRANSMAN M.: *Technology and Economic Development*. Sussex; Wheatsheaf Books limited, 1986.

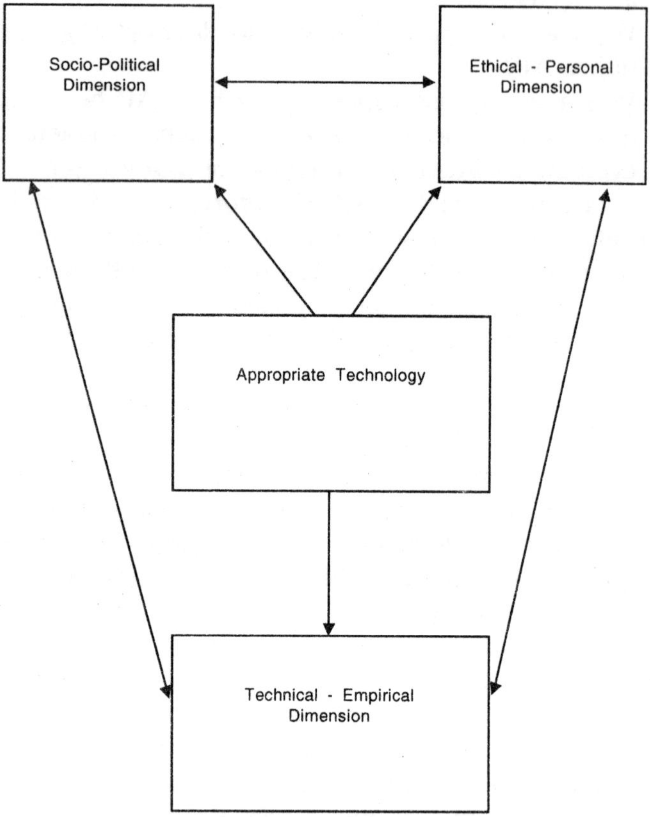

Source: Adapted from M. Fransman, *technology and economic development,* Sussex, Wheatsheaf Books Limited, 1986

Fig. 2 : Three Dimensions of Technology - Practice for Evolving Appropriate Technology

Whatever the choice of technology and whatever circumstances prevail in any region, these dimensions are involved to varying degrees. The necessary and distinctive feature of appropriate technology is that these three dimensions should be integrated harmoniously. A cogent theory of AP becomes feasible if AP is understood as consisting of these mutually interdependent dimensions of technology-practice.

Bankura Experiment

Technology can be applied in a given situation in two ways: either transferring it in its original form by bringing about changes in social and economic conditions which are compatible with the requirements imposed by it or modifying it to the extent that it could be easily adapted to match the local conditions. The present-day technological development has a strong relationship with urban economy, which can be explained through the objectives and functions of new economic structures. The main challenge facing the society is to create new economic opportunities in rural areas through the application of science and technology. There is, thus, a need to evolve a third way involving a practical conceptual approach aimed at applying science and technology for the economic upliftment of people in rural areas. A participatory approach provides ideal conditions for delivery of technology and its proper management in rural areas. The central theme of this approach is to set up an interactive mechanism involving the technology receiver, the technology producer and several other stakeholders in technology management. At the practical level, a technology delivery system compatible with the local conditions is to be established locally in a region for upgradation of traditional technology by blending it with modern technology.

The National Institute of Science, Technology and Development Studies (NISTADS) set up an S&T Field Station at district Bankura (West Bengal) in 1982. The station has undertaken a project on "Village Artisans and Science", an experiment aimed at evolving a technology delivery system for rural artisans in the district[8].

[8]SEN Uma, MUKHOPADHYAY A. & JAIN A.: *An Experiment in Developing Technology Delivery System - Bankura Project*. New Delhi : National Institute of Science, Technology and Development Studies (NISTADS), 1989.

The traditional crafts covered in this experiment include : fishing hook, brass and bell metal, pottery, leather work, rural wool and handloom. The Bankura experiment constitutes an approach pursued for choice of technology and the manner in which artisanal technology has been successfully upgraded and commercialized. To demonstrate the utility of this experiment, a case study on leather work has been conducted with the following objectives: (i) application of modern technology for economic use of bones and flesh of dead animal, (ii) modernizing the tanning process for value addition, and (iii) transferring the upgraded technology to the leather workers.

The underlying experiment comprises the following four components :

(i) Assessment of techno-economic status and identification of needs and problems of people in the area where upgradation and commercialization of technology are to be attempted.

(ii) Understanding social and economic relations of the sections of people to be considered for technical skill upgradation with science and technology as input for the technology delivery process.

(iii) Establishing a mechanism of interaction among stakeholders for implementation of the upgraded technology.

(iv) Setting up an appropriate organizational and management structure for sustaining innovation, production and commercialization.

The first two components seek to understand the social, economic and technical constraints faced by people residing in the area. This required undertaking of a series of analytical and action research studies. The remaining two components are concerned with mobilization of resources, capabilities and socio-economic and political structures for upgradation and commercialization of technology. The prime mover of this experiment is the "change agent" who should be able to interact with artisans, scientists, financing agencies and local administrators for creating the requisite conditions for prompting the artisans to adopt the new technology.

A techno-economic survey carried out by NISTADS in Bankura district indicated that the economic condition of the artisan households has been deteriorating with the passage of time. People in almost every household expressed the desire to leave their traditional occupation for some better alternative profession. The artisans have lost their traditional outlet in agriculture and have engaged themselves largely in agricultural activities. The problems faced by them in sustaining their activities are related to scarcity of raw materials, lack of demand for their products, rising fuel prices, non-availability of finance and their lack of access to training and educational institutions for skill upgradation[9].

Social and Economic Relations

The relations of artisans' class with science and technology depend on various socio-economic factors. Firstly, the artisans belong to the socially and economically handicapped section of the society. They are not in a position to acquire S&T capabilities, generate technology or contribute financially to the generation of technology. The S&T potential is largely under the governmental authority. The elite sections of the society have a decisive influence at central and state levels. Above all, they direct the development of scientific and technological capabilities. Since the traditional technology is generally considered as backward and inefficient, these elite by and large keep the artisans' production activities outside the framework of modern science and technology[10].

Secondly, the artisans in general lack the capability to apply R&D results for upgrading their traditional techniques and raising the scale of their production. Nor can they influence government policies in favour of creating institutions and instruments for the application of science and technology for the promotion of

[9]NATH Pradosh: *Techno-economic Survey of the Craft-based Rural Industries of Bankura.* New Delhi; NISTADS, 1985.

[10]QURESHI M. A.: Social Linkages of Artisans with Technology-Upgradation of Village Pottery Craft. *Economic and Political Weekly* (31 March 1990) p. 638-688

traditional crafts. At the most, traditional technology operates at the periphery of the policy and decision-making structures of government-supported science and technology. There is much criticism about the low contribution of indigenous R&D to the GNP. The basis of support for science and technology was provided mainly by the need felt for applying science and technology to the development of agriculture, industry, defence and health. As a result, there is a stiff competition for resources for R&D in these sectors. Usually, what is done for the production sector in rural areas, excluding the modern industry, is considered as an act of benevolence, the national programmes of R&D in rural technologies being considered more as a humanitarian gesture by the elite rather than a firm commitment of the government to apply S&T for the development of rural industries[11]. The technological attitude is intimately related to property and is not an outcome of enlightenment. Many case studies have shown that the degree of innovation in agriculture is closely related to the level of income and the size of land-holding. The artisans are no exception to this rule if their income and assets are taken into account.

Thirdly, the artisans, to a large extent, show a rational behaviour when they are exposed to techniques new to them. Once workability and profitability of new techniques are demonstrated to them, they become interested in adopting these techniques. They are also aware, while adopting innovations, of the difficulties which they may face from government machinery, and in matters concerning financial assistance and marketing of new products. However, due to their current low social and economic status, a majority of the artisans are hesitant to take initiatives for benefiting from the existing facilities[12]. The

---

[11]UNITED NATIONS EDUCATIONAL SCIENTIFIC AND CULTURAL ORGANIZATION - NATIONAL INSTITUTE OF SCIENCE, TECHNOLOGY AND DEVELOPMENT STUDIES: *International Seminar on Science and Technology for Social Development*, 12-14 December 1994, Final Report. New Delhi, Paris; NISTADS, UNESCO, 1995.

[12]QURESHI M. A. & KHAN S.: *NISTADS and Rural Development*. In (Jain, Ashok, Qureshi, M.A., and Khan, Subhan, Eds.), CSIR and Rural Development. New Delhi; Deep Publications, 1995. p.19-34.

level of motivation of the artisans can be raised by providing them support for upgrading their skills, diversifying production and expanding the market for their products.

Interactive Mechanism

The only way to improve the economic status of rural artisans is to draw them closer to science and technology. Their deficiencies in this respect can be made up by initiating a dialogue between artisans and their clients, scientists, technologists, local development authorities and functionaries, financial institutions, Non-Government Organizations (NGO's) and local leaders. An interactive mechanism is established for holding such a dialogue, which involves different levels of interaction involving more than one stakeholder. The interactive mechanism is instrumental in:

(i) Making objective studies of artisans' traditional technologies;

(ii) Identifying the technological gaps in the artisan technology and industry;

(iii) Effecting the needed improvements in artisan technology;

(iv) Arranging demonstration of on-shelf technology and examining whether the technology is suitable for adoption by artisans or requires modification to suit local conditions or an alternative technology is to be evolved by blending it with traditional technology;

(v) Developing infrastructure for adaptive research, design and development work; and

vi) Making adequate arrangements for financial assistance, raw material procurement, skill upgradation and other types of institutional support.

A resourceful institution/NGO/ individual needs to act as a "change agent" to keep the interactive mechanism alive, so that the stakeholders may sustain the dialogue. It helps in making right choice of technology and creating the requisite conditions for availing of the technical and economic opportunities.

Organization and Management

At the S&T Field Station, Bankura, the primary objective is to upgrade the artisan technology by blending it with modern technology and promote its commercialization. The Project Manager stationed at Bankura administers and coordinates the activities in respect of various projects being carried out at different places in the rural areas of Bankura. NISTADS has been discharging the role of the "change agent". The project team is at the bottom level of the organizational matrix. The team collaborates with scientists and technologists of national laboratories and local S&T institutions for identifying technological gaps in the selected artisan industries as the first essential step in the direction of generating appropriate technology at laboratory scale. It works in tandem with the local administrative authorities and district agencies engaged in developmental work. It interacts with artisans in the tasks of arranging fields trials, demonstration, adaptation and skill formation in the upgraded technology. At the field level, investigators, technicians and economic analysts undertake planned activities relating to the project. For each project, a standing committee, chaired by an elected representative of the local political constituency specific to the project area, is constituted for facilitating implementation of the project.

The government of the State of West Bengal is fully involved in implementation of various projects. It has constituted a State Level Advisory Committee to coordinate activities under the programme " Village Artisans and Science". The main functions of the committee are to advise, oversee the progress of projects and provide assistance in the implementation of projects. The committee is made up of representatives of the State Department of Small Scale Industries, Directorate of Small Scale Industries, District Rural Development Agency, District Industrial Centre (DIC), concerned national laboratories, Council for Advancement of People's Action and Rural Technology (CAPART), Director of NISTADS and Project Manager and State Khadi and Village Industries Board (KVIB). The committee, chaired by the local Member of Parliament, plays a pivotal role in the mobilization of resources and in ensuring smooth running of projects. Presently, all the projects underway in Bankura have been sponsored by central agencies. The CSIR Headquarters and the State Government provide partial financial assistance, particularly for training of artisans and technology diffusion.

Creating Ground Conditions

NISTADS, while developing its role as a technology change agent, organized a series of workshops on "Village Artisans and Science" at Sewagram, Wardha and Bardoli for the purpose of gaining proper understanding about the problems of artisans and artisanal technology and the kind of institutional linkages needed for upgradation of artisan technology and commercialization of artisanal products. The workshops facilitated effective interaction among the village artisans, the scientific community of national laboratories/institutes, government and non-government agencies and R&D institutions engaged in development work, in order to make them familiar with the problems of rural artisans-based industries and induce them to generate a process of thinking for tackling identified problems by working out enduring solutions for them. District Bankura was chosen for carrying out the programme on "Village Artisans and Science" with stress on upgradation of artisan technology. The above mentioned workshops served to make the task of motivating the scientists in the programme activities easier. Participation of scientists from various laboratories created the needed awareness and orientation. Initially, it was a difficult task to involve the artisans. They were apprehensive of the purpose of the city folk coming to their doorsteps asking various questions about the problems faced by them. The problems concerning their crafts relate primarily to the techniques which they follow, procurement of raw materials and fuel, marketing of products, and financial and other social and economic issues. Some of the artisans thought that the project team was visiting them for making assessment in respect of some loan or subsidy to be given to them by the government. This section of the artisans cooperated in taking the team members to various household-work sites and provided information about the problems they were facing in purusing their traditional crafts. Those artisans and the officials who had a negative attitude towards the project objectives and the methodology of technology delivery did not play a positive role in respect of the ground work. A few of them were not only sceptical but also tried to avoid extending initial help in the work of the team. However, this phase did not last long and once the technology demonstration programme was taken up, the artisans gradually developed interest and their initial resistance and attitude of non-cooperation diminished to a considerable extent. Despite their illiteracy, the artisans seemed to be quite aware of the environment in which they were struggling for their survival. The younger members of the community were more informed than the elders about the social and economic conditions in which they were pursuing traditional crafts.

At the outset, it was difficult to get authentic information regarding the artisans of the district, places of their concentration, nature and types of crafts they were engaged in, the level of technology and their economic conditions. The DIC was also unable to provide much information, except some basic facts about well known crafts like those practised by the famous terra cotta potters of Panchmurha village, silk weavers and bell and brass metal and Baluchari Saree craft workers of Bishnupur, stone carvers of Chhandapathar, and Dokra craft workers of Bikna village. However, statistics about the number of artisan clusters and their spread in the district provided to the team by the district agencies were later found to be adhoc and unrealistic. In subsequent meetings with the district development agencies, necessary clarifications were secured regarding the nature and extent of assistance required from the district and state level agencies. Also, the areas of cooperation were identified. The project team also got acquainted with the problems faced by these agencies. These problems were concerned with artisan cooperatives, which had become non-functional due to debt burdens, failure to distribute loans and lack of market demand for artisanal products. The need for bringing about product diversification and product improvement through technology intervention in the artisan production system was appreciated. It provided a new dimension to the programme, necessitating the involvement of various state agencies in the advisory committee. The State KVIB evinced interest in associating itself with the activities relating to village pottery, ceramic and leather industries.

Case Study of Leather Work

Bankura is the third largest district of West Bengal. It has a long tradition in leather trade due to its proximity to Calcutta. The estimated livestock population in Bankura is around 200 thousand and on an average two thousand hides are available in different weekly markets, which form the conduit for the transmittal of hides to the organized tanneries in Calcutta. Normally, when a cattle of a farmer dies in a village, the persons in the village belonging to scheduled caste community take the dead animal to the nearby dump yard/field where it is flayed by them for the hide; the body is left behind for vultures and other meat-eating animals and birds. The farmer pays Rs 20-30 to the flayers for disposing off the dead animal. In some villages, the animals are buried under the ground due to non-availability of flayers. The flayers then pass on the hide to local leather worker at a nominal price ranging from Rs 50 to 60, depending upon the quality and size of the hide. The leather worker gives salt treatment to the hide for preserving it for a day or two. It is then sold by him to the local collector, who is in reality a petty contractor, at a price ranging between Rs 70 and 80.

The role of a flayer is limited primarily to flaying of the dead animals for hides and passing on the hides to petty contractors. Flayers are not always the tanners. Sometimes the petty contractors carry out local tanning to improve the quality of raw hides, thereby increasing their life. The traders operate through petty contractors or local agents covering 20-25 villages. Once a truckload of hides is collected through the network of agents/collectors, the hides are taken to the organized tanneries in Calcutta.

The bones of the dead animal are collected from the field mostly by women folk. The bones thus collected are carried by the women to the godowns of local agents. Such a godown is generally situated in every village where the livestock population is high. The bones are sold to the local agents who are operating in an area covering 10-15 villages. When slaughtering is done in a slaughterhouse, the bones become the property of the slaughterer who sells them to the local agent directly from the slaughterhouse. The intestine and other parts of the dead animal are hardly processed for further utilization. In a few villages, the flayers also take out intestine of the dead animal and process it to produce special types of ropes used for tying bullock cart components. Only tallow is collected for local consumption. On an average about 3 kg of tallow per cattle is obtained.

In another situation, usually the same leather artisan carries out all the three functions of flaying, tanning and crafting the leather into footwear. In relation to the emerging market economy, this situation has undergone rapid change. Now only flaying is done at the village level, the hides are taken to towns for tanning and the finished leather comes back from towns to the village artisans for producing leather products. The practice of tanning in the villages has almost totally disappeared. Considerable number of leather workers have moved out of traditional professions completely. Those who have remained in the profession are either confined to flaying operation or are engaged only in footwear making. Tanning has become a large scale centralized industry away from the rural setting. Two important factors have contributed to such a situation. Firstly, at the social level, leather work is considered a dirty occupation. Unless economic returns are high enough to offset the social stigma associated with leather work, the traditional social worker would prefer to opt out of this low income-generating occupation. Secondly, most of the modern technology inputs specially required in the leather finishing process are capital intensive and beyond the reach of poor leather workers.

There are several possibilities for introducing technological innovations in the form of upgradation of leather workers' skills and capabilities. After the body of the dead/slaughtered animal has been flayed, the following innovations could be implemented stepwise, depending upon the local conditions.

(i) After recovery of the carcass, the flesh to be converted into poultry feed and the bones ground for use as fertilizer.
(ii) The hide to be processed through vegetable tanning and the leather made available to cobblers for footwear manufacture.

(iii) The hide to be subjected to wet-blue tanning, so that it can be preserved for a few months for bargaining for a better price with the urban traders.

(iv) Hair-on-tanning to be attempted for value addition to leather production for sophisticated market segment. In the present project, good quality hair-on-leather could not be produced, as the length of hair of the sheep was too small to be acceptable in the market for fashionable products. This innovation was abandoned, as the product could not withstand competition with the products available in the market.

There are other possibilities. These include chrome-tanning and diversification of leather products. The former can lead to production of quality leather for sale in both rural and urban markets. The result will be additional employment and income generating opportunities. While upgrading the technology, it is important to keep in mind that technical innovation is implemented on incremental basis, so that it is easily assimilated by leather artisans. They will not be able to cope with several innovations at a time. They will also face constraints simultaneously in relation to their technical and economic capabilities. They may not be able to digest all levels of technical inputs at one go. Such efforts are doomed to fail. Thus, upgradation of skill should be gradual. Further, introduction of innovation must not drive the artisans out of their traditional occupations and related activities. They have to be trained to retain their employment and to be able to raise their income level. Once the contacts are established and the interactive mechanism is in place, the upgradation effort must follow a certain sequential order depending upon local conditions. The upgradation and its benefits are briefly mentioned below:

(i) The bones collected by the leather workers are ground manually and are generally sold to the farmers at the rate of Rs. 0.50 per kg. Now a bone crushing centre has been set up in a village of Bankura. The centre has a small electrically operated bone crushing machine costing Rs. 3.25 lakh. The funds for setting up the centre were provided by the local government. The centre is managed by Zila Parishad. At the bone crushing centre, training programmes were conducted on operation of the machine. The bone collecters are offered two options : either they can sell the bones to the centre at the rate of Rs 1.5 per kg or they can get the bones ground by paying in kind. The ground bones are then sold to the farmers @ Rs 3 per kg. The amount spent on setting up the centre and the operational expenses are being recovered by setting aside one-eighth portion of the ground bones as grinding charges. The earnings from the sale of the ground bones would gradually offset the capital cost in a few years. Simultaneously, it takes care of the operational expenses of the centre. The mechanized bone grinding facility has permitted considerable increase in the income of the bone collecters as well as value of the ground bones used as a fertilizer.

(ii) Earlier, leather workers were salting the hides to preserve them for a day or two. Most of the time the leather worker had to sell the hide at a price offered to him by the local agent fearing that if it was not disposed of quickly, it would get spoiled and he would suffer a loss. Now a half-tanning method, called wet-blue tanning, has been introduced, which enables the leather worker to keep the half tanned hide for about two months. The leather workers have been trained in wet-blue tanning process. They are very happy with this innovation. It has increased the value of the hide as well as bargaining power of the leather worker while negotiating the sale price.

All these innovations have been accepted by the leather workers and implemented successfully, resulting in enhancement of their income level. These innovations have evolved through interactions and interactive processes. Training of leather workers was carried out vigorously. Follow up action is being continued for augmenting their capabilities by providing assistance in technical, financial and marketing aspects. The focus of the study has been on upgradation of skills of leather artisans for value-added production thereby reducing their exploitation by middlemen and increasing their income level. There is considerable scope for further upgradation of technology through product quality improvement and diversification. This requires sustained efforts for innovation. The key issue to be kept in view is the need to make appropriate choice of technology. Thus, choice of innovation is crucial in the context of the conditions prevailing in the rural areas. Ultimately, innovation must contribute to happiness of the people who are targeted to benefit from it.

13. Introduction and Utilization of Industrial Technologies in Rural Areas

Wang Wenyong

Abstract

This study aims to analyze and document available data and information on the experiences of China with the introduction and diffusion of industrial technologies in rural areas, with special reference to economic growth, employment, income generation and alleviation of poverty in rural areas. Introduction and Utilization of Industrial Technologies in Rural Areas (IUITRA) is one of the important policy measures in China - a country largely based on agricultural economy. The most important effort is the establishment of Township and Village Enterprises(TVEs) in the last decade.

Introduction

At present, China has more than 20 million Township and Village Enterprises (TVEs). In 1995, their contribution to GNP was 25 per cent. However, during 1986-95, the output value of the TVEs grew even much faster, i.e. with an annual increase of 15 per cent on an average. For a long time in China, scientific research and production were divorced from each other and this has cost China dearly. Since 1985, China launched reforms on science and technology management systems, the central idea of which was to make scientific research directly serve the national development objectives, creating motives as well as pressures for suppliers and recipients of technology to come closer to each other. Research institutes are forced to go to the market for research projects or research sponsors and partners. On the other hand, business enterprises are also putting in all efforts looking for new scientific achievements by the R&D Institutions.

With a large rural population, China's rural area is a huge potential market for industrial products. However, if rural economy is not developed, it would still be a mere potential market. The fast growth of rural enterprises has helped in gradually materializing this potential market. Peasants' income increases with the development of rural enterprises. The increased purchasing power, in return, creates a large market for industrial products. Since the reforms and opening up in late 1970's, the economic progress has met the fundamental needs of a majority of peasants. The living standard has improved considerably. Statistics for 1991 showed that food costs 56.8 per cent of a peasant's total income, while the demand for better food, clothing, and housing is increasing at a fast rate. It showed that demand for electric fans, motor cars, and cameras was 20 per cent higher than that in 1990. It is still more important that improvement in living standard eventually leads to demand for products of higher quality. Demand for larger quantity and higher quality thus makes better introduction and utilization of industrial technologies in rural areas necessary and obligatory.

Social Needs

1. Poverty Alleviation

Out of its total population of 1.2 billion, 70 million Chinese are still living under the poverty line, with an annual income below 200 RMB yuan (30 US$) per person. The most important factors that cause poverty are: (a) lack of access to the outside world; (b) poor infrastructure and natural resources; (c) backward education and technological level; (d) decision-making mistakes; and (e) the usually adverse climate and land conditions. The fact that 70 million Chinese live below the poverty line has prompted the Central Government to pledge to solve all food and clothing problems in rural areas by the year 2000. Despite the rapid decline of poor people—down from 250 million people in 1978 – China still accounts for 10 per cent of all people living in poverty across the globe. From 1980 to 1994, the Central Government allocated 16.14 billion yuan to 1410 counties and 9876 TVEs. 80 per cent of this fund was allocated to the 592 targeted priority counties in Central and West China. The efforts are mainly aimed at technological capacity building and related activities. With experiences, both positive and negatives, accumulated, it has become one of the priority considerations in the Chinese policy to introduce S&T in rural, particularly, poverty areas. The government has organized and implemented various kinds of programs for poverty alleviation, especially by helping the poor areas to develop rural enterprises and improving their management and techniques. As rightly stated in a State Council Circular recently, "Advancement of science and technology is an important booster to economic growth. It is more so to develop economy in the poverty–stricken areas. For the development of such areas, it needs favourable treatment in the nation's development endeavour, information and technological support. Of the three, technological support is the most important at present. The Spark Program that is under implementation should target the poor areas as its priority. The governments at different levels should vigorously promote the dissemination of adaptable technologies, and mobilize technicians and skilled labour to go to the poor areas and transfer technologies to the local peasants.

2. *Increase in Income and Employment*

For about 30 years, the average income of a Chinese peasant was below 200 RMB yuan. During this period, not much attention was given to the development of industry in rural areas, which was confined mainly to cities. Thus, urban development took place at the cost of rural progress. However, this proved to be misleading. Due to a comprehensive set of factors, China's urban industries faced adverse conditions. Unemployment increased in cities. On the other hand, with technological progress, though slow in agriculture, more and more excessive labour emerged on the farmland too. Under these circumstances, an attempt was made to develop rural enterprises. This ultimately provided the way out and helped in increasing income and employment in rural areas.

The total income generated by the rural industries increased from 66.91 billion RMB yuan in 1980, when China just started reform and opening to the outside world, to 1161.35 billion yuan in 1990 and further to 1765.95 billion yuan in 1992. At the same time, there was very fast increase in the per capita annual income of Chinese peasants. The figure was 191.3 yuan in 1980, increased to 686.3 yuan in 1990 and further to 784 yuan in 1992. This is largely attributed to the fast increase in the number of rural industries and average payment to their employees. The number of township industries was 1.4246 million in 1980, 18.504 million in 1990, and 20.916 million in 1992 while the per capita economic returns and average payment to employees in the rural industries were:

(RMB yuan)

	1980	1990	1992
Per capita economic returns:	2230	10342	16622
Average payment to employees:	480	1092	1668

3. Population Control

It was more or less a tradition in China that a family should have as many children, especially sons, as possible. As the old Chinese saying goes, "To raise children is to prepare (safety assurance) for the old-age days," and "more offsprings means more happiness" for a family. However, with the advancement of science and technology, manpower input, especially unskilled manpower input, has greatly lost its significance. Against this background, peasants may use machinery, better seeds and more sophisticated technologies to reduce manpower input to improve yield. Now many Chinese peasants, instead of producing more children, are trying to seek employment at a nearby township enterprise. Often, under the township enterprise management system, employees also have, what we call in China "an old-age safety guarantee."

4. Prevention of Rural or Urban Migration

China's rural labour force is estimated at around 400 million. With the introduction of advanced technological inputs and management system, coupled with liberation of production forces in an ever-deepening economic reform in China, the countryside now has more surplus labour. One estimate by an agricultural official puts such labour force at 100-170 million. The tolerance and accommodating capacity of the Chinese cities is, of course, very limited. Immigrated workers seem to have created a problem already beyond the cities to handle. The development of rural industries accelerates construction and expansion of small towns, which could be the strategic choice of China to accommodate surplus rural labour in the predictable future.

Development and Utilization of Local Resources

Poverty and backwardness are often associated with shortage of resources. With the third largest land area in the world, China is rich in various resources. However, there are numerous examples where resources are available, but there is lack of ability to utilize and develop them and as a result people are still living in poverty conditions.

From the national perspective, China's economic development was and is still restricted by financial shortage. Under the centrally planned economic system, the government was not able to afford enough investments to exploit many local resources. In recent years, Chinese government has paid increasing attention to find a way out to develop and utilize local resources and giving incentives to local economies. In this respect, rural enterprise development, which does not demand much investment, has been given special attention. The contribution of rural enterprises in development and utilization of local resources is indicated in the following table:

Industry	*Enterprises No.*	*Production (Million ton)*	*Output value RMB Yuan*
Coal mining	23,903	513.43	21.67 b
Black metal mineral	6034	–	7.81 b
Nonferrous metal mineral	5,169	–	8.1 b
Iron	4,245	118.92	5.98 b
Gas and oil	12	–	13.43 m
Chemicals	661	–	553.83 m

Note : *b–billion; m–million*

To have better knowledge in this regard, a specific case analysis on utilization and development of local resources is provided here.

CASE : Fujian Experience

Fujian province is a mountaineous area in south east China with a variety of fruits growing and marine creatures living in its coastal part. It has abundant bamboo resources. Due to the complicated and mountainous topography, the communication system was poor and development of the province's economy was restricted. People were poor and were not allowed to exploit local resources by themselves. When they were permitted to do this, they lacked necessary technologies to develop them. In recent years, after thorough analysis of the advantages of exploiting local resources, the provincial government has launched a comprehensive plan to encourage rural enterprises to exploit the local resources, especially edible mushrooms raising, bamboo planting, marine products breeding, production of various fruits and vegetables, and processing of the items mentioned above. A typical example is Tantou town of Fu'an city, an appropriate place to grow lotus plum, which is a very delicious indigenous fruit. The local government succeeded in applying Spark Program and supported the project "Lotus Plum planting and processing". Since implementation of the project in 1993, over 3,300 hectares land was put under lotus plum producing 6, 500 tons fruits. The local people also established a processing plant to make lotus plum into candied fruits. 80 per cent of the candied fruits were exported. Planting and processing as well as sale of lotus plum have become the backbone industry in this town. Its output value covers 31 per cent of the town's total output value. With the development of this project, 70 per cent of the local peasants have got rid of the poverty situation and some have even become rich.

Under the rigid centrally planned economy, the decision on location of enterprises, was taken by the government instead of being profit driven. What is more, for a long time, China practised an equal wage system in urban and rural areas. That means, if enterprises locate their industries anywhere, they will have to pay the same wages to their employees.

With the development of China's reform process, urban industries have been given increasing freedom in management. Driven by search for cheaper production cost, big enterprises in urban areas began to establish some collectively-owned industries, often in their rural neighbourhood, where labour is much cheaper than in urban areas, to produce components of their products.

A more common case is the location of industries from overseas in coastal areas. By 1992, China had 12,060 TVEs that had some kind of international linkage, such as joint venture and joint operation. The South China Province of Guangdong, close to Hongkong and Macau, alone had 9,317 such ventures or 77.25per cent of the total. In 1992, the yearly income of the TVEs employees averaged 2,597 yuan, among the highest in mainland China. However, it is still believed to be one of the lowest in the world and certainly much lower than the average payment in Hongkong.

Status of Rural Industrialization in China

After the beginning of economic reforms in 1979, rural enterprises have received considerable attention. From the national perspective, rural enterprises are mainly concentrated in coastal areas, especially in the eastern and southern China, which are the most dynamic regions. The fast growth of rural enterprises has become an engine of growth for the entire economic development of the two regions. However, 85 per cent of the rural enterprises are scattered in villages far away from the cities, markets, raw material sources, and other infrastructural facilities like electricity, water, transportation and communication. This leads to low benefits and extensive wastes. As these are scattered in large areas, it makes it difficult for the government to monitor and deal with pollution.

The major ownership of China's rural enterprises is collective. Under the collective ownership, the property of a rural enterprise is owned by all the peasants of the village or the town which establishes it. Owners of rural enterprises shall have the authority to decide the management through the Peasant Conference.

China started to allow family cooperative and private

ownership as early as 1978, when the Third Plenary Meeting of the Central Committee of the Chinese Communist Party's Eleventh Congress passed a resolution on economic construction. The Central Government declared that private ownership was legitimate and began to establish a new environment to encourage the development of the two ownerships. From 1985 to 1993, the number of private enterprises doubled, the output value increased by 18.8 times, the fixed asset increased 13 times, tax increased 38 times. Among the total output value of private enterprises, industry contributed 53.6 per cent, construction 10.2 per cent, transportation 18.3 per cent, commerce and restaurants 17.9per cent.

Generally speaking, private enterprises are tiny and few, with an average of 2.4 persons for each enterprise and have backward equipment and untrained manpower. Further, workers' interests are not under valid legal protection.

Of late, the following two types of ownership are emerging rapidly and are arousing stronger interest from both government and enterprises nation - wide.

a. Joint Operational Enterprises

These enterprises, in compliance with collective enterprise management and operational system, have joint investments, and are jointly managed by rural enterprises, enterprises of other ownership, and institutions as well. All the partners share the benefit, loss as well as risk. On 3 January 1992, the Ministry of Agriculture issued the regulation on rural joint operational enterprise in order to increase cooperation among rural enterprises, to optimize economic and technological structure, and protect the legitimate right of those joint operational enterprises. There were 4,010,000 joint operational rural enterprises in 1993.

b. Share Holding Cooperative

It is an economic entity which has two or more labourers or investors, who, in compliance with laws in China, have shares in capital, goods, technology and the right to use the land. They cooperate with each other to engage in various production activities in accordance with agreements or regulations. The share holding cooperatives have a democratic management system and a distribution system according to both work contribution and share holding and draw public accumulated funds.

	Number	*Employment*	*Output value*
All Rural Enterprises	23,210,000	112,740,000	2,902.2 b
Share Holding Cooperatives	130,000	488,000	212.2 b
Percentage	0.56%	0.43%	7.31%

Covering only 0.56per cent of the total rural enterprises with only 0.43 per cent of the total rural employment, the share holding cooperatives contribute about 7.31 per cent of the total output value. The output value per capita of share holding cooperatives is almost 17 times as much as that of the average rural enterprises. Encouraged by the obvious economic benefits, on 24 December 1992, the Ministry of Agriculture decided to popularize and improve this kind of ownership. The Ministry also issued a policy aimed at enhancing the restructuring of rural enterprises which have given priority to expanding share holding and share holding cooperative ownership.

Product Variety and Quality

Rural enterprises cover almost all traditional industries, which could be roughly classified into the following categories:

1. Light Industry

Shoe making, leather and fur manufacturing, paper making, production of plastic products, timber processing, bamboo, cane, palm and grass made products, furniture manufacturing, toys making, industrial art manufacturing (Sculpture, crafts made from materials like bamboo, cane, palm, grass, embroidery, blanket, earthen ware and china clay, fireworks).

In 1992, there were 545, 000 rural enterprises in light industry which had an employment of 18,286,000 workers and an output value of 550.41 billion yuan.

2. Construction Industry

The growth of this industry has played an excellent role in

absorbing the surplus rural population and alleviating the pressures thereof. In 1992, the construction industry employed 15,407,000 persons.

3. *Food Industry*

Corn processing, vegetable oil processing, milk food processing, aquatic product processing, canned food processing, sugar manufacturing, meat processing, drink manufacturing, white alcohol, yellow alcohol, beer, aerated water, tea and tobacco manufacturing.

Food industry is a weak field in rural enterprises, accounting for 7.4 per cent of the total output value of rural enterprises. The growth rate is 16 point lower than the overall growth of rural enterprises. This has, in turn, affected agricultural production. On the basis of a comprehensive survey, the State Council has taken a decision to promote food industry in rural area.

4. *Machinery and Electronic Industry*

In 1992, there existed 225,000 machinery and electronic rural enterprises all over China, 7.4 per cent more than in the previous year, with an employment of 8, 994,000. The output value of these rural enterprises already covers 35 per cent of the total national machinery and electronic industry. The main products are electrical fans covering 63 per cent of the total market, agricultural machinery covering 95 per cent of the nation's market, components of cars, environmental protection equipment, electric wire, cables, and common devices and instruments. A lot of problems exist in this field.

5. *Building Material Industry*

This industry had 171,000 enterprises with an employment of 7, 836, 000 and an output value of 118,640 million yuan in 1992. The major products are:

Products	*Annual production*	*per cent of total national production*
Cement	100 m ton	33
Lime	210 m ton	85
Brick	495.48 b	90
China & Earthenware	11.74 m	75

Other products include cement prefabricated components, marble, plaster.

6. *Mining Industry*

This industry includes coal mining, oil recovery, natural gas extraction, ferrous metal mining, non-ferrous metal mining, apatite, aluminium, raw salt, marble, clay mineral.

7. *Cosmetics Industry*

Cosmetic industry is mainly established in Jiangsu, Guangdong, Zhejiang, Shanghai, Shandong. In 1992, there were a total of 595 enterprises with a total employment of 31,000 and output value of 2.67 billion yuan, which was 44.5 per cent of the total national production.

8. *Service Industry*

Transportation, commerce, storage, supplying and sale service, tourism, estate, culture centre, information consultancy, entertainment centre, monetary service, water resource service. In 1992, about 28,860,000 people, 23.1 per cent of the total rural enterprise population, were engaged in service industry.

As seen from the above table, a majority of the rural industries fall within traditional categories. On the other hand, electronics, information, and biotechnology and those based on advanced technology have not emerged in China's rural enterprises. Among the traditional industries, building material, mining and manufacturing cover a large portion in China's rural industries in terms both employment and output value.

Technological Level (including incorporating new technology)

A major part of the staff in the TVEs are peasants who are less educated. Untrained staff, shortage of funds, tiny scale are the main factors keeping TVEs at a low technological level. In a general sense, the technological level of China's rural enterprises is still very low. The level could be judged from the number of trained staff consumption of resources, and scale of fixed assets.

The Educational Background of Rural Enterprise Staff:

Total Employment	57,843,125
College and Above	532,978 (0.92per cent)
High and Vocational School	11,527, 016 (20per cent)
Junior Middle School	30, 534,582 (52.8per cent)
Certificate Holding	
Engineers	607, 116 (1.05per cent)
Technicians	1, 493, 759 (2.6per cent)

The total output value of rural enterprises was 3155 billion yuan in 1993. The average energy consumption for a thousand yuan output value was about 100 kg crude coal, 37.7 kg gasoline, and 380 KWh of electricity.

In 1993, the total investment on fixed assets was 217,627 million yuan, of which 102,733 million yuan was used for purchase of equipment. 272,765 projects were under implementation. The average investment per project was about 797,855 yuan.

From the perspective of human resources, with a small percentage of technicians and well - educated staff, the technological level of China's rural enterprises is relatively low. Energy consumption for every 1,000 yuan output value is much higher than that in developed countries. On the other hand, the input into fixed asset (especially equipment) increased a lot, which implies tremendous growth in capital equipment of rural enterprises.

Incorporating New Technology

Rural enterprises are required by the competent authority to engage in development of new technology and technology innovation. The key point is to encourage rural enterprises to attach priority to application of adaptable advanced technologies. Projects which are identified as incorporating new technology are granted favourable policies so as to promote rural enterprises to attach importance to technological innovation. The criteria for adaptation of technology are:

(1) It can improve the production structure as well as product quality
(2) It can improve productivity;
(3) It can make full use of resources and raw materials
(4) It can reduce or avoid pollution and damage to ecological environment.

Skill Development

As the working staff of rural enterprise comes from peasants, skill development has been one of the most urgent tasks for these enterprises. Concerned authorities have issued many regulations on staff training for enterprises.

(1) Until 1992, there were 500 formal programs set up by the universities to train students from rural enterprises. Among them, quite a number of rural enterprise colleges were established under the universities. 49 colleges and 235 vocational schools specialized in agriculture related fields were established. In 1992, 26,679 students graduated from agricultural colleges, and 40,076 from vocational schools. At the same time, a lot of television schools and correspondence schools were also established aiming at training personnel for rural enterprises.
(2) A national authority has been formed to be in charge of Rural Enterprise-oriented Education Affairs. Under the direction of these authorities, a national network of training bases for rural enterprises have been established. Statistics for 1992 showed that above 4,000 training bases at all levels (state, provincial, prefectural, and county level) were available to rural enterprises. Every year, thousands of peasants were trained in various programs. In 1993, 194,534 personnel were trained in courses at college level (including evening, correspondent, and television school), and 734,920 people were trained at the vocational schools.

Through these efforts, the quality of TVEs staff has greatly improved during these years. Until the end of 1992, the number of personnel holding a college diploma and technical personnel increased at the annual rate of 58.7 per cent and 69.5 per cent respectively compared to 1983.

Stimulation of Agricultural Production and Mechanization

Scores of years ago, Chinese government tried to launch mechanization in rural areas. It has established a very good but still premature ground for agricultural mechanization. In many areas, the efforts did not succeed because of large agricultural population with small pieces of farmland. During 1980s, the arable farmland per peasant was about 0.12 hectare. Since the implementation of 'Family Contract Responsibility System' in rural economy, every family became an independent production unit. They were granted a small piece of land, which was generally scattered in different places for the sake of land quality and divided equally among all members of a village. Mechanization makes no sense in this situation which is characterized by small farming. However, fast rural industrialization has created much employment and transferred a lot of excessive population out of agriculture, which makes agricultural mechanization possible. For example, in southern Jiangsu Province, population distribution has been as follows:

Year	*Agriculture*	*Manufacturing Industry*	*Service Industry*
1952	80.9%	8.5%	10.6%
1978	60.0%	28.9%	11.1%
1990	28.9%	52.2%	18.9%

Since 1978, as peasants were allowed to purchase agricultural machinery, there was fast development of rural enterprises specializing in agricultural machinery to meet peasants' demand. The output value of agricultural machinery industry in 1992 was 43.86 billion RMB yuan, which was 24 per cent more than that in 1991. In 1992, mechanization was done mainly through introduction of 755,054 (large) + 7,504,000 (small) tractors, 9,123,600 irrigation machines, and 3,371,100 harvesting machines.

During the primary development stage of rural enterprises, due to lack of proper techniques, serious pollution was caused. Statistics show that:

1. About 7.7 per cent of the total number of rural enterprises caused pollution, which had an output value of 35.3 per cent of the total.
2. The waste water piped out by rural enterprises covered 6.7 per cent of the nation's total waste water. Paper manufacturing industry contributed 44 per cent of the total waste water, chemical industry 8.7 per cent, and printing and dying industry, 6.7 per cent.
3. Tile and brick manufacturing, coking, and casting are the three major energy consumers and polluters. In 1989, the extent of the amount, output value, energy consumed and waste gas for these three industries as part of the entire industry was as follows:

Enterprises 1.9%
Output value 6.8%
Energy consumed 45.3%
Waste gas piped out 45.6%

The situation is serious in chemical industries too.

The government of the People's Republic of China has formulated a series of regulations and laws for environment protection. In order to save energy and protect the environment, the Ministry of Agriculture, with World Bank aid, has launched a joint program on, "efficient use of energy and pollution control for China's rural enterprises", in an attempt to find a comprehensive solution for the problems in question. In 1990, the Ministry of Agriculture, the State Environment Protection Bureau and the State Statistics Bureau have established an environment protection database, so as to have better macro-control system. This situation is getting better, but still much remains to be done to solve the pollution problem.

Economic Policy Reforms

Very recently, the Agricultural Development Bank of China burst into life with the breakup of the Agricultural Bank of China into separate policy-and commercial-lending banks. A major function of the Bank is to grant low-interest, long-term loans to most-needed agricultural projects, many of which involve farm-produce retail, storage, wholesale, trade processing, irrigation and technology development.

Favourable taxation policies were also practised towards township enterprises, which include:

(1) Low state tax levies or tax exemption for township enterprises;
(2) Low tax on agricultural products and the individual income of the farmers.
(3) Low tax on certain products of rural industries, namely, glassware, canned food, porcelain and farming tools.

Public and Private Financing of Rural Industrialization.

Statistics for 1992 show that actual investment in rural enterprises was 104,684 million yuan, which was used in new construction, expansion, technical innovation and other matters. The major part of investment in rural enterprises came from banks and the enterprises themselves. Foreign investment covered only a small part of the total investment. Foreign investment mainly went the Guangdong and Jinagsu provinces; it was 2,093 million and 1,739 million yuan covering about 32.7 per cent and 27.2 per cent respectively of the total foreign investment in China's rural enterprises.

Supportive Institutional Framework for Technological Development of Rural Industries

In the current development, China has suffered a lot from technologically backward industrial enterprises in rural areas. The prosperity of the rural enterprises is largely attributed to the priority attached to relying on science and technology development both by the government and the rural enterprises themselves. Thus, the further development and survival of rural enterprises in the rural areas, in the existing highly competitive market, will remain dependent upon progress in science and technology supported by these two agencies.

Spark Program and Demonstration

a. Objectives

It is the first ever program chalked out by the Chinese government to promote the development of rural economy by means of science and technology. Through the organization and implementation of "Spark", projects, it imports, digests and spreads technologies in the hope of promoting commercialization of scientific achievements. By introducing advanced appropriate technologies, it makes in-depth development and full use of the resources of rural areas, gradually forming backbone industries to urge development of other economic sectors, thus promoting regional economic development in rural areas. Spark Program, as part of the governmental reform, has made a bold experiment in transition from complete administrative plan towards mutual coordination of administrative plan and market orientation, and has gained preliminary achievements.

b. Management

The management of Spark Program is mainly concentrated on project application and approval, program formulation, supervision of implementation, management of funds and final examination of projects.

The authority in charge of Spark Program is the Science and Technology Commissions at all levels. The Spark Program Office of the State Science and Technology Commission, renamed in 1998 as Ministry of Science and Technology is the highest body, mainly responsible for the formulation of related guidelines and policies. Provincial (Ministerial) Spark Program managing agencies are responsible for formulating and implementing provincial policies and strategies, organization, supervision and implementation of state-level Spark Projects, and guiding and coordinating spark activities in the province in question. Prefectural and county-level Spark Program managing agencies take the responsibility for specific guidance and supervision of the enterprises undertaking Spark projects, formulation and management of local Spark projects and make applications to the Provincial Commission of Science and Technology for Spark Projects.

c. Stipulations to Select a Spark Project in the Program and Criteria for Establishment of "Spark" Projects

Projects to be included in Spark Program should be in conformity with the following requirements:

(1) Be in line with the nation's industrial policy, technological policy and relevant stipulations of the government, and able to function as a demonstration as well as an engine to improve the local and even the national industrial structure.
(2) Be in line with local overall plans for economic and technological development, focused on S&T progress so as to form Spark demonstration industry;
(3) Be conducive to improvement of rural productivity and living standard of peasants and development of regional key industries:
(4) Be able to expand export, incorporate high and new technologies to make better utilization of local resources.

d. Implementation of the Spark Program

Implementation of local Spark Projects is organized by the local S&T commissions, and the local S&T commissions are to supervise and follow the implementation. The project undertaking enterprises must follow strictly the program stipulations and report the project implementation situation punctually and periodically to local S&T commissions. "Spark" projects management departments, S&T commissions at all levels and relevant departments should provide necessary supporting conditions for the implementation of projects so as to ensure that the projects could be finished on time and meet the requirements.

Upon conclusion of Spark Program, local S&T commissions should examine the results and report the documentation to SSTC. The success rate of Spark projects is above 95 per cent, although a few projects end with failure. Those enterprises which achieve great success serve as demonstrations. Local governments are requested to disseminate their successful experiences and organize study programs with the successful enterprises for other economic entities. Demonstration is a fundamental requirement of a "Spark" project.

The Spark Program, started at a small scale in 1986, is being implemented at a larger and larger scale since then. Till the end of 1990, it completed 22,212 demonstration projects out of a planned total of 42,389. The total investment of 9.378 billion RMB yuan has led to a total output value of 63 billion RMB yuan, including 14.6 billion RMB in tax and benefits and training of 5.04 million people.

d.1 Provision of Technology Information, and Access to R&D

A variety of consultant services have emerged through various channels, including governmental, institutional, and commercial ones.

Governmental Organizations

China has established its State Information Centre. All ministries have their own Information Institutes. At the same time, a large number of specialized S&T institutions are available, including: Chinese Academy of Sciences, Chinese Academy of Engineering, Chinese Academy of Agricultural Sciences and so on. They have established branches in many provinces with institutes specialized in various fields. China's higher education institutions are also a very important science force in the country engaged in research. Before the economic reform, their activities were planned by government and they were not allowed to develop their own business. Since the reform policy, all these institutions are encouraged to serve the cause of economic construction. Now almost all of them are open to public. They have begun to cooperate with and provide services to enterprises in various forms. Until the end of 1993, there were 5440 science organizations located all over China, with 707,053 scientists engaged in research activities.

Institutions

Even taking full advantage of science institutions, these as yet cannot meet the vast and increasing demand of rural economy. Thus, many technology associations, to meet the demands of rural development, were established by local peasants to promote experiences, exchanges and information consultation. It is a great invention of Chinese peasants, and its dynamism gives incentives to rural economy. Until the end of 1991, there were 120,000 Rural Technology Associations with 4,510,000 members in the whole country.

Marketing Support

All production ministries of the central government have established marketing support agencies, e.g. Electronics Product Market Promotion Centre under the Ministry of Electronic Industry, Textile Market Promotion Agency under Textile Ministry (now the Textile Industry Council). These agencies have their network of marketing support system. A detailed example is cited below :

> Under the State Science and Technology Commission, an institution 'China Technology Market Promotion and Management Centre' has been established to provide technology marketing support. At the provincial and in some cases at prefectural level, there have also been established 'Technology Market Development and Management Centres, which together form a national network for technological information collection and circulation. It publishes several newspapers and monthly journals to disseminate technological achievements and information to enterprises. At the same time, they annually hold more than 10 international technology fairs or exhibitions in different cities in China and several Chinese Technology Fairs abroad. Their activities in promotion of circulation of technological information as well as marketing are popular among entrepreneurs, including rural and urban ones.

Standardization and Quality Control

Rural enterprises have made great contribution to China's national economy; however, a lot of problems arise in respect of product quality. In 1992, the Technology Superintendent of China published the year's product quality survey results. The rates of qualified 'rural-enterprise-made products' in the year's four quarters were respectively 56 per cent, 52.9 per cent, 49.4 per cent and 51 per cent which were lower than the average level. The GB/T19000-IS019000 (quality control and guarantee) Standard Series are the regulatory files for enterprises to establish a quality system. In order to get the standard better understood and abided by, the Rural Enterprises Administrations developed programs to train concerned personnel on the standards.

During the past years, the Ministry of Agriculture has issued many documents requiring rural enterprises to carry out quality awareness education among employees. The Department of Township and Village Enterprises under the ministry has done comprehensive planning, and developed the syllabus, and has been in charge of examining implementation of the program. The programs are undertaken in the form of correspondence courses, tele-courses and so on. The department has also established a strict examination system to issue certificates to those who pass the examination.

The Department of Rural Enterprises under the Ministry of Agriculture has accelerated the construction of product quality monitoring and testing authorities. There are now 21 authorities all over China which are responsible for supervising the quality of food, building materials, fireworks, leather, tea and cosmetics.

d 2. Human Resources Development

Traditional education in China was aimed at training students to pass various examinations, and sending more students to higher education. Under the centrally planned system, the situation got exacerbated. The system paid little attention to the actual social demands. The consequence was that many graduates could not find satisfactory jobs, while enterprises could not find proper employees.

Now more and more vocational schools have been opened up. Many colleges have established joint programs with rural enterprises to train students who can meet the rural requirements. Some enterprises attract competent personnel to work with rural enterprises by offering them better emoluments. Some establish corporation with S&T and educational institutions, thus obtaining help through technical consultancy, or ever technology transfer. Through these channels, some scientists and engineers get out of the ivory tower and go directly into business.

Training Centres

Those rural enterprises which are selected to be sponsored under the Spark Program are granted opportunities to train their staff in Spark Program Training Centres. Under the State Science and Technology Commission, there are about 60 training centres. Under provincial, prefectural and county S&T commissions, hundreds of training centres providing service to rural enterprises have been established.

Enterprises have also established their own training programs, which are characterized by flexibility. There is no fixed model for the training programs. Some programs invite college teachers and some require only the engineers and technicians working in the rural enterprises who could train peasant workers. Some Rural Enterprise Colleges have been established. These colleges have achieved major success and have acted as demonstration centres for the nation's education and enterprises.

The course design in the training programs is geared to the demand of rural enterprises. The Rural Enterprises Managing Personnel's College functions under the Ministry of Agriculture to meet the requirements of foreign market - oriented enterprises, develop International Trade, Foreign Languages for Foreign Trade, GATT, Western Accounting System, Accounting Principle, and popularization of high and new technology.

d 3. Mechanisms of Interaction and Communication among Institutions and Rural Industries

Interaction and communication among institutions and rural industries are conducted in the following ways:

a. Meetings on Rural Issues at National or Provincial Levels

These meetings are held on an irregular basis to address issues of common concern or issues of urgency and at the time of making major decisions. Participants in the meetings are mainly people in charge of rural development in the provinces or autonomous regions. State leaders are also invited to attend or even chair the meetings. With the decision-making power of the participants, these meetings have proved to be very effective in solving problems, especially those of high magnitude. For example, at the end of 1992, the Chinese Premier participated in a meeting on rural situation, which was chaired by a vice-premier, and ten major strategies were adopted.

b. Communication Through News Media Like Newspaper, Journals, Radio and TV

In the media mentioned above, special columns or forums are set aside for the institutions and rural industries to publicize and exchange information, and consultancies are also available. In addition, offers and demands of technology and personnel are also available, and this greatly facilitates transfer of technology and talents in rural areas.

c. Organized Activities

Organized activities for interaction and communication among institutes and rural industries include technology trade fairs at both national and provincial levels, and also regular technology consultancies organized by local S&T commissions. Many technology fairs make it possible for the technology developers and producers to meet and communicate, and this is deemed to be a very efficient way for the technology developers to get information on technology needs and for the producers to access the better technology they desired.

d. Extension Services

For rural industrial technology extension services, China has the following institutions: governmental departments, academic institutions, technical departments of the enterprises, and non-governmental technological associations/societies.

Government Departments:

There include the State Science and Technology Commission which shoulders the main responsibility. They help chalk out extension plans and specific programs.

Rural Technological Associations/Societies

In 1979, China introduced a management reform in rural areas, namely, family contracting system to replace the previous people's communes. As a result, China's peasants had much more freedom, than before. One natural outcome of this was the development of sideline occupations other than the traditional agriculture focused on food production. For more benefits, demands on technologies and extension services became much higher. This led to the formation of rural technological associations/societies. More than 130,000 such associations have been established since 1980, with a membership of over five million rural households. The associations are fully self-operational, and function without any government financial support. They cover the whole territory of China, and more than 140 industries in all the rural production sectors. 24 associations are at the provincial level, and 40 associations are inter-provincial.

The associations are divided into three categories. Those in the first category engage themselves in introducing and disseminating appropriate technologies, technological training and demonstration, and exchange of information. The second category normally go deeper so as to help in purchase of raw materials for its members and provide technical services and consultancy before and during production. Associations of the third category may be considered as economic entities. They provide their members a "through" service, that is, production-related services before and during production, and in addition after-production service, i.e. marketing. Marketing is at present one of the most serious headaches faced by the rural entrepreneurs.

Case 1: The Water Melon Association of Xincheng County, Hebei Province

The Association was founded in August 1985 in Xincheng County, Hubei Province in north China, where water melon cultivation has a long history of hundreds of years. At present, it is a full-fledged economic entity specializing in water melon technology dissemination, extension and services of different nature. It has at its disposal 6 trucks, 11 agricultural extension machines, and a real estate worth 300,000 yuan, with an annual net income of 400,000 yuan. Headquartered in Xincheng, the Association's 16,000 members are from 31 counties and 12 provinces across the country.

Though Xincheng excelled in China in water melon cultivation for its unique techniques, its varieties were few and fairly outdated. Development of new varieties was both expensive and beyond individual cultivators. In December 1985, soon after the Association was founded, it invested more than 140,000 yuan to import 120 fine and new varieties. A number of leading agricultural and biotechnology university professors were invited as advisors on payment basis. Today, China's water melons are represented by a fine variety of new breeds and have been disseminated throughout the country. Among them we find popular brands such as "Gold Flower", "Zhengzhou Hybrid Five", "Zhengzhou Hybrid Seven," "Shandong Red," and "Beijing Pleasure." More than 20,000 hectares of melon fields are growing these new varieties and an additional 300 million yuan were generated as a result in the past six years.

The Association has been growing fast, partially because of its excellent services. It has divided its membership into seven geographical regions. Professional experts sitting at the Association Council are assigned one of the seven regions to take up professional leadership responsibilities. They are responsible for technology dissemination, information exchange, material supply and marketing services in their regions. Each of the seven regions has designated a technician and a liaison officer to aid the Council experts and conduct day-to-day work, including technical consultancy, seeds dissemination, and product marketing. The seven regions convene two leadership meetings of Council experts, technician and liaison officer on the 13th and 28th of every month.

Since its inception, the Association has attached great importance to human resources development in the following forms:

— Training courses: More than 90 courses have been conducted to train over 5,000 peasants as master trainers in each of the 31 membership counties.

— Technical book: A book, "Technical Guidelines on Water Melon Cultivation", is provided to every individual household member.

— News letter: 7,500 regular newsletters have been circulated.

— Water melon calendar: A first ever calendar of this kind in China, it is, as a matter of fact, a small encyclopaedia on all the technical points to be taken care of during the whole phase of water melon growth.

— Demonstration: In the poor Tang Xian County, Hebei Province, the Association entered into contract with a group of poor peasants to make them as its members - each household was provided with seeds, technical services through the whole growth period and help to market the product. The contract stipulates that any gain shall be of the member growers, and any losses shall be born by the Association. This was done in 1987 and within that year, all the families contracted made gains. So far, 1,000 poor families have been selected as contracted demonstrators and they have become rich. Farmers in their neighbourhood followed suit rapidly to become Association members.

Case 2: The Long-hair Rabbit Raising Society of Oxford Township, Laiwu County, Shandong Province

Founded in 1984, the Society now has 780 individual members and 15 corporal members. It combines rabbit raising with scientific research and marketing of rabbit hair. The Society has a rabbit breeding farm, a sales centre, a fodder plant, a research institute, a veterinary pharmacy, and a rabbit hair processing mill. The Society distinguishes itself among the rural technical associations/ societies through its scientific endeavours.

With the help of professors from Shandong Agricultural University, the Society launched a number of fruitful research programmes and made remarkable achievements. The major ones are: Artificial insemination technology, rabbit sex determination technology, parturition timing technology, rabbit immunity technology, fodder additives, and fast raising technology.

The Association has also attached great importance to human resource development. More than 470 training courses have been conducted for its members and outsiders in the past years. Among the trainees 1,200 have become master trainers, and 152 were sent to Shandong Agricultural University and other institutions of higher learning to receive further training.

In Oxford and its neighbourhood, up to 135 villages are specialized in rabbit raising, involving 60,000 households. The 3.6 million rabbits raised every year alone help generate 14 million yuan for the town.

China's 10 Best Rural Technology Associations/Societies

In a recent national appraisal on rural technological associations/societies, the following 10 were named the best:

1. The Long hair Rabbit Raising Society of Oxford Township, Laiwu County, Shandong Province
2. The Water Melon Association of Xincheng County, Hebei Province
3. The Planting Association of Nanbu County, Sichuan Province
4. The Frog Association of Yixing County, Jiangsu Province
5. The Vegetable Society of Luomen Township, Wushan County, Gansu Province .
6. The Shedded Vegetable Cultivation Society of Beizheng County, Liaoning Province
7. The Green Pepper Society of Rural Cultivators, Heilongjiang Province
8. The Wax Society of Zangjiang County, Hunan Province
9. The Tea growers' Association of Shaoxing County, Zhejiang Province
10. The Courtyard Cultivation Association of Shangcai County, Henan Province.

d 4. Technology Linakges and Transfer among Large and Small Industries in Urban and Rural Areas and International Cooperation

Initially, during the 1960's, big enterprises in urban areas began to setup collectively-owned rural industries, often in their neighbourhood, as subcontractors to produce components for their products. The mother enterprises usually provided to the subcontracting rural industries, among others, raw materials, machinery, technological advisory services, and manpower training. In addition, they usually provided product marketing and after sale services to their clients. As a result, the rural neighbourhood areas of the big enterprises built up dozens of subcontracting rural businesses. Later, some of the contracting partners grew into competitors of the mother companies. Such development mode of rural enterprises is still very much seen today, particularly in central and west China. A study showed that in 1992, China had a total of 18,085 rural industrial enterprises having partnership in different forms with their big brothers in the urban areas. They employed a labour force of 2.16 million and had fixed assets of 28.6 billion yuan. Their total output value in 1992 reached 89.3 billion yuan including 4.64 billion yuan as profit and 3.02 billion yuan to be handed over to the state as tax.

International Cooperation: Since the early 1980s, soon after China started opening up to the outside world, foreign investments flowed in trickles, gradually in streams and now in rivers. These foreign investments normally were used for international purchase of advanced facilities, technologies, consultancy, technical services, or technical training. By 1992, China had a total of 134,505 joint ventures with overseas companies employing 8,248,888 workers and yield a total of 2,561 billion yuan.

Success Factors, Current Problems and Constraints:

Rural enterprises grew up mainly relying upon themselves. Governmental assistance was not directly involved in the development of rural enterprises which obtained capital, human resources, technologies, equipment and source material at relatively high prices in the market. Their success and contribution are basically attributed to, apart from their own efforts, the national environment of reform and opening up. The emergence and growth of rural enterprises were not protected by the "paternal love" from the government. From the very beginning of their growth, they were granted no share in the national planning system. They were forced to go to market and develop in a highly competitive environment. Further, they are more independent from government interference than state owned enterprises. Market environment is highly competitive and demanding for rural participants. To cope with the changing market, an enterprise must have the power to take decisions of investment, management, pricing, employment, and distribution. Compared to the state - owned enterprises, peasants in rural enterprises collect the investment, borrow money from banks and return it by themselves. Therefore, they cherish the property very much. On the other hand, a rural enterprise does not have as many competent authorities as state - owned enterprises, which results in less of bureaucratic hinderance in the way of their development. No governmental interference intruded in their management, no planning system to protect their weakness. There are thousands of rural enterprises which go bankrupt and thousands of new rural enterprises emerge.

However a number of problems still exist. The first problem is the striking disparity—the extent of development of rural enterprises in different parts of China. Eastern China is regarded as a booming and rich area, while western and central parts of China face poverty and backwardness. The conception is largely attributed to the differences in state of development of their rural enterprises.

In 1991, of the total output value of rural enterprises, Eastern China's contribution amounted to 65.7 per cent. However, with two thirds of the nation's total population, the central and western areas' contributions in the output value are 30.1 per cent and 4.2 per cent respectively. Contrastive analysis is shown in the following table:

	East	*Centre*	*West*
Rural Enterprise Output Value	65.7%	30.1%	4.2%
Average Income of a Peasant	927	596	566
Rural Enterprise Contribution	24.4%	18.3%	11.6%

The rural per capita Gross Social Product in eastern area is 1,858 yuan higher than that in central part of China and 2,247 yuan higher than that in the western part. Many social and economic problems arise from this disparity.

The second, problem is that the development of rural enterprises is seriously handicapped by shortage of monetary support. In 1992, the gross output value of rural enterprises covered 32.3 per cent of the year's GNP, while the loans obtained by rural enterprises this year covered only 3 per cent of the nations total bank loans, including loans obtained from credit cooperatives; rural enterprises shared only 8 per cent of the this year's total loans. Especially, in western area, it is not easy for rural enterprises to obtain loans.

The third problem is the lack of power with the competent authority, i.e. Rural Enterprises Administrations at all levels. They are not given enough power to fulfil their responsibility of dealing with and especially, promoting the development of rural enterprises. These authorities still lack effective management and cannot provide necessary services to rural enterprises.

The fourth problem is that a complete regulation and law system concerning rural enterprises has not been established yet. Village Collective Ownership Enterprise Regulation Law of People's Republic of China has played a very important role in stabilizing development, protecting the legitimate interests as well as normalizing management of rural enterprises. However, there are still many problems in abiding by the regulations, for example, illegal change of ownership or imposing many unreasonable charges upon the rural enterprises in some areas.

With rapid growth of national economy, several serious problems have emerged inspite of the high rate of growth. Although a large number of rural enterprises have been established, a lot of them are generally tiny-sized enterprises, suffering from sub-standard products and not well known brands. Many of them cannot compete in the market.

Conclusions

- The national government should incorporate the programme of Introduction of Industrial Technology in Rural Areas into the national development plans, so that adequate policy and other supportive mechanisms may be built up.
- Regular meetings should be held at national level to review progress.
- Annual award meetings should be convened to create public awareness.
- Seminars should be conducted regularly on the introduction of industrial technology in rural areas. These could also be conducted at regional levels, to strengthen regional cooperation.
- Data bank on industrial technology should be created.

References

1. *1993 China Yearbook on Rural Enterprises,* Agricultural Press, December 1993.
2. *The Statistical Data of Rural Enterprises* (1993), Ministry of Agriculture, Department of Rural Enterprises, May 1994.
3. *1993 China Statistical Yearbook on Science and Technology.* The State Statistical Bureau, the State Science and Technology Commission, February, 1994.
4. *Rural Enterprise Policies: Volumes 11 and 12.* Ministry of Agriculture, Department of Rural Enterprises March 1994.
5. State Science and Technology Commission: *Guide to China's Science and Technology Policy - The White Paper on Science and Technology No. 1, 2, 3, 4, 5.* International Academic Publishers, 1988,1989,1990,1991,1992.
6. *Spark Program Yearly Reports.* The State Science and Technology Commission, 1988-1993.
7. *The Collection of Economic Development Projects in Poverty Area.* Ministry of Finance, October 1994.
8. SONG Lingei: *The Social and Economic Mechanism of "Southern Jiangsu Province Pattern".* Nanjian University, may 1994.
9. *The Report on Poverty Alleviation Institution No. 1 and No. 2,* The Study Group for Poverty Alleviation Institution of the State Council Development Research Centre.
10. *Final Report of The People's Republic of China on Transfer of Small Scale Productive Technologies and Encouragement of Entrepreneurship among Rural Women,* P-12/E/90 of APCTT
11. *Document of the Seminar on experiences of Technology Associations,* China Association of Science and Technology, Department of Science Popularization, October 1992.
12. ALCORTA, Ludovico: *The Impact of New Technologies on Scale in Manufacturing Industry: Issues and Evidence,* UNU/INTECH Working Paper No. 3 The United National University, Institute for New Technologies, June 1992.
13. *China Environmental Strategy.* World Bank, April 1992.
14. *Today's Women Farmers—The Situation of Chinese Women.* Women of China Magazine, June 1994.

PART V : DEVELOPMENT AND PROBLEMS OF SCIENCE AND TECHNOLOGY MANPOWER

14. Management, Cultivation and Employment of S&T Manpower Resources in China

Chen Hao

Abstract

In view of the chronic shortage of S&T manpower in China, it is imperative to formulate a rational policy of work force management in a bid to effectively cultivate, employ and protect the existing manpower resources. The author here aims to delineate a rough picture of S&T manpower management in China, with emphasis on higher educational institutions and their interactions with other R&D institutions and enterprises. Irrational distribution of Chinese R&D manpower is epitomized by the fact that a great majority of them (63.7per cent) are concentrated in universities and R&D institutions, while enterprises account for only 27.5 per cent. Discussing the existing problems and issues, the author highlights that both internal and external braindrain pose a great threat to development of S&T in China. The root cause of this situation is irrational income structure and unhealthy work conditions in Chinese S&T institutions. This calls for some concrete steps to effectively improve the situation.

Introduction

For a populous and developing country like China, it is self-evident that we have to stress on the effective management of S&T manpower resources. This is also because ours is an economically underdeveloped country marching towards national modernization from a state of "poverty and backwardness". As a matter of fact, we are extremely deficient in S&T talent. What is more urgent is that China must achieve advancement in large scale capital construction and that too at a fast rate, if we have to get rid of our long-standing poverty as soon as possible. Such a situation aggravates the chronic shortage of S&T manpower in the country. Thus, it is imperative to formulate a rational policy of workforce management in a bid to effectively cultivate, employ and protect the existing manpower resources in an appropriate way. It would never be an overstatement to emphasize this if we take a panoramic view of its vital importance for S&T progress and socio-economic development across the country.

In the wake of the burgeoning S&T development, in particular with the unprecedented hi-tech upsurge since the beginning of 1990s, the international rivalry is focussed on S&T strength competition among different countries, where the most intense and most influential aspect of the face-to-face confrontation lies in the contention for S&T talent. Therefore, rational management of S&T manpower becomes more and more vital and complicated with each passing day.

Based on the author's own research, this article aims at delineating a rough picture of S&T manpower management in contemporary China, emphasizing on higher learning institutions and categorized distribution. The term 'China' in this article refers to the Chinese mainland excluding the areas of Taiwan Province, Hong Kong and Macao. Related data and materials in this article are all collected from the Chinese mainland.

State Policy Concerning S&T Talent Cultivation and its Effect

Since the founding of the People's Republic in 1949, cultivation of S&T talent has been highly stressed by the national authorities. Starting from the First Five-Year Plan period, the authorities have emphatically pointed out that the developmental focus of the Chinese educational system lies in polytechnic colleges and the departments of natural and engineering sciences in the universities. At the same time, other kinds of collegiate communities specializing in agronomics, pedagogy and medicine should also undergo appropriate development. In the installation of specialities, our efforts have to be focused on machine-building, civil architecture, geological prospecting, mineral mining, power generation and metallurgy.[1] Statistics show that such a guideline has been put into practice without reservation. Table 1 clearly indicates that, after the establishment of the new China, normal,

---

[1]*LIANG Guizi & MENG Huili: "Chronicled Annals of Academic Degree Conferment & Post-graduate Education in the PRC* (1949.10. . .1993.3)". Xian Traffic University Press, 1994.

medical and polytechnic colleges underwent rapid development, while collegiate institutions majoring in politics, law, the humanities and other disciplines grew at a lower speed.

The System of Higher Learning

As a result of repeated organizational restructuring, a nationwide system of higher learning with Chinese characteristics has been formed. The State Education Commission (SEC) under the State Council is the highest national administrative body in charge of educational affairs in the country. Its subordinate units include various educational administrative bodies at provincial (autonomous regional or municipal) , city and county levels, responsible for administrative management of institutions of higher learning, the middle and the primary schools. Attached to all the departments or commissions under the State Council, there are independent agencies specializing in educational work of specific departments, and their professional work is under the leadership of SEC.[2]

At present, Chinese institutions of higher learning are of two types. One type are the ordinary and full-time universities or colleges and the other, adult universities. In 1993, there were 1,065 full-time universities and 1,183 for adult education across the country. According to their administrative affiliations, 34 are administered by the State Education Commission (SEC), 670 by provincial (autonomous regions or municipal) authorities and 361 by departments or commissions under the State Council.

Since 1980s, the status of Chinese universities has undergone a great change. First of all, the college-run forms have been diversified. Besides the ordinary collegiate communities of full-time schooling, a lot of new-type universities have made their debut, such as broadcast and TV universities, in-service or part-time universities, management colleges or normal colleges, correspondence courses and evening courses given or hosted by ordinary universities or colleges, SEC-sponsored program for self study examination and non-governmental colleges. All of the latter are termed adult education universities, and their number has surpassed that of the ordinary universities, while they have been at par with the ordinary universities in total quantity of enrollment (Table 2), but the adult education universities do not emphasize on specialities of natural or engineering sciences, as their aim is to provide collegiate schooling to their enrolees.[2] Another enormous change is seen in the distributive pattern of their graduates. Quite different from the past over-rigid practice of mandatory employment of all graduates under the national system of planned economy, now a system of independent employment under State's guidance is adopted for all collegiate graduates.

Table 1: The Total Number of Chinese Graduates from Different Faculties of Institutions of Higher Learning

(in 1,000)

	Pre Liberation graduates (A)	*Total* (B)	*Graduates from 1949 to 1993* 1949-65	*1966-78*	*1979-93*	*The ratio of B/A*
Total	185	9170.8	1554.4	1392	6224.4	49.57
Engineering Sc.	32	3071	530.6	543.1	1997.3	95.97
Agronomics & Forestry	13	635.7	147.1	126.1	362.5	48.9
Medical Sc.	9	877.5	166.6	207.9	503	97.5
Pedagogy	21	2594.6	392	244.6	1958	123.55
Humanities	24	551.6	82.6	105.1	363.9	22.98
Natural Sc.	16	537.6	112.4	111	314.1	33.6
Finance & Eco	19	626.2	71.3	27.9	527	32.96
Politics & Law	51	133.4	23.1	4.8	105.5	2.62
Physical Culture	—	73.7	14.1	12.1	47.5	—
Arts	—	69.6	14.6	9.4	45.6	—

Note: The pre-liberation years refer to the 20 years between 1928 and 1947.

Source: *The Statistics Yearbook of Chinese Educational System* Beijing; The State Education Commission, 1994.

[2]*Statistics Yearbook of Chinese Education System(1994)* compiled by the State Education Commission. Beijing; The Press of Higher Learning, 1994.

***Table 2:* Basic Data on comparison between China's ordinary and adult education universities—1993**

	Number	*Number of graduates (1000)*			*Enrolment (1000)*			*Students on campuses (1000)*		
		Total	*Regular**	*Special**	*Total*	*Regular*	*Special*	*Total*	*Regular*	*Special*
A**	1,065	570.7	298.9	271.8	923.9	386.5	537.4	253.5	141.7	111.8
B***	1,183	441.2	48.1	393.1	952.6	76.0	876.6	1,863	181.3	1,681.7
Ratio of B/A (per cent)	111.1	77.3	16.1	144.7	103.1	19.7	163.1	735	128	1504

* Regular graduates in Chinese colleges refer to student graduates after a campus life lasting 4-5 years or long er, while special graduates have a campus schooling of 1-3 years, receiving a simplified version of the normal curicular courses set for regular students.

** Ordinary full-time universities or college

*** Various institutions of higher learning for adults.

In February 1980, the 13th session of the Fifth National People's Congress (NPC) passed a resolution on the Regulations of Academic Degrees in the PRC, introducing a tripartite system of BS, MS and doctorate degrees and relevant stipulations on qualified norms for their recipients. Since then, China has had its own system of academic degrees and its enforcement in past years has proved to be of critical significance in modernizing institutions of higher learning, facilitating international exchange with foreign universities.

Training System for S&T Professionals

Ordinary collegiate education is the main form for cultivating S&T professionals in China. According to the decree "A Decision on Reshaping the Educational System" issued in 1951 by the Central People's Government, institutions of higher learning are universities, colleges, polytechnics and special-purpose schools. In line with "Some Views on the Problems in the Teaching Practice Now Cropping up among Majors in Natural Sciences in Universities and Colleges", a document decreed by the then Ministry of Education, institutions of higher learning mainly aim at cultivating research professionals engaged in natural sciences, majors in natural sciences in our higher normal colleges are to breed teachers for middle schools, while Chinese polytechnics predominantly aim at cultivating senior engineers and high-ranking technicians. The regular faculties of natural and engineering sciences at universities usually have four-year schooling.[1]

So, the system of higher education for incubating S&T professionals in China consists of higher professional schools, mono-disciplinary schools attached to universities, polytechnics, and faculties of natural and engineering sciences in universities. In 1993, for example, among the 570,000 regular graduates from collegiate communities in the country, majors in natural sciences, engineering, medicine, agronomy and forestry were 15,000; 127,000; 33,000; 13,000; 3,000 in number respectively, accounting for 2.6per cent, 22.3per cent, 5.7per cent, 2.3per cent and 0.6per cent of the national total. In sharp contrast, among the 440,000 graduates from adult education universities, there were only 668 majors in engineering sciences, accounting for 0.15per cent of the total number of adult graduates.[2] From this we may see what a critical role our ordinary universities have played in breeding Chinese S&T workers! This is not a solitary or exceptional case, however, as in most cases, since the founding of the People's Republic in 1949, the majors in engineering faculties have always been quantitatively unmatched in terms of annual enrolment, the number of campus students and the number of graduates each year leaving from Chinese full-time collegiate communities.[3]

Graduate education has been an important means for China to cultivate senior scientists, particularly during the early years of new China. Almost at the same time as universities began enroling freshmen, post-graduates were also admitted. According to the 1951 governmental decree "A Decision on Reshaping the Educational System", universities and polytechnics are qualified to enrol collegiate graduates and youngsters with equal schooling in a training system with two years of schooling or longer. In cooperation with other research units, the Chinese Academy of Sciences(CAS) cultivates collegiate lecturers and research professionals in a graduate program. At present, the Chinese

---

[3]GUO Yugui: *A Comparative Study of the Academic Degree Systems of the US and the Former Soviet Union-Some Additional Considerations of Ours.* Fudan University Press, 1991.

program has two parts set for MS students and PhD students.

In August 1955, a resolution on Provisional Regulations for Post-graduate Education in the CAS System was passed at the 17th Plenary Session of the State Council, institutionalizing the program for cultivating post-graduate students. For this, the *People's Daily* carried an editorial on its first page. We may say with certainty that under the guidance of the Regulations, our post-graduate work was proving fruitful and rewarding before the outbreak of the Cultural Revolution(CR) in 1966.

During the period 1966-1977, however, the work was at a standstill.

In October 1977, the State Council ratified the circular "A proposal on the Restoration of Post-graduate Enrolment in Universities and Colleges" drafted and submitted by the Ministry of Education, requiring all campuses, if the conditions permitted, to do their best in making a good job of the post-graduate program. Thus, the admissions and cultivation of post-graduate students were reinstated. In 1992, there were 248 collegiate and research units across the land authorized to confer doctorate degrees, and 588 units to confer MS degrees. From 1981 to 1993, a total of 258,000 post-graduates received MS degrees, including 52,000 majors in natural sciences, 120,000 in engineering sciences, 12,000 in medical sciences, and 30,000 in agronomics. At the same time, more than 14,000 obtained their PhD degrees, including 4,000 majors in natural sciences, 5,000 in engineering, 500 in agronomics and 400 in medicine.[2]

The actual situation in recent years has indicated that postgraduate schooling has become a necessity when applying for a position as a lecturer or researcher in a collegiate community or research unit. Also, a nation-wide survey on research and higher learning circles shows that post-graduates are playing a more and more important role in these fields.

The post-doctorate program is an effective means for training senior scientists. As a fecund incubator of high-ranking specialists, it has a life span of only about one decade in China. Because of its unique advantages in scientific creation, prodigy training and talent protection, it has achieved much success and bright prospects in S&T undertakings across the land. Not only supported by governments at various levels, it is also welcomed and appropriated by the broad masses of Chinese S&T workers.

By the end of 1993, 299 post-doctorate stations were inaugurated in 165 campuses or research institutes throughout the country, with 2,800 post-doctorate students working in them. The research scope involving the students covered 47 primary disciplines and 163 secondary disciplines in eight academic categories.

Sending youngsters to study abroad is an important State-decreed policy for breeding senior S&T professionals in China. The policy has gone through three stages since the founding of New China. The first stage covered the whole of the 1950s, when the new born People's Republic was at the high tide of national reconstruction and badly in need of various qualified personnel. So, the students sent abroad were mostly majors in science and engineering. Because of the historical conditions created by the "cold war", they were all dispatched to the Soviet Union and eastern European countries, totalling some 90,000 people in number.[1] During the succeeding years of the 1960s, the outflow slackened and got phased out due to convulsions in international politics. The second phase was from the mid-1960s to mid-1970s when China was tortured by the chaotic decade of CR without any large-scale groups of students being sent abroad. The third stage started at the end of the 1970s and has kept going on up to now, when a massive exodus is being seen in an unprecedented upsurge of youngsters flocking abroad for advanced study programs. By 1994, as many as 230,000 students were crossing the borderline to study in foreign countries.[4] Meanwhile, the involved disciplines, admittance forms and residing countries became greatly diversified. In addition, the faculties and specialities selected by the Chinese students are not limited to natural or engineering sciences.

[4]From in-house materials issued by the Education Bureau of the CAS.

Composition and Distributive Layout of the Chinese S&T Contingent

Cultivation of S&T professionals is a key aspect of S&T personnel management. In the final analysis, however, the ultimate motivation of the management lies in better employment of S&T talents, enabling them to make maximum contribution to R&D progress and growth of the national economy. So, employment of S&T personnel is one of the main contents of the management of S&T manpower resources.

Structural composition of the Chinese S&T contingent includes technical professionals, S&T activists and R&D personnel.[5]

The technical professionals refer to majors in science, engineering, medicine and agronomics who have obtained State-conferred professional titles or have graduated from universities or secondary polytechnic schools. In working practice, they have been promoted in various departments of the national economy and governmental organs, being engaged in research, teaching, production and other works. They include technicians, agronomists, medical workers, researchers and lecturers. Totaling 18.123 million in number, they consist mainly of teachers, accounting for 49.7per cent of the total. The industrial technicians account for 29.6per cent, medicos for 16.1per cent, agro-technicians for 2.7per cent and researchers for 1.8per cent.[5]

The S&T activists refer to workers immediately engaged in S&T activities or offering direct service to S&T activities. By 1993, China had 2.426 million S&T activists, including 1.48 million scientists and engineers in R&D units under the administration of governmental departments, universities and State-owned large or medium-sized enterprises. Among their employers, enterprises take the lion's share, accounting for 37.9per cent. Next to them, institutions of higher learning hired 32.2per cent of S&T activists.[5]

R&D personnel refer to specialists in research and development. By 1993, there were 917, 000 R&D personnel in China. Among them, 598,000 were scientists and engineers, accounting for 65.3per cent of the total number. They were predominantly working

[5]*China's S&T targets (in the Year of 1994)*, compiled by the State S&T Commission of the PRC. The Chinese Press of Personnel Affairs, 1994.

for government-run R&D units, large or medium-scale enterprises and universities. Specifically, 344,000 were employed by State-owned R&D units, accounting for 37.5per cent, 252,000 and 241,000 by firms and universities, accounting for 26.2per cent and 27.5per cent respectively.

Some studies have shown that during the Eighth Five-Year Plan period (1991-1995) the national volume of S&T manpower resources went up from 23.147 million in 1989 to 31.82 million in 1994, a net increase of 8.04 million, averaging 6.94per cent annually, going up from 4.2per cent in 1989 to 5.1per cent in 1994.[5]

Problems and Issues

Currently the management of China' s S&T manpower resources is being challenged by the following problems and issues:

(a) *Irrational distribution of talent composition.* According to statistics for 1992, for example,[5] Chinese S&T workers were concentrated in industrial undertakings (up to 3.41 million accounting for 19.36per cent), educational, cultural and arts and broadcasting units (up to 8.487 million, accounting for 48.23%), and public hygiene, sports and social welfare (up to 2.253 million accounting for 12.8per cent). The three categories totaled 80.4per cent of the Chinese S&T contingent, while other professionals specializing in agronomics, forestry, fisheries and water conservancy numbered only 888,000, accounting for 5.1per cent of the national total. Another example: there were 532,000 agronomists and 334,000 researchers across the land in 1989, accounting for 5.14per cent and 2.87per cent respectively. By 1993, the figures were down to 496,000 and 297,000 accounting for 2.74per cent and 1.84per cent of the year's national total and down by 2.45per cent and 1.03per cent respectively. As the latter decrease was caused by some researchers who had been shifted to the social market as a result of the current restructuring drive of the S&T institutions, the phenomenon may be said to be normal. The former figure was slashed to half, quantitatively equivalent to some 40,000 professionals in absolute value.

The big-margin drop really puzzles us. As a rule, the burgeoning development of the national economy or S&T activity always gives rise to more demand for S&T workers. The fact posing a challenge for us is: instead of a proportionate growth seen in the number of S&T professionals in these occupations, the figure has been decimated. So, has the S&T level in these occupations become so high that the S&T workers have become a "surplus" workforce?

Irrational distribution of Chinese R&D manpower is epitomized by the fact that a great majority of S&T workers are concentrated in government-run R&D institutions (344,000 in number) and universities (241,0000) according to 1993 statistics. The two account for 63.7per cent of the national total while only 252,000 scientists are working for enterprises, accounting for 27.5per cent of the national total.[5] Furthermore, the former two have poor links with enterprises. Our research professionals are hardly throwing themselves into the grassroots units, nor are they taking part in the latter's R&D activities. On the other hand, it seems that our enterprises are far from being urgently in need of an S&T-driving force introduced from universities and research units, which are less in consideration of the former's demands when designing their own research projects. So, the three links of the manufacturer, the university and the research unit are practically disengaged from one another. Governmental interference may prove helpful. To critically change the situation, it seems that we have to overcome some major problems including long-standing institutional snags, conceptional predicaments and the habitual inertia.

(b) *Poor treatment and "brain drain"*: These two problems are interactive, as in most cases "brain drain" is caused by low remunerations paid to S&T workers. In China, the poor treatment received by the intelligentsia has two dimensions. One is the occupational gap in personal income. For example, scientists at universities and State-run research units (such as the CAS) have on the average lower salaries compared to employees in banking, foreign trade, tourism, film & TV and other trades. Some of them are paid less than even a taxi driver in terms of monthly income (which is roughly equivalent to a third or quarter of the latter) let alone the enviably high payments made to employees of foreign-funded enterprises. On the other hand, the gap in personal remuneration between two workers at different administrative levels in the same occupation is trivial. Probably, a professor may be paid no more than his assistant and the director of a bank has the same income as his driver. This is only one aspect of the problem: relatively low payment. Another more serious aspect of the problem is the extremely low absolute value of personal income. Restricted by the length of this article, we cannot deliberate on this point in detail.

Low remuneration leads to a "brain drain" as one of its direct outcomes. The exodus of Chinese scientists has two outlets. One is permanent residence in foreign countries as a result of their participation in overseas study or cooperative programs. According to relevant statistical data, students returning home on time hitherto account for only 58per cent of the youngsters who have been dispatched abroad to study by the State. Still fewer number of self-funded students are seen coming back from overseas. In the CAS system, the figure is 3per cent.[4] Most of them have completed their schooling higher than the regular collegiate level at home and some of them have been bred as backbone researchers at domestic units. At first, they went to the US, western Europe and other developed countries or regions. Now, some of them have joined the flow to the newly emerging countries, including those of Southeast Asia. Another form is the trans-occupational drain at home. The current trend in this aspect is directed to joint ventures, foreign-funded firms and highly paid trades, such as banking, real estate, foreign trade and securities.

(c) *Institutional Forms and Teaching:* The current institutional forms and teaching contents in China cannot meet the demands posed by the development of the national economy and S&T advancement. The system of higher learning in China is copied from the stereotyped pattern of the former Soviet Union, characterised by too meticulous division of disciplines and specialities, excessively narrow coverage of student knowledge, over-emphasis on book learning, lack of hand-working capability and divorce between theory and practice, so as to strangle the students' creativity. The defect is especially seen in the breeding of engineering students. This leads to less R&D activities at Chinese enterprises both in contriving new technological circuits and in developing new products. So, our manufacturing is notorious for its modelling and imitation in design practice and lack of technical innovation. Besides irrationality in our existing S&T system, institutional flaws in the on-going educational system and curricular contents are also to be blamed.

(d) *Imbalance in Talent Cultivation:* Over-emphasis on cultivation of scientists and engineers leads to an imbalance in proportionate layout of various talents needed in capital construction across the land. As shown by data in Table 1, during the four decades and more since the founding of the People's Republic, China has bred some five million professionals specializing in science, engineering, medicine and agronomics, while there were in all 130,000 experts in politics and law and some 600,000 financiers and economists. Such a miscalculation of talent cultivation leads to extreme scarcity of legal and economic managerial experts in China today.

Prospects

In line with the "Outlines for Reshaping the Educational System of Current China and Related Developments" decreed by the Party Central Committee and the State Council, the present targets for Chinese education during the 1990's are: collegiate cultivation of professionals has to be geared to the requirements of socio-economic development and S&T circles across the land; our efforts have to be focused at superbly running a group of key universities and key disciplines, so that high-ranking experts can be brought up at home. Chinese institutions of higher learning have to explore new approaches in a bid to expand their scope, rationalize their structure, increase the enrolment of post-graduates, stabilize the build-up of basic disciplines and properly develop new and frontier disciplines and selectively promote some applicable disciplines. In this way, an effective system of governmental macro-control over educational undertakings can be set up step by step, leading to the establishment of society-oriented and independent operation of various schools and universities. The current systems of university enrolment and graduate employment have to be renovated, while post-graduate training and the system of academic degrees have to be perfected and properly developed if necessary.

With the inauguration and perfection of the socialist market economy in this country, the State should gradually adopt measures to rationalise the deployment of S&T talents across the country in line with market demands. As the overall structure of the nation-wide S&T manpower resources becomes balanced and rationalised, talent cultivation will be tilting towards meeting social demands. Some long-neglected kinds of talents will be stressed and revived. So, the disciplinary layout and speciality arrangement will be adjusted and relocated so as to satisfy the requirements posed by the national economy and S&T development. According to a prediction made by the State Planning Commission, by the year 2,000, the maximum volume of social employment in China will be 640.42 million (i.e. the working population of up to 781 million x labour participation rate 82per cent). If the critical value of S&T manpower density reaches 7per cent of the national population, for ushering in an economic boom by 2,000, the S&T workforce in this country must be at least 44.8 million.[6]

It would be the most vital and difficult problem for managerial experts in charge of S&T personnel administration to maintain, stabilize and develop the Chinese S&T contingent at present and in the years to come. For enhancing the remuneration paid to S&T workers, we must take some effective steps. Otherwise,

[6]ZHEN Yuantai & WU Wenfang: Analysis on Current Situation of S&T Human Resources. *Forum on Science and Technology in China* 1996 (3) Sum.64.

the contingent's stability will be affected, further posing a potential threat to the cause of China's S&T. So far, however, neither the apparent initiative nor the policy realignment advocated for effecting improvement in a big way is in sight.

15. Planning, Development and Problems of Science and Technology Manpower in India

A.R. Rajeswari

Abstract

The paper traces since independence India's policies, thrusts, planning and development relating directly or indirectly to science and technology manpower as a resource. The mismatch between the higher education system (the supplier) and the industry (the user) is highlighted. The analysis includes the status of S&T manpower in terms of availability, demand, utilization, training and associated problems/issues, besides providing some relevant socio-economic indicators.

Introduction

Science and Technology (S&T) manpower is believed to be a valuable resource and an important input for economic growth and social development. The strength of S&T personnel is being seen as one of the indicators of progress of a country. In recent years, study of S&T manpower has been receiving considerable attention in the belief that it may guide planning bodies to make appropriate allocation of resources to the S&T sector and thereby gain maximum benefit.

In India, a major development problem is the generation of employment, in both rural and urban sectors. This is intimately connected with industrial and economic growth, where appropriate use of S&T can play a crucial role. In recognition of this fact, India adopted the Scientific Policy Resolution (SPR) in 1958, National Policy on Education (1968), revised Education Policy (1986) and Technology Policy Statement (1983) on the assumption that investments in S&T education and in scientific research are prerequisites for industrial and economic growth and for ensuring abundant supply of highly qualified personnel through a formal system of education.

Policies

Policies of the government relevant to S & T manpower production, development and utilization are as under:

1. Education Policy
2. Science Policy
3. Technology Policy

The essential extracts of the policies relevant to S&T education and manpower development are:

1. National Policy on Education

1.1 Education Policy, 1968: "This involves a transformation of the system to relate it more closely to the life of the people, a continuous effort to expand opportunities, a sustained and intensive effort to raise the quality of education at all stages, an emphasis on the development of S&T and the cultivation of moral and social values".

1.2 Revised Education Policy, 1986: The National Policy on Education (NPE), formulated in 1986, envisages the preparation of a blueprint for complete revamping of the country's educational system. It pointed out that disparities among the urban, semi-urban, rural and regional-level educational institutions have grown out of proportion, with a few elitist institutions dominating the scene. While centers of excellence are few, a large number of institutions - schools, colleges and even universities-dispersed throughout the country, lack basic facilities and remain a lingering formality in the domain of national educational activity. The policy spells out that there should be a commitment to pull these institutions out of their prevailing situation and to induct in them an acceptable level of competence. To achieve this, "Courses and programmes will be redesigned to meet the demands of specialization better. There will be increasing flexibility in the combination of courses."

2. Scientific Policy Resolution, 1958

"To ensure an adequate supply within the country of research scientists of the highest quality and to recognize their work as an important component of the strength of the nation; to encourage and initiate, with all possible speed, programmes for the training of S&T personnel on a scale adequate to fulfil the country's needs in science and education, agriculture and industry and defence; to encourage individual initiative for the acquisition and dissemination of knowledge and for the discovery of new knowledge in an atmosphere of academic freedom; to ensure that the creative talent of men and women is encouraged and finds full scope in scientific activity".

3. Technology Policy Statement, 1983

"Human resources constitute our richest endowment. Conditions will be created for the fullest expression and utilization of scientific talent; skills and skilled workers will be accorded special recognition".

Five Year Plan Thrusts

Thrust in different five Year Plans relating to S&T education, manpower and programmes/projects have been as under :

THRUST IN FIVE YEAR PLANS

First Five Year Plan (1951-1956)	Giving importance to establishment of infrastructure for science (Chain of National Laboratories was established).
Second Five Year Plan (1956-1961)	Strengthening of research facilities, coordinate research between agencies, link research at national level, create sufficient S&T manpower (1958–Scientific Policy Resolution was adopted)
Third Five Year Plan (1961-1966)	Encouraging basic research in universities, enlarging research scholarships and fellowships, improving farming methods and giving thrust to industrial research, application and commercial exploitation of scientific results.
Fourth Five Year Plan (1969-1974)	Integrating research with programmes of industrial development, achieving greater coordination among laboratories, institutions and universities and avoiding duplication.
Fifth Five Year Plan (1974-1979)	Formulation of science and technology plan was initiated, "Approach to Fifth Five Year Plan" was prepared and widely circulated and integrated with the plan document.

	Emphasized strengthening and coordinating existing facilities of S&T infrastructure rather than creating new organisations, undertaking goal-oriented research programmes and basic research in new and emerging areas of research.
Sixth Five Year Plan (1980-1985)	Draft S&T Plan was prepared and discussed extensively. This was integrated in the Plan. (1983-Technology Policy Statement announced). Emphasised achievement of technological self reliance and creation of appropriate instruments relating to policy formulation and planning for S&T.
Seventh Five Year Plan (1985-1990)	A Steering Group for Science, Technology and Environment was set up and report prepared. It incorporated the recommendations of various S&T working groups. Objective was to make S&T an essential and integral part of all major socio-economic sectors and to undertake S&T missions.
Eighth Five Year Plan (1992-1997)	The procedure of Seventh Five Year Plan repeated. The objective was to utilise S&T in enhancing the pattern of utilisation of natural resources, to propagate scientific temper and S&T popularisation.

High Level Committees for Implementation

High level committees are set up from time to time for implementation of the science policy and planning for national development; promoting effective utilisation of the existing and available capabilities by coordinating research efforts over a number of institutions/agencies; dealing with problems of international collaboration; correcting imbalances; revamping of scientific, technological and industrial activities for civil and defence requirements; advising and assisting the government in preparing a science plan; achieving self reliance, fuller utilization of indigenous research etc.

High Level Committees Set up for Coordination/ Implementation of Science Policy and Planning

Name of Committee	*Year*
Science Advisory Committee to Prime Minister (SAC)	1948
Scientific Advisory Committee to the Cabinet (SACC)	1956
Committee on Science and Technology (COST)	1968
National Committee on Science and Technology (NCST)	1971
Science Advisory Committee to the Cabinet (SACC)	

and also Cabinet Committee on Science and Technology (CCST)	1981
Science Advisory Council to the Prime Minister (SAC to PM)	1986
National Council on Science and Technology (NCST)	1993 ??

The mandate of the above high level committees was to advise and make viable recommendations to achieve the aims, objectives and thrusts of science and technology policies and planning.

Infrastructure for Formal Education

Due to the objectives laid down in the enlightened S&T policies, post-independent years saw enormous growth in the number of universities, colleges, polytechnics, schools, industrial training institutes, professional colleges and training institutes, offering a variety of courses at various levels in different disciplines. Increase in the number of educational institutions has resulted in expending considerable amounts of national resources on education and training, both formal and informal.

Education and training are usually imparted to the S&T personnel through a formal system of education based on regular hours of study in schools and colleges and sometimes through a non-formal system of education through correspondence courses, evening courses, sandwich courses, etc. Till recently, emphasis in India has been placed predominantly on the former rather than on the latter.

After the attainment of independence, the major goal set by the national government was to achieve economic progress so as to improve the quality of life of its people. Its policies were, therefore, aimed at transformation of the nation's traditional economy to that based on S&T. Such a transformation necessitated abundant supply of qualified S&T personnel of the highest order. This resulted in the expansion of universities, engineering and medical colleges and enormous growth of enrolment in various courses. The development of S&T education in India has been largely attributed to the social demand rather than economic or industrial demand. This situation is being examined now to make S&T education relevant to industrial and economic development and also to vocationalise education at the school leaving level to reduce what is popularly called ',the diploma disease". Having said that, it may perhaps be useful to explain the Indian educational/ training infrastructure for development of S&T manpower and the quantum of scientific and technical personnel available.

The development of university education has been guided by the University Education Commission (1948-49), Secondary Education Commission (1952-53) and the Education Commission (1964-66). Presently, there are lot of discussions and deliberations on revamping university education, especially enabling S&T education to meet the needs of industry, bringing excellence in S&T education and making it more need-based. These were the main thrusts of the revised Education Policy announced in 1986.

The University Education Commission drew up the blueprint for university education in free India. One of its major recommendations was the establishment of a University Grants Commission. The Commission was to support the universities, particularly in their research programmes and to guide their future development. The Secondary Education Commission's major recommendation was with regard to the structure of the degree course. The Education Commission (1964-66) surveyed the developments since independence, identified new needs and made specific suggestions to meet the requirements, including the scale of investment in education. It made specific recommendations with regard to promotion of science. As a result of the above commissions/reports/recommendations, there has been a phenomenal increase in the number of universities and professional colleges since independence.

The Union Government (Central Government or Federal Government) is directly responsible for coordination and determination of standards in research, scientific, technical and higher educational institutions. For discharge of these responsibilities, the Ministry of Human Resources Development has set up statutory organisations like the University Grants Commission, All India Council of Technical Education (AICTE), Indian Institutes of Technology and other autonomous organisations at the national level like the National Council of

Educational Research and Training. The government has also set up four Institutes of Management and four Technical Teachers Training Institutes. In the three decades of planned development, the number of educational institutions has grown manifold. The University Grants Commission has established centres of advanced study with areas of specialisation under different departments of various universities to undertake research and to train manpower to meet the needs of the country in specialised and emerging fields. The specialised areas of work of some of the centres are chemistry of textile fibres and dyes, chemistry of natural products, cell biology, proteins and vitamins, chromosome research, radiophysics, electronics, etc. The Institutes of Technology, and Science, and colleges of engineering are engaged in the development of high level technological skills in all branches of engineering, technology and science.

Over the years, in the areas of S&T, a variety of incentive schemes for scientific and technical manpower have been introduced. These schemes were designed to attract students towards scientific and technical education, to encourage young scientists to take up research as a career, to promote the welfare of professional scientists, to accord due recognition and honour to scientists and engineers for their R&D performance, to remunerate scientists and technologists adequately for the efforts put in by them and to check the flight of trained personnel to foreign countries. Under the Assessment Scheme, the scientific and technical personnel are upgraded after satisfactory completion of a specific period of service.

Issues and Problems

As a result of these policies and the major efforts since independence to increase the availability of qualified S&T manpower, the number of universities has increased from 27 in 1950-51 to about 210 in 1993-94 with 5000 science and 850 technical and professional colleges and this required spending of considerable amount of resources on S&T education. As a consequence, the supply base of S&T manpower increased considerably and the estimated stock of S&T manpower by end of 1990 was around 4 million. The yearly outturn of S&T manpower was of the order of 200,000.

The twin objectives of S&T manpower production and utilisation are to make S&T education relevant to national needs and to ensure optimal utilisation of S&T manpower. But the question still to be answered is whether these objectives have been achieved and whether expansion in S&T manpower has helped to achieve the national objective of economic and industrial growth and employment generation in the urban and rural sectors. It is not easy to answer this question especially since unemployment and poverty continue to be chronic problems despite more than four decades of development planning in India. It is indeed difficult to quantify the deficient employment situation in the rural sector, where it is generally agreed that there is much under-employment or disguised unemployment which are difficult to measure. Even in the organised sector, the extent of under - utilisation of labour force is considerable. Moreover, it is quite distressing to note that the overall productivity of the economy is amongst the lowest in the world. It can, therefore, be said that the explosion in the strength of S&T personnel has not yet eliminated or even reduced umeployment. Moreover, the phenomenal growth in the supply of S&T personnel has resulted in employment difficulties within the S&T sector itself, such as unemployment, under-employment, malutilisation and decline in standards. The problems/issues specific to S&T manpower are explained under different heads in the ensuing paragraphs.

Education (production or supply side): There exists an improper mix of S&T personnel production, as the data on institutions, enrolment, outturn and stock show. Curriculum/ course structures do not match the demands of employers and markets. Outdated syllabi/courses have outlived the relevance and need. These resulted in oversupply of some categories of S&T personnel whose skills are either obsolete or are becoming obsolescent, as also undersupply of skills which are in shortage or in zero supply. This may be seen from the data presented at the end of this paper on the lopsided share of different fields of S&T in enrolment, outturn and stock, as outturn is a function of enrolment and stock is a function of outturn. The pure science undergraduates (bachelor's degree holders) constitute nearly 60 per cent of the total production. The reasons for higher output (or oversupply) in pure sciences in India could be:

(i) Large capacity for admissions.

(ii) The cost of pure science degrees is far less than that of a degree in engineering or medicine.

(iii) The number of years required to obtain a degree in science is less as compared to engineering and medicine.

Linkage between education planning (comprising curriculum planning and course planning) and employment planning is lacking. Facilities for vocational education are also lacking.

Education is expanding due to socio-political reasons, but not due to relevant economic considerations and developmental needs.

Employment (demand or utilisation) : Though roughly 66 per cent of S&T manpower are employed, there is sizeable underemployment and misplaced employment amongst them. Besides, unemployment, migration abroad (otherwise called brain drain), wastage (out of labour force) as also unfilled vacancies exist side by side. Employers, especially industry, are hesitant to employ highly qualified scientific personnel, as curriculum/ course lacks relevance. Research and development (R&D) system though supposed to be one of the major employers of highly qualified S&T personnel, absorbs only 5per cent of total S&T personnel stock. Since industrial R&D activity is low, highly qualified S&T perosnnel do not find job opportunities in industrial R&D. Moreover, the shares of technology-intensive industrial output or exports in respective totals are low. There is a growing trend among prospective S&T personnel towards banking and industrial management jobs which do not require S&T input. In fact, a large percentage of pure science graduates prefer clerical jobs in banking and financial institutions. These preferences are mainly due to the expectation of higher income and better job prospects, as low income and poor job prospects are generally the characteristics of scientific jobs. So, the shift of S&T personnel to non-scientific jobs seems to act against the objective of increasing industrial growth through appropriate deployment of S&T manpower. Unemployment, poor salary structure of S&T manpower and slow growth (2-3per cent per annum) of S&T manpower employment in R&D sector, which is supposed to be one of the largest employers of S&T personnel, are the other causes which need examination and corrective action. Could one argue that S&T personnel might respond to the tight employment situation by setting up small scale industries, i.e. resort to self-employment as they possess technical/scientific skills? But, this is not true. The structure of Indian society is such that it attaches a great deal of importance to the secured earnings of an individual, which are often used as an indicator of social rank. Income deflated or inflated by some "coefficient" of security of earnings is the manner in which status is determined. This situation is further aggravated by the fact that there is little evidence to show that potential for entrepreneurship qualities exists among scientists and technologists in the government, university and service sectors. In addition, little encouragement can be derived from the fact that there are currently very few self - employed scientists and technologists in the country.

The above discussion on the supply, unemployment and "misemployment" of the S&T manpower may lead one to conclude that there is an excess supply of S&T personnel in India in relation to the level of economic and industrial development. The data provided at the end indicate that there has been no appreciable increase in manufacturing activity, in gross national product, in manufacturing employment or in non-traditional exports to absorb the available supply of S&T manpower. Moreover, the available stock of S&T manpower is a sizeable percentage of the persons employed in manufacturing activity, while the growth of employment in manufacturing is marginal.

Towards Some Corrective Measures

To provide increasing employment to S&T personnel, to streamline S&T education and to increase industrial growth, some measures are discussed under three broad heads: Education, Industry and Others.

Education Related Measures

A question arises as to the need for producing such a sizeable number of science graduates when the economy of the country is unable to utilise them. There are reasons to believe that barring those who pursue higher education in science or work as laboratory technicians/teachers, a major chunk of them go in for clerical jobs which do not require scientific knowledge or skills. In order to reduce the dimension of the problem, it is necessary to adopt the following approach to the problem:

(i) Vocationalization of education after the school stage;
(ii) Selective enrolment through tests of aptitude and examination results in BSc courses;
(iii) Limited admission in MSc/PhD courses according to manpower needs;
(iv) Serious gaps have appeared in technical education in creating medium level skills. The courses of engineering colleges and polytechnics need to be restructured to fill these gaps.

It would be a step in the right direction if along with delinking of academic degrees from job specifications where such qualifications are not required, a number of existing colleges are made to vocationalize college education. To begin with, they can concentrate on such courses as publishing, printing, tourism, insurance, accounting, public health and hygiene, commercial publicity, photography, stage craft, script writing, industrial documentation, electronics, musiology, etc. The case for introducing courses in public health and hygiene should be considered seriously. Vocationalization in rural colleges should centre round the needs of rural development-cottage industries and programmes designed for the development of backward areas. UGC should set up an expert group to examine the development of new vocational courses. A high level body of experts should work out the requirements of various categories of jobs which are likely to be created in the near future and to prepare job profiles with the levels of training at each point.

There is a widespread view that technical education, especially at the university level, should be need - based and job oriented. There is a general feeling among the employers that the products from the university possess some amount of theoretical knowledge, but they are not sufficient in skills and, therefore, are not able to handle problems of the industries. This may partly be attributed to the fact that the teachers themselves do not have operational knowledge which is necessary to handle the day-to-day problems of industry. There is, therefore, a need for frequent professional contacts between faculty members and technologists in the industry, so that the teachers may update their practical knowledge and the industrial technologists become more fully acquainted with the current developments. In this context, interaction between industry and technical education has to be made compulsory for industries. Some of the steps which may be taken in this regard are as follows:

(i) Members of the teaching faculty should be encouraged to take-up consultancy work and this could be treated as part of the working load.

(ii) Industrial technologists who have good record of performance should be appointed to suitable visiting positions in educational institutions and they should be permitted to direct students for PhD degrees in applied topics and in other research programmes. They should be encouraged to serve on curriculum committes of the universities and other institutions.

(iii) There should be courses on continuing education for engineers working in industries.

(iv) working engineers in industries should be encouraged to take up part-time courses in universities leading to higher degrees.

(v) Industrial management courses, specially for the middle level should be introduced in technical institutions.

(vi) Scientists from universities and research laboratories should occasionally change positions with technologists in industries. Necessary changes in regulations may have to be made to facilitate its implementation.

The universities should be encouraged to introduce nonformal education in science and technology. Evening/vacation courses in science and technology, which are not necessarily experimental-oriented, can be introduced. This may add to employment opportunities without additional capital input. Introductions of evening courses will relieve pressure on admission in the day colleges. Further, it will motivate students to seek employment (including self-employment).

The present system of education has locked itself in rigid discipline-oriented departments. It is, therefore, necessary that both universities and IITs should have greater flexibility in deciding both curricula and courses, so that the new emerging areas can be exploited quickly with efficiency.

Industry-related Measures

The second major requirement for easing the S&T manpower employment situation is increasing the rate of industrial growth.

However, the way industrial growth rate has fluctuated of late rules out the feasibility of such containment of the number of educated unemployed. So far, we could not achieve more than 5per cent growth in spite of pious intentions. It is hardly possible to find employment unless industrial growth is stepped up to at least 12-14per cent to take care of most of the existing unemployed and the future outturn of S&T personnel during that period. A quantitative growth rate here is specified as an indicator, as it was observed that industrial growth is highly related to S&T employment in the organized sector. The pattern of industrial investment so far rules out such a growth rate within the prevailing framework of resource availability. The best that is generally hoped for is that India will manage with a backlog of educated (matriculates and above) unemployed of under three million. Therefore, something massive has to be done to achieve 12-14per cent growth rate. In addition, special schemes must continue to be implemented and expanded in self-employment enterprises.

Eradication of unemployment is closely linked with increasing the income of the people and development of the economy and as such means to solve the evils should be found out. The true solution, both long-term and short term is to build up the economic infrastructure of the country and lay a sound base for the growth of industries, both small (cottage, village and agriculture) and large scale.

Training

Since humanware is the key element of any transformation, training to enhance/update operating abilities, setting up abilities, adapting abilities, reproducing abilities and improving innovative abilities, besides managerial, interpersonal and financial skills, is of utmost importance. In this section, the in-plant training to upgrade skills is dealt with in detail, at the same time bringing out the importance of training and the lack of it in the country.

The importance of In-plant Training (IPT) as a source of skilled human resource development is becoming apparent now, especially its value in enhancing national competitiveness.

IPT typically aims to achieve the following objectives:

— induction/orientation of new employees;

— updating of knowledge and theoretical concepts to enable employees to appreciate fully the nature of their current job responsibilities;

— acquisition of technical skills by existing employees, as demanded by their current job profile;

— enhancement of technical skills and/or learning of new skills, developing the capabilities required to take up advanced technical jobs and/or higher-level positions; and

— acquisition of managerial and attitudinal skills for planning and control, as well as for inculcating a positive disposition towards work among employees.

From the point of view of the employee, IPT is an avenue to enhance efficiency and dexterity on the job, while it eases job conversion and facilitates handling of multi-skill jobs. IPT is also expected to facilitate promotion of employees to higher positions and towards employment in occupations requiring enhanced skills.

In India, for example, no direct institutional support for IPT is available from the government. Nevertheless, government initiatives have an indirect influence on IPT. For instance, the new national economic policy of the government has made enterprises more open to international competition, making it necessary for them to enhance their productivity and cost effectiveness. This creates the need for well-designed IPT in enterprises to enable them to meet demands for new skills resulting from the updating of technology. On the whole, however, it appears that IPT has not been the focus of government policymakers in India. The governing philosophy appears to be that IPT is best left to individual enterprises.

In India, many large enterprises, in both manufacturing and services sectors, have good training infrastructure and expend considerable effort in planning for IPT. Smaller enterprises usually perceive IPT as a non-critical function, and financial constraints often limit resources committed to IPT, in any case.

In India, there is no policy document on vocational training that specifies a government strategy for skilled manpower development in different fields, including the relative emphasis it would like to place on different sources of skilled manpower development. In fact, IPT is probably not even recognized as a source of skilled manpower development which needs to be exploited by the government. Second, the new national economic policy seeks to trim the labour force in enterprises, and employers have had to initiate voluntary retirement schemes. The government has instituted a national renewal fund aimed at providing financial support for rehabilitation of the voluntarily retired employees. Till the time of this writing, however, no part of the fund has been made available for retraining of such employees, which would enable them to develop new skills and be absorbed into other enterprises. Third, there is no evidence that technology management at national level has influenced macro-level IPT systems or policy. It appears that there is insufficient coordination among government departments responsible for technology scanning, forecasting and transfer at national level on the one hand, and those responsible for macro-level planning of training activities on the other.

Studies conducted in India indicate that IPT is not considered as a critical activity in many enterprises. There is often no company policy on human resources development, and provisions for the administration of IPT are usually inadequate. Investment in training is usually influenced by the labour market situation, resulting in insufficient financial resources for training. Enterprises, especially the smaller ones, lack upto-date equipment, machinery and facilities. They are unable to release workers for training due to the small size of their workforce and the shortage of workers.

These studies have also revealed that smaller enterprises depend more on other sources of skilled manpower - recruiting skilled manpower from other organizations-and, therefore, give lesser weight to IPT. Other enterprises employ predominantly unskilled workers, and do not perceive IPT as a critical function.

IPT also takes a back seat when profits fall and the enterprise faces severe constraints on resources. In fact, some employers have argued that IPT results in the dissipation of organizational resources, since many employees, having undergone skill development programmes as part of IPT, leave the enterprise for better employment elsewhere. Thus, planning for IPT, far from being pro-active, is usually incremental and bears little relatioship to future skill needs.

In so far as training infrastructure goes, many enterprises in India provide training machines and equipment of relatively inferior quality and features, since it is expensive to provide the latest models for training. Generally speaking, larger organizations have a better training infrastructure compared to the smaller ones. Financial constraints are often given as the reason that enterprises cannot provide high standards of training infrastructure. Further, sharing of such infrastructure among similar enterprises is uncommon, though many enterprises are aware of the potential mutual advantage.

Other Measures

In the context of generating employment through more investment, the following alternative could also be considered. The investments by non-resident persons of Indian origin in shares of Indian companies on a non-repatriation basis should be viewed in a larger perspective of national interest as an instrument to generate economic benefits to the country and it should be imaginatively interpreted and applied, if need be, after suitable amendment. The unfortunate and negative tendency to treat it as a set of punitive regulations needs to be discouraged and a fresh view is required to be taken of the situation. Already, under liberalisation and economic reforms adopted by the government since 1990, some corrective measures have been adopted.

Overseas Indians have now acquired the capacity to locate and negotiate for new and difficult technologies and they can import sophisticated technologies into India with much greater ease and facility and assist in filling up the technlogical gap of our country.

In addition to improving our educational system to be more need-based and increasing investment in both small and large industry, India should go in for large scale industrial extension services, on similar lines as agricultural extension services. In order to bring in more collaboration between industry and institutions in the utilisation of technologies developed in our institutions, such industrial extension services would be very useful. Moreover, these extension services would absorb a large part of our S&T personnel.

Further, there is an urgent need to establish a national S&T manpower information system for creating databases on supply and demand of S&T manpower to study and analyse the whole issue of education and employment planning for S&T manpower. As of now, no reliable data are available and whatever are available are incomplete, incompatible, outdated and unreliable. Due to this, sound, rational analyses/studies are not done or are done only as pilot studies on sample basis. Even to establish an information system and collect data, standardisation of concepts and definitions of terminologies associated with S&T manpower and education are prerequisites.

Conclusions and Perspective

The foregoing analysis has clearly shown that there is excess enrolment, output and stock of S&T personnel in India, compared with the conventional indicators of her level of economic development. To put it differently, India's objective of contributing to the elimination of poverty through appropriate use of science and technology required, it was believed, increased availability of qualified manpower through expansion of the educational infrastructure, which has not been fulfilled to any appreciable extent. Moreover, there has been no real change in the high(er) technology content of India's manufactured goods exports, nor has India succeeded in exporting proportionately more high technology goods. This is coupled with unemployment and underutilisation of S&T personnel. This raises doubts as to whether the S&T content of the present educational system suffers from lack of relevance to the needs of industry or whether industry lacks dynamism and professional management, or both. To check unemployment of

S&T manpower, a two-pronged attack is required-streamlining S&T education and increasing industrial growth and exports and, in turn, overall economic growth. Some remedial initiatives were already discussed under three broad headings: education, industry and other measures. They consist of both long-and short-term measures. These are only broad suggestions and the details will have to be worked out case by case. The spectrum of skills available in the country in the form of S&T manpower should be codified into : obsolete skills, obsolescent skills, surplus skills, transferable skills, shortage skills and zero stock skills and they should be continuously monitored to correct the imbalance between supply and demand of S&T manpower. Such a dynamic exercise will lead to education planning (comprising both curriculum and course planning) and employment planning leading to appropriate manpower planning and development.

Some Useful Indicators

Total employment (main workers) (1991) (number)	285930000
Total employment (main workers) growth rate (%) (1991 over 1981)	2.5%
Manufacturing employment (main workers) (1991) (number)	28679000
Manufacturing employment (main workers) growth rate (%) (1991 over 1981)	1.3%
Total labour force (1991) (number)	346900000
Estimates of unemployment (1987-88) (number)	12400000
Unemployment as a % of labour force (1987-88)	4%
Total stock of S&T personnel (number) (1990)	3809200
Economically active stock of S&T personnel (number) (1990)	3110700
Economically active stock as % of total stock (1990)	81.7%
Total stock as a % of total labour force (1990)	1.1%
Economically active stock as % of total labour force (1990)	0.9%
Growth rate per annum of S&T stock	9.0%
Total stock of S&T personnel in total employed in manufacturing	13.3%
S&T personnel employed in R&D (1990)	202673
S&T personnel employed in R&D (1992)	193688

S&T personnel in R&D as per cent of total stock (1990)	5.3%
S&T personnel in R&D as per cent of Economically Active Stock (1990)	6.5%

Growth rate of S&T personnel employment in R&D
(i) 1990 over 1980 = 5.1%
(ii) 1992 over 1990 = -2.0% (negative growth)
(iii) 1992 over 1980 = 3.9%
(iv) 1990 over 1984 = 5.1%
(v) 1992 over 1984 = 3.2%

Stock of S&T manpower (1990)	3809200
Employed (estimated)	58%
Self-employed (estimated)	8%
Unemployed, students and not seeking employment (estimated)	34%

Composition of the total stock of 3.81 millions S&T personnel

Degree holders	3.07 million (100%)
Science Graduate	55%
Science post-graduate	14%
Engineering & Technology (graduate and above)	15%
Medical (graduate and above)	10%
Agriculture & Veterinary (graduate and above)	6%
Diploma and paramedical personnel	0.74 million
Total stock of S&T personnel	3.81 million (100%
Economically active stock, i.e. available for employment or already employed	3.11 million (82%)
Wastage (not available for or seeking employment)	0.70 million (18%)

Total S&T outturn (1989)	232000
S&T outturn as % of total outturn in all disciplines	21%
Total S&T outturn	100%
Science	70%
Engineering	15%
Medical	10%
Agriculture	5%
Total S&T enrolment (1991-92)	1347000
S&T enrolment as % of total enrolment in all disciplines	29%

Total S&T enrolment	100%
Science	67%
Engineering	17%
Medical	12%
Agriculture	4%
Background of students in MBA Programs	
All MBA students	100%
Science background	11%
Engineering background	66%
Arts, Commerce, etc. backbround	23%
Composition of Imports (1990-91)	
(i) Technology Intensive* (%)	69.6%
(ii)Non-technology Intensive (%)	30.4%
Composition of Exports (1990-91)	
(i) Technology Intensive (%)	27.8%
(ii) Non-technology Intensive (%)	72.2%
Composition of Industrial Output	
A. Gross Value of Output (1990-91)	
(i) Technology Intensive (%)	17.1%
(ii) Non-technology Intensive (%)	82.9%
B. Gross Value Added (1990-91)	
(i) Technology Intensive (%)	8.8%
(ii) Non technology Intensive (%)	91.2%

* Technology intensive industries are the following:
Chemical and Allied Products
Machinery
Electrical Equipment
Aircraft and Missiles
Professional and Scientific Equipment
Primary Metals
Fabricated Metal Products
Motor Vehicles
Other Transportation Equipment

Source: Various published sources.

References

1. *Research and Development Statistics,* 1992-93. New Delhi; Department of Science and Technology, 1994
2. *Science and Technology Pocket Data Book,* 1993. New Delhi; Department of Science and Technology. 1994
3. *Various Issues of Fifth Five Year Plan Documents.* New Delhi Planning Commission
4. *Statistical Year Book.* Paris; UNESCO, 1993.
5. *Economic Survey,* 1994-95, Ministry of Finance.
6. *Monthly Abstract of Statistics,* January 1995. New Delhi; Central Statistical Organisation.
7. *Manpower Profile. India Year Book* 1993-94. New Delhi; Institute of Applied Manpower Research.

PART VI : BIOTECHNOLOGIES

16. Chinese Biotechnology at the Dawn of the New Millennium

WANG Qinnan

Abstract

Describing China's development programme for biotechnology, it is pointed out that Phase 1 of the programme was primarily aimed at basic research and R&D for product development. Some of the achievements made during this phase have been described. In Phase II, the biotechnology programme is to be concentrated on national resources in a few well-chosen promising areas which are likely to have a major impact in the 21st century. When selecting major areas, novelty is accorded paramount importance, in order to increase the competitive edge of the Chinese biotechnology industry. In terms of technology, the interdisciplinary aspect should be emphasized. The first biotechnology drug to be approved by the Chinese Government for entering the market is human interferon. Concludes that in the 21st century, R&D in biotechnology is of vital importance for the Chinese economy, for raising the standards of living of people and for providing them adequate health care. Phase II is aimed at strengthening the concept of property rights, protect Chinese patent rights, and at the same time, training a large number of advanced biomedical scientists. To accomplish these objectives, it will be important to stress on basic and interdisciplinary research, and to encourage international collaborations.

Introduction

Biotechnology has been on the National Science and Technology Plan since 1982, with emphasis on R&D in genetic engineering. It has also been in the National High-Technology Development Programme ('638 Programme') since 1986. These two programmes are divided into three sub-programmes:

(1) new crop strains with high yield, good quality and drought-resistance;
(2) new drugs, vaccines and gene therapy;
(3) protein engineering.

During the last 10 years, large scale research has been initiated in the fields of agriculture, animal husbandry, medicine and pharmacy. This forms Phase I of China's biotechnology development programme.

The Progress

During Phase I programme, emphasis was on basic research and product R&D. The following are the notable developments:

1. Cloning of 400 new genes was done in animals and plants, and the DNA sequences were deposited in NCBI. Ten of these new genes have been elucidated in terms of their potential functions.
2. 20 novel gene carriers have been developed and used routinely.
3. Total synthesis of rice rbcl and rbcs structural genes and their over-expression in E. coli (Rubisco sub-units) has been achieved.
4. Elucidation of the mechanisms of the cis - and trans-response elements of the yeast acid phosphatase.
5. Sequencing of the smallpox virus genome (Tiantan strain) (196kb)
6. Over-expression of the thermostable basic phosphates and reverse transcriptase.
7. Isolation and purification of 7 membrane proteins from sperm, production of antibodies and monoclonal antibodies against these proteins, and cloning of their cDNA.
8. Completion of a software for hammer-head ribozyme design, construction of auto-splicing ribozyme plasmid for *in vivo* and *in vitro* expression.
9. Construction of stable strains of bacteria for the production of antibiotics.
10. Discovery of a few new genes related to liver cancer; isolation of three new cDNA clones from esophagus cancer cell lines whose expression can be activated by retinoic

acid; discovery of tumor suppressor role of the 3′ UTR of IL-6 transcription factor.

11. New techniques : Establishment of primer-mediated high-resolution chromosomal microdissection and gene microcloning; construction of DNA library for 24 specific regions in human chromosome 17, and human chromosome Xp II. 2-21, Xp 11.3 YAC library; 3-D structure of proteins and molecular graphics; effective treatment of the immune status of chronic hepatitis.
12. Completion of physical maps of rice genome.

During the Phase I Biotechnology Programme, over 3500 papers were published, 42 patents were taken, 421 PhD students graduated, and 73 postdoctoral fellows received further training.

Overall Guidelines for the Biotechnology Programme

After careful assessment of Phase I programme, it was decided that the central theme for Phase II Biotechnology Programme would concentrate national resources in a few well chosen promising areas which are likely to have a major impact in the 21st century. It was decided that this national effort should be highly organized, aiming to transform basic research into manufacturing capabilities, to facilitate maturation of Chinese biotechnology industry, and to support sustained growth of agricultural and health-care industries, and other industrial bases, such as chemical engineering, energy, and environmental protection. When selecting major areas, novelty is of paramount importance, so as to increase the competitive edge of the Chinese biotechnology industry. One promising area of work would be utilization of genetic resources in China. In terms of technology, the interdisciplinary aspect would be emphasized.

The Programmes

Under the above guidelines, the programme will continue to have molecular and cellular biology as its foundation, to make use of cell, organ and tissue culture and DNA recombination techniques. To make gradual transition to Phase II Programme, which will begin in the 21st century, the on-going programmes will be focused on the following major fields:

1. Crops giving high yields of good quality and resistant to severe conditions

a. application of two-line hybrid rice technology
b. basic research in two-line hybrid rice technology
c. pest-resistant cotton and other quality transgenic crops
d. saline-resistant genes in plants
e. symbiosis and N_2 fixation

2. Animal Husbandry

a. genetic engineering of lean-meat pig
b. genetic engineering against swine fever
c. transgenic fish
d. animal embryo technology
e. ES cell research

3. Research on Major Diseases

a. gene therapy of malignant tumors
b. diagnosis and gene therapy of genetic diseases
c. human tumor suppressor genes
d. viral hepatitis (molecular cloning of the viral genome, blockade of viral gene expression)
e. cardiovascular diseases (bioactive peptides, CGRP expression in *E.coli*)

4. High-tech Vaccines, New Drugs and Gene Therapy

a. high-tech vaccines against major infectious diseases, for birth control and others, new dosage forms (— adjuvants, slow releasing pumps, etc.)
b. high-tech peptide drugs (interferon, interleukins, natriuretic hormone and derivatives, CSF)
c. antibodies and related molecular biology research (rat-human chimeras, reshaped ab, small ab — especially scfv, ab library technology)
d. new expression systems for high-tech new drugs and their regulation: *E.coli* vectors and related research, stable fermentation yeast vectors and over-expression of acidic phosphatase, fungal gene expression systems, YACs, isolation and purification of target proteins)

5. *Protein Engineering*

a. medicinal proteins
b. protein products and molecular design of new drugs
c. 3-D reconstruction of biological macromolecules

6. *Other Areas*

a. genome plan
b. crops genome and organ-specific expression
c. transplantation

Organizational Structure

The programme in biotechnology is operated under the aegis of the Chinese Centre for Biotechnology Development. The total number of staff working under this programme is 15,000; they are from both academia and the related industrial bases. Some of the main laboratories and institutes are as follows:

CAS—	National Laboratory for Molecular Biology
	National Laboratory for Plant Molecular Genetics
	National Laboratory for Fresh Water Ecology and Biotechnology
	National Laboratory for Plant Biotechnology
CAAS —	National Laboratory for Plant Pest Biology
CAMS —	National Laboratory for Medical Molecular Biology
	National Laboratory for Molecular Oncology
CAPM —	National Laboratory for Viral Genetic Biotechnology
Others—	National Laboratory for Oncogenes and Other Related Genes
	National Laboratory for Medical Genetics
	National Laboratory for Crop Genetics and Breeding
	National Laboratory for Family Planning and Reproductive Biology
	National Laboratory for Protein Engineering and Plant Genetic Engineering (Peking University)
	National Laboratory for Genetic Engineering (Fudan University)
	National Laboratory for Agrobiotechnology (China Agricultural University)
	Shanghai Medical University Laboratory for Medical Molecular Virology

The Second Shanghai Medical University Laboratory for Human Genome Research
Academy of Military Medical Sciences (AMMS)
Beijing Medical University.
Academy of Agricultural Sciences (Hubei Province)
Centre for Hybrid Rice Research (Hubei Province)
Beijing Institute for Bio-Pharmaceuticals
Shanghai Institute for Bio-Pharmaceuticals
Changchung Institute for Bio-Pharmaceuticals
Lanzhou Institute for Bio-Pharmaceuticals
Huabei Pharmaceutical Factory
Centre for Genetic Engineering of Vaccines
Centre for Genetic Engineering of Drugs
The Genome Laboratory

For assesment of any biotechnology product from safety angle and freedom from adverse effects, the concerned institutes are required to apply to special governmental bodies attached to the Ministry of Health, Section for Drugs Administration and the Ministry of Agriculture, Section for Science and Technology.

Commercialization and Economic Gains

Chinese biotechnology today may be slightly different from that in other countries in that its major impact is still on its social aspect. But economic aspects are becoming more and more obvious, particularly in regard to agricultural production, high-tech new drugs, gene therapy and environmental improvement.

Agriculture

(1) *Hybrid Rice:* The use of two-line hybrid rice technology resulted in an average yield increase of 10-15 per cent, in a total area of 520,000 hectares. The two-line method offers significant advantages over the older three-line method, such as simplified production procedure, shorter time course, higher hybrid advantages, and much higher yield per hectare.

(2) *Transgenic Crops*

Thirteen strains of Bt transgenic cotton with pest-resistance are at the stage of field testing.

Transgenic wheat strains with disease-and stress-resistant transgenes from wild wheats have been selectively bred; they have strong disease resistance and other advantageous characteristics.

From 1000 transgenic blight-resistant potatoes, three strains were selected, which have 1-3 grades higher blight-resistance; strains of disease-resistant tomato and tobacco have also been successfully bred; complete sequencing of all 12 genes of the RDV virus and partial elucidation of their functions has been achieved.

(3) Recombinant Microbes

Significant progress has been achieved in respect of recombinant N_2 - fixation microbes and other disease-resistant microbes. Molecular genetics and engineering studies on the common rice and maize N_2-fixing microbe, with large scale field testing resulted in significant increase in crop yield.

Symbiotic N_2-fixation microbe in soybean and peanut is being used in the field and significant increase in total yield has been achieved.

Genetically-altered microbes against various diseases in rice and wheat are now being field tested and good effects have been demonstrated.

(4) Animal Biotechnology

Test tube cattle and sheep (*in vitro* nuclear transplantation, embryo separation, artificial insemination) are being tested for large scale production.

Transgenic fish [with growth hormone (GH) gene] showed increase of > 20 per cent in yield. It is being tested for large scale production.

Success is expected to be achieved in human GH production in the mammary gland of transgenic sheep.

Drugs and Gene Therapy

(1) Drugs

The first biotechnologically produced drug to be approved by the Chinese Government to enter the market is human interferon. It comes in different dosage forms: Topical $IF_{\propto}lb$, $IF_{\propto}lb$ and $_{\propto}2a$ for IV use. Approval was given after clinical trials in 2400 patients in 20 hospitals, to treat 30 different viral diseases with this drug. IF∞lb is a Chinese invention in that it is a new structural form.

EPO, GM-CSF and others are in different phases of manufacturing trials.

Biotechnologically Produced Vaccines: The first approved biotechnologically produced vaccine to pass clinical trials is the surface vaccine against hepatitis B, which conforms to international standards in 20 parameters. It was first manufactured in Changchun and Beijing biopharmaceutical factories in 1991. All children injected with this vaccines tested positive for antibodies against surface antigen. Now five biopharmaceutical factories have been selected for routine production of this vaccine. Besides this vaccine, vaccines against dysentery, cholera and polyvalent or multiple vaccines and other vaccines are also in different phases of clinical trials. Several other biotechnologically produced vaccines have passed animal testing stage.

IL-2 has been approved for clinical testing. Recombinant IL-2 can give as high as 35-40per cent yield of total yeast protein. For 10 Kg fermentation solution, 1 g IL-2 can be isolated.

Three vaccines against diarrhoea in infant domestic animals have passed clinical testing. Tests in 100,000 mothers and 1,000,000 infants have indicated that these vaccines are highly effective in reducing the incidence of death caused by diarrhoea in infants.

(2) Antibodies

Iodinated monoclonal antibody (Hab 18) against liver cancer has been approved for use in the treatment of liver cancer. A human antibody library has been established. Twelve diagnostic mab kits are being made and four have been approved for production. Some progress has also been achieveid towards biotechnological production of drugs for major infectious diseases and birth planning (Table 1).

(3) Gene Therapy

Genetic diagnosis and gene therapy have been performed in 1000 patients. Complete success was achieved in two cases of hemophiliacs. Reconstruction of thymidine kinase (TK) and cytosine deaminase (CD) gene in viral carrier has been done successfuly and their use for treatment of malignant tumors has been attempted in animals. TK gene therapy is currently at clinical trial stage.

***Table I*: Biotechnology Vaccines and New Drugs**

Name	*Diseases*	*Year of Testing*	*Institutions*
IF∞lb	tumor, hepatitis, B,C	1989	CAPM
IF∞2a	tumor, viral diseases	1991	CAPM
IFγ	tumor, viral diseases	1992	CAPM
IL-2	tumor	1990	CAPM, Institute for Virus Research
IL-3	tumor	1993	BMU
GM-CSF	anemia, leukopenia	1993	CAPM, Institute for Virus Research
EPO	anemia, erythropenia	1994	CAPM Institute for Virus Research
Insulin	diabetes	1994	CAS, Shanghai Biochemical Institute
Urokinase	thrombosis	1994	AMMS
TNF	tumor	1994	Fudan University
Vaccine for hepatitis B	hepatitis B	1989	CAPM, Institute for Virus Research
Vaccine for dysentery	dysentery	1995	AMMS
Vaccine for cholera	cholera	1995	CAPM, Institute for Infectious Diseases

Monoclonal antibodies against 38 leukocyte differentiation antigens have been made which find use in targeted therapy against human leukemia.

Immunotoxin conjugates of these antibodies have been used in a number of leukemia patients before bone marrow transplantation.

(4) Protein Engineering

Insulin and urokinase have passed the manufacturing stage. Five insulin mutants with high efficacy, and long-or short-lasting effects have been made. A recombinant urokinase with thrombolytic effect was made. Pharmacological testings indicate that they all have a high therapeutic ratio.

The 3-D structure of glucose isomerase was elucidated at 0.25 nm resolution. It can be over-expressed and reproduced and its mutants for manufacturing process are being tested.

Molecular design of drugs (antiarthritic, anti-tumor, painkiller) have been explored and a number of promising guides have been obtained.

International Collaboration

During the course of the last 10 years, a number of international collaborations were established to successful conclusion:

1. Hepatitis B vaccines, recombinant interferon y, GM-CSF, EPO — Pharmacia
2. Cardiovascular molecular biology — University of Toronto, Station for Biotechnology Collaboration
3. Rice genome project — Rockefeller Foundation
4. Symposiums and training — EU
5. Biotechnology training — ICGEB

Perspectives

In the 21st century, R&D in biotechnology would be of vital importance for the Chinese economy, for raising the standards of living and for proviing sufficient health care facilities to people.

In the future, large investments will be made in the areas of grains and cotton research, with emphasis on high yield, good quality, and stress resistant hybrids, on selection of special strains suitable for different ecological zones of China, on basic genetics and molecular biology research on hybrids, on practical technology for growing grains and economic crops, and on developing high efficiency biological fertilizers.

Another area of major research effort will be production of new drugs and vaccines. Gene knock-out and gene-targeting technology will be used to study mamallian expression systems in transgenic animals and plants, and to develop recombinant vaccines against hepatitis, AIDS, and for birth control.

Therapy of malignant tumors and cardiovascular diseases will be stressed. Related research on relevant genes and cancer cell autocrine mechanisms will be encouraged. A comprehensive approach involving drug targeting, tumor vaccines, cell therapy, and cytokinins will be tried.

Strategic research will include exploration of genetic resources, transplantation and gene therapy.

Phase II of the Biotechnology Programmes will be aimed at strengthening the concept of property rights, protecting Chinese patent rights, and at the same time, training a large number of advanced biomedical scientists. To accomplish these objectives, it will be important to lay stress on basic and interdisciplinary research, and to encourage international collaborations.

References

1 *The Ten Splendid Years—the fruits of 863 Plan.* The Scientific and Technology Committee of China, Scientific and Technology Publishing House of China.
2. JIA Shirong: *The Studies Advancing and Future Prospects about Agrobiotechnology.* Shirong USTC Publishing House. 1995
3. LU Liangshu: *The Hope of Agricultural Modernization in China.* The Meeting of Agricultural Science and Technology, 1993.
4. MEI Qiming & ZHU Yingguo: *Chinese Journal of Genetics* 1991,18(2).
5. ZHANG Ziguo & ZENG Hanlai, *International Rice Research Notes* (IRRN) 1992, 17(6)
6. GU Minghong *et al* : *Rice Genetics* (IRRI) 1991, p.766-769.
7. GUO D. & HONG Z.Y.: Agricultural Biotechnology P.A.P.C. 1992, p233-234.
8. HE Z.L. *et al: Chinese Science Bulletin* 1994, 39(7).
9. SCHEU A.K. & ECONOMIOU A. *et al*: *Molecular Microbiology (U.K.)* 1992, 6(2).
10. LI C.J. & FAN B.Q.: *Archives of STD/HIV Research* 1995, 9(19).
11. WANG Xiao-Quan & SHEN Ge-zhi *et al: Plant Science* 1993 (94).
12. CHEN Tianmi *et al: Vaccine,* 1992, 10(4).
13. ZHANG Zhaoshan: *Chinese Journal of Biotechnology* 1993, 9(1).
14. GUO Fu Su *et al: Infection and Immunity* 1992, 60(5).
15. CAI Z. H. & SONG G. C. *et al: Southeast Asian J Trop Med Pub Hlth* 1993, 24(1).
16. WEI Tang *et al: Tetrahydron Lett* 1994, 35(35).
17. ZHU De-xu *et al: Biol. Chem, Hoppe-Seyler.* 1993, 374, p.903-908.
18. PING Qiu & YI DingQin *et al: Biol. Chem.* 1994, 375, p.413-418.
19. TANG Zhao-you *et al: Antibody, Immuno conjugates, and Radio pharmaceuticals.* 1993, 6(3).
20. CHEN Zhi-Nan *et al: J.Med Coll PLA* 1993, 8(2).
21. LI Side *et al: Chinese J. Cancer Res.* 1991, 3, p.56.
22. GUO Ning *et al, Acta Microscopica* 1993, 2(2) p.196.
23. HUA Liu & XU Zuo-Liang: *Hum, Antibod, Hybridomas* 1993, 4 (January).
24. WANG Yan, XU Jianjun *et al: Chinese J.Cancer Res,* 1993, 5(1).
25. GU Jiangren *et al: Chinese Cancer Res.* 1992, 4(3).
26. CHANG J. C. & LIU D. P. *et al: Proc. Natl. Acad. Sci. USA,* 1992, 89, p.3107-3110.
27. ZENG F. Y. & HUANG S. Z.: *Haemoglobin,* 1992, 16(6).

28. ZENG Y. T. *et al: Am. J. Hum. Genet* 1991, 49(suppl) p.202.
29. WANG Yu & TAO Qi-Min *et al: J. of Hepatology* 1994, 21(3).
30. ZHANG R. E. & LIU G. Q. *et al: J.of Chromatogr* 1993, 646, p.45-52.
31. TU Hongmin & SHAO Xiaoxia: *Chinese Biotechnol* 1992, 8, p.219-225.
32. HE Fuchu *et al: Hepatology,* 1993, 17(2).
33. CHAI Jianhua & G.U. Yanghong: *Chinese J. of Genetics* 1993, 20(3).
34. DENG Han-xiang & H E Xiao-xuan *et al: Prog Natural Sci* 1994, 4(1).
35. G.U. Shuyan et al: *Colloque Inserm*/John Libbey Eurotext Ltd, 1993.
36. XIONG Jian Ping & XIA Zong-Xiang *et al: Nature Structure Biology* 1994, 1(10).
37. LI Yidi & HUANG Qichen *et al: J. Biochem* (Tokyo) 1994, 116(1), p.18-25.
38. LIU Hai Yan & SHI Yun Yu: J. *Comp. Chem.* 1994, 15(11).
39. DING Dafu & TANG Haixu: *Biomath* 1994, 9(2).
40. LAI Luhua & WANG Yanli *et al: Science in China* (Series B) 1994, 37(1).
41. HOU Ke-ke *et al: Science in China* 1993, 36(5).
42. Li Daizong *et al: Carcinogenesis* 1993, 14(2)
43. GAO Wende & LING Jun *et al: FEBS lett* 1994, 347, p.257-260.
44. LIU Zheng & QIN Hong *et al: European Biophys,* 1995, 23, p.1.

17. Emergence of Biotechnology in India : Promises and Prospects in the 21st Century

Kavita Mehra

Abstract

Biotechnology is one of the high technology areas which are likely to have considerable impact on the national economies in the 21st century. In this context, the present paper studies the efforts being made in India to be at the forefront of this high-tech area. Starting with the policy framework, and the organizational structure, the major research programs being pursued have been identified. Efforts at creating capabilities in indigenous R&D and its commercialisation have been examined. Points out that in India, the existing R&D capabilities have not been fully utilised at industrial level. Patenting of innovations is also not widely popular among the Indian scientists. Although most of the biotech companies in India are producing a few diagnostics, antibodies, enzymes, vaccines and agricultural products, as yet no company has produced a hightech biotechnology product. Concludes that biotechnology research in India is primarily funded by government and the role of private sector in the promotion of biotechnology has been negligible. There is an urgent need to increase research-industry relationship to acquire the capability to face international competition in the 21st century.

Introduction

Biotechnology is one of the new technologies which promise to revolutionarise the world economy. The present-day biotechnology (New Biotechnology) has evolved through three phases. In 1920s, commercial utility of biological process was realised in the production of butanol and alcohol from starch through *Clostidium acetobutylicum*. The process made a significant contribution in the field of explosives during World War I. In second phase of its development, during World War II, biotechnology found its way in the production of antibiotics (pencillin) due to advances in the field of microbiology. The new biotechnology is based on a great discovery of this century, the DNA structure and the parallel advances in the fields associated with genetic technologies so as to lead as a recombinant DNA revolution. The importance of biotechnology greatly increased in the mid- seventies with the possibilities of altering or manipulating the life form by the use of recombinant - DNA techniques.

Development of Biotechnology in India

In India, biotechnology followed the route of classical biology to modern biology and finally took the shape of present-day biotechnology. However, the pace of development of modern biology remained very slow.[1] The development of basic modern biology included the disciplines such as biochemistry, biophysics, molecular biology, cell biology, immunology, genetics, virology, microbiology, etc. In India, modern biology had its beginning (basically through researches in biochemistry) in late 1950s and early 1960s. Initially, there were a few groups pursuing research in biochemistry. Later, departments of biochemistry started coming up in universities and medical colleges and the spread of basic modern biology took place. In 1977, the first institution specially for research in modern biology - Centre for Cellular and Molecular Biology (CCMB)-was set up in India at Hyderabad. A few years later, in the early 1980s, National Institute of Immunology (NII), New Delhi and Institute of Microbial Technology (IMTECH), Chandigarh, were set up. Thus, in the late 1970s onwards, India's entry into the field of biotechnology became evident through the creation and strengthening of institutional structure for research. Programmes/projects in biotechnology as such began to be pursued from 1980 onwards.

---

[1]BHARGAVA, P.M. & CHAKRABARTY, C.: The role and present status of biotechnology in India. *Curr.Sci.*1991,60, p.513-517.

Under the research funding scheme, Science and Engineering Research Council (SERC), of the Department of Science and Technology (DST) , during 1975-80, an exploratory approach was adopted to identify thrust areas of research in life, physical, chemical and engineering sciences. On that basis, thrust areas in life sciences were identified by DST in 1980. An explicit mention was made of certain areas of biotechnology, such as genetic engineering, immunology, molecular and cellular biology, etc. (Annexure-I). That was the way research in basic modern biology was getting transformed into biotechnology research. SERC in DST initiated another scheme-Intensification of Research in High Priority Areas (IRHPA) aimed at setting up core groups or units around outstanding scientists, formulating national programmes and setting up infrastructures to carry out research in identified areas. Activities of IRHPA gave a boost to the follow up of programmes generated out of Thrust Area mechanism of SERC. During the Sixth Plan period (1980-85), five units in genetic engineering were set up in the country.

At the recommendation of the then Science Advisory Committee, National Biotechnology Board (NBTB) was set up during March 1982 in DST. NBTB was an inter-agency body assigned the responsibility to coordinate biotechnology-related R&D done under several departments, namely, DST, Council of Scientific & Industrial Research (CSIR), Indian Council of Medical Research (ICMR), Indian Council of Agriculture Research (ICAR), Department of Atomic Energy and University Grants Commission. NBTB prepared a long term-plan wherein some priority areas were identified (Annexure II). As per the decision of CSIR to opt for biotechnology research, it set up two institutes CCMB and IMTECH. In the plan document, CSIR made special mention of tissue culture applications in the field of molecular biology.[2] Apart from the research conducted in these institutes, financial support for biotechnology research in the universities

[2]MEHROTRA N.N. & TANDON Poonam: *Development of biotechnology in India - Issues for effective diffusion to harvest its potential.* Paper presented at International Symposium on Biotechnology for Sustainable Development, UNEP, Kenya, 1989.

and in other agencies was provided through the projects sponsored by SERC in DST.

In the Seventh Five Year Plan (1985-90), NBTB's emphasis was on manpower development and initiation of a few multidisciplinary and multi-institutional programmes in biotechnology and many other activities (Annexure III). In a way, in the Seventh Plan period biotechnology was given considerable importance. A separate Department of Biotechnology was set up under Ministry of Science and Technology during February 1986. The responsibilities assigned to DBT include planning, promotion, and co-ordination of biotechnology research in India.

Setting up of DBT was a part of government of India's efforts towards achieving the targets set in the Seventh Plan document. As a follow up of NBTB's activities, DBT concentrated mainly in two areas, manpower development and creation and strengthening of infrastructural facilities along with its commitments for other activities (Annexure IV). A Product and Industry Development Division was created in DBT in 1988 for the development of biotechnologies through government funded projects and transfer of institutionally developed technologies to the industry.[3]

For the Eighth Five Year Plan, DBT undertook the task of identifying thrust areas in biotechnology. Thirteen task forces for various sub-areas (Annexure V) were set up and these task forces, through brain storming sessions, meetings, interactions, workshops, etc identified the thrust areas. Emphasis was on implementing projects in the gap areas covering upstream research and development, product-based technology development, demonstration of proven technologies, setting up of pilot plants, etc. During the same period, realizing the potential of scientific manpower and the R&D infrastructure existing in the country for the production of economically feasible products, processes and technologies in the area of biotechnology, the government thought of establishing an institution-industry

[3]GHOSH P.K.: Indian Experience in commercializing Institutionally Developed Biotechnologies. *J. Scientific & Industrial Research* 1996, 55, p.860-872.

interactive group. An institutional mechanism was evolved in the form of Biotech Consortium India Ltd. (BCIL) which was set up in September 1990. It was promoted by a consortium of financial institutions led by Industrial Development Bank of India (IDBI) in association with DBT. The main objective of BCIL is to promote production and process development and improve technology development and technology transfer with the support of financial and industrial institutions.

Organisational Structure of Biotechnology in the Framework of Science and Technology in India

Science & technology in India is organised in a planned manner through provisions made in the Five Year Plans. In India, significant part of R&D is conducted by the laboratories of government agencies. In contrast with United States, private sector investments in R&D in biotechnology in India had been negligible till recently. Research is carried out in the university departments and in the national laboratories functioning under various government ministries or councils. In India, since there is a separate Department of Biotechnology under the Ministry of Science and Technology, responsibility for national level planning and promotion of biotechnology activities lies almost exclusively with DBT. It coordinates the biotechnology-related R&D activities of several government agencies. Some university departments too conduct biotechnology research. Applied research aimed at product and process development forms an insignificant part of Indian university research system. Several laboratories under Council of Scientific & Industrial Research, Indian Council of Medical Research, Indian Council of Agricultural Research and laboratories of autonomous or other institutions under many ministries are working on basic and applied (technology development) aspects of biotechnology.

DBT, in its working, is assisted by the Scientific Advisory Committee consisting of representatives of major organisations like universities, national laboratories, some other agencies, industries, etc. concerned with biotechnology research and also by a Standing Advisory Committee (overseas), whose members are eminent Indian scientists working abroad. DBT also derives the benefit of advice from members of international peer committees in review of its major programmes. For example, the committee in the field of fertility control was headed by Nobel Laureate, Roger Grillman, of Salk Institute of USA.

DBT has a staff of nearly 200 constituting scientific and administrative personnel. The budget of DBT during 1995-96 was Rs. 965.2 million, whereas for the period 1992-97 , it was Rs. 4 billion.

National Network of Linkages of DBT

DBT has the responsibility to coordinate biotechnology-related activities in the country. For that, DBT has linkages with:

- Planning Commission
- Ministries
  - — Ministry of Human Resource Development
  - — Ministry of Agriculture
  - — Ministry of Health and Family Welfare
  - — Ministry of Commerce
  - — Ministry of Food and Supplies
  - — Ministry of Textiles
  - — Ministry of Environment and Forests
  - — Ministry of Rural Development and Employment
- Science departments and agencies
- State level S&T Councils
- Autonomous institutions and universities
- Financial institutions
- Industries

DBT interacts with the above institutions through:

— Joint Committees/Expert Task Forces
— Memoranda of Understanding (MOUs)
— Meetings, Seminars, Workshops and Training Programmes
— Individual scientist level

Biotechnology Programme

One of the missions of the Government is to convert genetic wealth into economic wealth through biotechnology. It is believed that it can create wealth for the nation, provide rural employment, improve health care, ameliorate environment and provide better methods for limiting population growth. The impact of biotechnology has been felt in the areas of agriculture, health, environment and industry in terms of availability of biotechnology-based processes and products. In the past 10 years or so, a number of projects supported by DBT have achieved a high degree of success and have resulted in product development. A few of the technologies developed have reached the consumers. Several projects in genetic engineering, agribiotech and animal health care have been initiated.

Maximum investments in the past years have been in the area of agriculture and allied sectors covering crop biotechnology, plant tissue culture, medicinal plants, sericulture, biofertilizers and biocontrol. Other areas covered include medical biotechnology, animal and aquaculture, basic research, including human genetics, industrial biotechnology, environment and biodiversity conservation and biotechnology-based programmes for weaker sections.

Programmes Initiated by DBT and Achievements

(1). Human Resource Development

The need for adequately trained/highly skilled manpower essential for meeting the expanding R&D and manufacturing requirements in the multidisciplinary area of biotechnology was felt, right in the beginning, when DBT was set up. Therefore, DBT initiated post-graduate and post-doctoral teaching programmes in biotechnology. DBT has provided funds and faculty positions for M.Sc./MTech./Post-doctoral training programmes in 29 selected universities and other institutions. There are special courses in agriculture, medical, marine, animal, biochemical engineering and bioprocess technology. MSc programmes supported cover the areas of general microbiology, biochemistry, biophysics, biostatistics, developmental biology, basic genetics, pollution control, biodegradation, molecular biology, genetic engineering, applied microbiology, advanced immunology, plant tissue culture and economically important plant products. Annual intake of students in such courses is 350. Government of India, through DBT, is also sponsoring short-term training programmes of 2 – 4 weeks duration in high-tech specialised areas in institutions where expertise and infrastructure required for the conduct of such courses exist. Under this programme, so far more than two thousand persons have been trained. A certain number of biotechnology associateships - both national and overseas - are awarded every year for advanced research or specialised training in identified priority areas of biotechnology. Nearly 20 associateships in each category are provided by DBT every year. In addition, there are programmes for training of industrial R&D personnel and visiting scientists from abroad.[4]

It has been revealed through a study on availability and requirement of manpower in plant tissue culture, hybridoma, protein engineering and enzyme engineering technologies, that trained manpower is available at all levels — research, production, marketing and extension services. It is estimated that the manpower available at post- graduate and doctoral levels by 2000 A.D. would be around 10,000 and 5,500 respectively.[5]

***(2)** Creation of Infrastructural Facilities*

Development and strengthening of infrastructural facilities in biotechnology had been another important mandate of DBT. In this regard, DBT supports a network of service-oriented infrastructral facilities.

(a) *The Infrastructural Facilities*

The infrastructural facilities set up by the government provide services to various scientific institutions and individual scientists in the country. The following are the main facilities set up in the country:

1. Microbial Type Culture Collection (MTCC), at IMTECH, Chandigarh.
2. National Facility for Blue Green Algal Collection, Indian Agricultural Research Institute (IARI), New Delhi.

---

[4]RAMACHANDRAN S.: Government Funding and Support- The Department of Biotechnology. *Current Science* 1991, 60, p.518-523.

[5]VISALAKSHI S. JAIN A., MEHRA Kavita & KUMAR Suresh: *An interim report on manpower projections in biotechnology by 2000 A.D.* (Submitted to DBT) 1992.

3. National Facility for Marine Cyanobacteria, Bharathidasan University, Tiruchirapalli.
4. National Facility for Plant Tissue Culture Repository (NFPTCR), New Delhi.
5. Animal House Facilities.
6. Biochemical Engineering Research and Process Development Centre, IMTECH, Chandigarh.
7. Genetic Engineering Units.
8. National Facility for Structure and Functional Studies of Biological Macro molecules through X-ray Crystallography, BARC, Bombay.
9. National Facility for High Field NMR, TIFR, Bombay.

(b) Autonomous Institutions

1. Centre for DNA Finger Printing and Diagnostics, Hyderabad.
2. National Centre for Cell Science, Pune.
3. National Institute of Immunology, New Delhi.

(3) R&D Cum-Production Programmes

Financial support is given for proposals holding promise of generation of products, processes and services for ultimate commercial exploitation[4,6,7]. The following are some of the R&D programmes/projects sponsored by DBT:

— Biological control of pests and diseases.
— Extraction of copper through bacterial leaching.
— Crop biotechnology : Crop-based research programmes on molecular aspects related to plant tissue culture, RFLP mapping, chloroplast and mitochondrial research, basis of cytoplasmic sterility and transformation and regeneration systems.
— Tissue-culture propagation of bamboo, coconut and cardamon, cocoa, blackpepper.
— Transgenic cotton

---

[6]DEPARTMENT OF BIOTECHNOLGY (Ministry of Science and Technology, India): Annual Reports 1987-88 to 1995-96.

[7]DEPARTMENT OF BIOTECHNOLGY (Ministry of Science and Technology, India): Ninth Plan Document submitted by the Department, 1996.

— Biomass production
— Biotechnology of silkworm and host plant
— Enhancement of alcohol production through the use of improved yeast strain
— Production of sex hormones(FSH and hCG)
— Production of insulin
— Prawn aquaculture
— Hormone treatment for increased production of major carps and production of transgenic varieties
— Animal birth control
— Development of biodetection kit for identification of biological toxins in the environment.
— Oil palm demonstration project.
— Establishment of pilot plant facilities for raising elite varieties of identified forest trees.
— Vaccine production and development.
— Embryo transfer technology.

The following are the salient sector-wise findings from R&D efforts:

(a) Health Sector

Vaccines: Effective control of communicable diseases is a national goal for providing health for all by 2000 AD. Vaccination has been considered as one of the most efficient and cost-effective preventive practices against communicable diseases of children. DBT has initiated several programmes in the area of research on and production of vaccines. An R&D-cum-production unit for Oral Polio Vaccine (OPV) is operational at Bulandshahar (BIBCOL) and is producing OPV from imported bulk from Belgium. So far, it has supplied 7.5 million doses of OPV. The company is planning to produce hundred million doses very shortly.

A major national R&D programme is going on in the country to develop new and improved vaccines and diagnostics against the major communicable diseases, *e.g.* tuberculosis, leprosy, diarrhoeal diseases, hepatitis, malaria, filariasis, typhoid etc. Success has been achieved in some cases.

Immunodiagnostics: Efforts have been made to develop filaria kit, procedure for detection of hepatitis B surface antigens, transfer technology for detection of falciparum malaria, amoebiasis and early pregnancy. Considerable progress has been made in respect of diagnosis of brucellosis, toxoplasmosis, tuberculosis, leprosy, typhoid fever and malaria.

Results of R&D in Health Sector :

— Specific DNA probe for diagnosis of tuberculosis developed
— A non-toxogenic *Vibrio cholerae* candidate strain for oral cholera vaccine developed.
— A genomic library of *Plasmodium vivax* antigen gene constructed.
— DNA libraries of *Mycobacterium tuberculosis* H37RV and *M. smegmatis* vectors have been generated.
— Diagnostic kit for hepatitis - A developed.
— An antibody detection kit for Aspergillosis developed.
— Genes responsible for increasing susceptibility to infectious diseases (leprosy and pulmonary tuberculosis) developed.
— Gene for chloroquin resistance in *Plasmodium falciparum* cloned. This may have role in reversal of chloroquin resistance.
— Monoclonal antibodies which recognise neuroectodermal tumors have been developed and their diagnostic potential is under evaluation.
— Hepatic cells from foetuses and adult livers have been cryopreserved/cultured. Such cells have been successfully transplanted in clinical subjects for recovery of hepatic failures.
— Use of liposome intercalated amphotericin B cured systematic fungal infections.
— Female anti-fertility vaccine (beta hCG) developed is at Phase II stage of clinical trials.
— Male anti-fertility vaccine (oFSH) entered Phase II of clinical trials.
— Diagnostic kits for HIV - I&II developed.
— Leprosy vaccine entered Phase III of clinical trials
— Immunotherapeutic developments for leprosy control.

— Six PCR-based kits developed for molecular diagnosis of Ducchane Muscular Dystrophy.
— Three - dimensional structure of an iron binding protein and lactotransferrin has been determined; the protein has strong antibacterial properties and has potential applications in medical and dairy industries.
— A genetically engineered strain has been constructed for the development of potential cholera vaccine.
— Techniques for pre-natal diagnosis for thalassemia based on DNA analysis have been developed.

(b) Agriculture Biotechnology

Crop Biotechnology

The new technique of plant breeding, involving genetic engineering, protoplast fusion and tissue culture makes it possible to quickly develop high yielding varieties of food crops and fruit trees and fast growing and stress resistant agricultural and silvicultural plants. Cropwise findings are:-

* Amaranthus — A gene encoding for protein of high lysine and sulphur containing amino acids cloned and sequenced.
* Tobacco — Genetically transformed for insect resistance; being field tested.
* Chickpea — A recalcitrant crop; A tissue culture regeneration system being developed; Success has been achieved in shoot regeneration.
* Rice — Successful regeneration from mesophyll protoplasts of Indica rice achieved.
* Latyrus sativus — A variety with lower neurotoxin levels (0.8-1.2 per cent level lowered to 0.06 - 0.08 per cent level) developed which will be high yielding. Five clones are under multilocational trials.
* Brassica — Transgenic plants with insect resistance gene have been field tested.
* Wheat — High frequency, reproducible embryogenic callus induction and plant regeneration have been achieved in several varieties.

Biopesticides

Biotechnology has been recognised as an important tool to exploit biological agents for efficient management of pests and diseases. In India, mass production technique, first on pilot scale and later on a large scale is being standardised. Viral, bacterial and fungal agents are slated to be produced for use in the control of several important insect pests.

Isolation and characterisation of insecticidal crystal protein genes of Lepidoptera have been done.

Sericulture

— Through integrated management, a successful demonstration to control uzifly at farmer's level achieved.
— Immunodiagnostic kits developed for early detection of pebrine and nuclear polyhedrosis disease of silkworm.
— Improvement in cocoon yield demonstrated.
— Successful introduction and expression of cloned foreign genes into silkworm through the application of recombinant baculovirus and microinjection techniques. Transgenic silkworm expressing luciferase gene from firefly.

Semi-intensive Aqua-culture

Prawn : Techno-economic viability of farming prawns demonstrated.

Carp : Technology package for 10-15 tonnes/ha/year evaluated and ready for commercialisation.

Biofertilizers

— Mass production of soil-based Blue Green Algae (BGA). Liquid polybag cultures of BGA and Azolla under production.
— Field demonstrations being carried out at farmer's fields.
— Increase in yield of paddy due to the supply of BGA as a fertilizer has been observed.
— With the application of Rhizobium, 3 – 21 per cent higher yield has been reported.

Field evaluation of tissue culture-raised cardamom

Product-based programme in cardamom to compare the yield of open pollinated clones vs tissue culture-raised plants from the identified elite planting material has revealed 40 per cent higher yield from tissue culture-raised plants. The average yield from tissue culture plants has been 363 kg/ha against 261 kg per ha in the case of pollinated seedlings.

Training programmes have been conducted to prove the value of tissue culture plants and to promote their adoption by farmers.

Multiplication of forest trees through tissue culture

Two tissue culture pilot plant facilities have been set up in the country. The two centres are working on technologies for the multiplication of elite trees using tissue culture methods. Tissue culture - raised plantlets are being evaluated at various locations in the country.

The following species of forest trees are under multiplication in the two pilot facilities set up by DBT

* At National Chemical Laboratory, Pune

— *Eucalyptus sps* (Nilgiri)

— *Tectona grandis* (Teak)

— *Dendrocalamus strictus* (Bamboo)

* At Tata Energy Research Institute, New Delhi

— *Populus* (Poplar)

— *Anogeissus pendula* (Desert Teak)

— *Eucalytus sp.* (Nilgiri)

— Bamboo

— *Leucaena hybrids* (Subabul)

So far, 2.5 million planlets have been produced and 2.0 million planted in an area of 1100 ha. The tissue culture raised plants showed 90 - 95 per cent survival rate and the resulting plants have greater height and better girth size than normal plants.

(c)Animal Health Care

Animal Birth Control Injection

An animal birth-control injection called Talsur has been developed. Besides being cheap, it is easy to use and has few side effects

Vaccines for Ranikhet disease in poultry and Herpes virus in cattle are ready for commercialisation.

Embryo Transfer Technology (ETT) for Cattle Herd Improvement

This technique offers scope for achieving faster multiplication of elite livestock and devising methods for conservation of valuable germplasm. In conjunction with artificial insemination, it offers a powerful tool for increasing the population of desired types of animals. Success has been achieved at down-stream level. Infrastructure facilities have been established at six laboratories and manpower is being trained. More than 500 calves and 100 buffaloes have been produced through ETT.

(d) Environmental Biotechnology

ELISA and PCR techniques and DNA probes have been developed for detection of enteric pathogens in drinking water. A microbial strain has also been developed for desulphurisation of fossil fuels.

(4) International Cooperation

As a result of DBT's efforts, international bilateral cooperation with USA, Germany, Switzerland, Sweden and UK, has been established. These cooperations have been beneficial from the point of view training of scientists, acquisition of equipment and exchange of visits and information through joint workshops and meetings. New programmes are being identified with Brazil, China, Egypt, Israel, Japan, Poland, Slovania, Srilanka and Tunisia. The thrust for collaborations is in high tech areas, specially from the point of view of biotech products. The projects are in the areas of reproductive biology, immunology, bioreactor designs, biosensors, production of industrial enzymes, development of products and processes, vaccines and immunodiagnostics for human health and animal use and enhanced agricultural productivity through genetic engineering.

G-15 Gene Banks for Medicinal and Aromatic Plants

In this project, multilateral cooperations have been established with SAARC and G-15 countries. India, as the overall coordinator, has pioneered the project on establishment of gene banks for medicinal and aromatic plants as part of South - South consultation and cooperation among G-15 countries. Inventories of important medicinal plants have been prepared and the programme to set up gene banks in all the G-15 countries is in progress.

Farmer- centred Agriculture Resource Management (FARM)

The FARM project is being supported by UNDP, FAO and UNIDO. All the eight participating countries of the region are exchanging technology, and information, organising training programmes and developing a good network arrangement in terms of information exchange.

(5) Recombinant-DNA Safety Guidelines

Advances in manipulation of genetic material had generated a sense of concern among scientists about safety of work involving pathogenic microorganisms and genes encoding virulence factors. In view of this, a document 'Recombinant - DNA Safety Guidelines has been prepared by DBT. The document covers areas of research involving genetically engineered organisms, genetic transformation of green plants and animals, recombinant-DNA (rDNA) technology in vaccine development and large scale production and handling of products obtained by r-DNA technology. It is also paying attention to matters relating to notification, containment, recognition of facility, competent authority, large scale experiments and manufacture, biologicals produced by r-DNA technology, release to the environment, field experiments, etc. Efforts are on in the country to strengthen and set up institutional Biosafety Committees in all institutes working in the field of biotechnology. Regular meetings of Review Committee on Genetic Manipulations are arranged and inter-ministerial interaction has been strengthened.

(6) Bioinformatics

A unique information system established by DBT spans a subject -oriented scientific grid covering many scientific institutions and providing a single reference source for biotechnology information in the country. The e-mail facility has been introduced through ERNET, which is facilitating access to INTERNET for worldwide biotechnology information. The BTIS Centres have developed several value-added databases and softwares for PC-based sequence analysis, etc.

Creation or upgradation of instrumentation facilities in some laboratories in the country either jointly or on their own, has been supported by the government. These can be particularly useful for structural biology work. A special effort has been made to evolve projects and activities which will use these facilities for drug designing and drug targeting, leading to the development of better products.

(7) Patenting

Scientists are encouraged to file patents. There is inter-agency (DBT, National Research and Development Corporation (NRDC) and BCIL) collaboration for technology transfer and patenting of innovations. So far, 15 applications have been filed. Six patent applications filed for recombinant products include plasmid expression vectors, gamma irradiation produced and salt inducible expression systems. A patent for gene process system is under processing.

(8) Indigenous R&D and its Commercialisation

The primary technologies involved in biotechnology are:

1. Recombinant DNA (r DNA) technology (Genetic Engineering)
2. Hybridoma technology
3. PCR - technology
4. Tissue and cell culture
5. Fermentation
6. Downstream processing
7. Protein purification
8. Synthetic nucleotide and peptide chemistry

Most of these technologies have already been established in several Indian laboratories, including some industrial R&D laboratories. One can specifically mention some developments for the following technologies :

(1) Hybridoma

Thirty nine main groups are involved in hybridoma research in India. Research interests have two motives (i) for use in basic research to understand the structure/function relationships of the biomolecules in human, animal and plant models, and (ii) to use the Monoclonal Antibodies (MCAs) to develop diagnostic tests. Use in therapeutics is a possibility for the future.[8]

The first Monoclonal Antibodies (MCAs) were produced against hCG (human chorionic gonadotropin) in India. Using hybridoma technology, several murine and human monoclonal antibodies have been produced against a variety of antigens. The focus of Indian efforts is towards evolving diagnostic aids based on MCA for (parasitic) tropical diseases, such as leprosy, tuberculosis, amoebiasis, filariasis, typhoid and leishmaniasis. Indian strength also lies in reproductive hormones. Monoclonals are used in immunocontraception. MCAs have been developed against a variety of hormones and antigens of the reproductive tract.[8]

(2) Plant Tissue Culture (PTC)

More than 100 groups, mostly located in universities and national laboratories are working on PTC. Research capabilities exist for micropropagation, disease-free cultivar via culture, somaclonal variation, anther culture, embryo rescue, etc. Efforts are needed to develop capabilities in protoplast fusion, secondary metabolises and genetic transformation. The most successful technique is micropropagation. Laboratory level micropropagation processes have been demonstrated for almost all categories of plants.[9]

A large number of protocol developments have taken place for horticultural crops, including flowering plants, ornamentals, small fruits, and plantation crops. Some of the protocols have been commercialised and a large number are available at laboratory level and can be up-scaled.[10] The following are some

---

[8]VISALAKSHI S.: R&D and Commercialisation Status of Hybridoma Technology in India: An international comparison. *Science and Public Policy* 1995, 22(4) p.231-238.

[9]MEHRA Kavita: Scope of plant biotechnology in developing countries- An Indian case study in plant tissue culture. *J. Scientific & Industrial Research* 1993, 52, p.795-802.

[10]MEHRA Kavita:'*Commercial applicability of Plant tissue Culture in India*'- Paper Published in proceedings of "Biotechnology and Human society." Edited by Kanan Nanda. New Delhi; 1998. p. 105-110.

of the specific plant species for which complete protocols have been developed under DBT-sponsored projects and can be taken up by the industry for production:

Populus deltoides (Poplar)
Alnus nepalensis (Utis)
Dalbergia latifolia (Rosewood)
Prosopis cineraria (Khejiri)
Tecomella undulata (Desert Teak)
Sapindus trifoliata (Soapnut)
Bambusa vulgaris (Golden Bamboo)
Cocoa
Pepper
Rubber
Apple
Soapnut
Rosewood

(3) *Recombinant DNA*

Several genes have been constructed using plasmid, viruses and phagmids. In some situations, synthetic genes have been made by cDNA and PCR techniques. Modified expressions of hosts have been created in *E.coli,* different species of yeast and cell lines of insect and vertebral origin. Laboratory level success has been achieved in producing transgenic plants, insects and fishes. Transgenic plants for some cash crops, cereals and fruits have been developed by incorporating bacterial genes to express toxic proteins to lepidopteran pests. Transgenic fishes have been made by incorporating growth hormone genes.[3]

Technology Transfer and Commercialisation

Based on indigenous knowhow, only a few diagnostics, antibiotics, enzymes and agricultural products, including tissue culture-raised elite plant material as well as some improved variants, have been produced in the country. No laboratory has succeeded in developing a recombinant molecule and bring it to the market place. Through hybridoma technology, diagnostic kits were developed which have been transferred to industry for production. 10 -15 companies in the Indian market import kits and repack and sell them.[8]

In the field of animal biotechnology, Embryo Transfer Technology has been developed and calves are being produced. 'Talsur', an animal birth control vaccine, has been developed and is being produced commercially.

Based on plant tissue culture research, a number of protocols have been developed by R&D laboratories in the public sector and they have been transferred to industry or other government agencies. Some micropropagation processes for horticultural plants have been developed by R&D laboratories of some private companies. Those products are being marketed in India and are being also exported.[11] Tissue culture - raised product range includes :

Domestic Market:	Cardamom, banana, grapes, papaya, sugarcane, vanilla, strawberry, orchids, lillies, gerbera, geranium, carnations, Diffenbachia, Syngonium, Philodendron, Cordyline, Dracena, Croton and eucalyptus,
Export Market	Foliage Ornamentals—Diffenbachia, Syngonium, Spathiphyllum, Cordyline, Philodendron, Ficus species, Gerbera, Gladiolus, Lilies, Carnation, Chrysanthemum, Alstroemeria, Anthurium, Rose, Orchid, Geranium, Strawberry and Banana.

Based on somaclonal variation techniques, a variant for lemon grass has been developed and released to the farmers.[10] Based on research carried out in public-funded laboratories, DBT has made efforts in some cases[3] to transfer some of them as a process or product to the industry or the user (Annexure VI). Some of them, viz. leishmaniasis detection kit, amoebic liver abscess kit, and 'Talsur' are under commercial production.

Biotech Industry

Most of the biotech companies in India are producing diagnostics, antibodies, enzymes, vaccines and agricultural

---

[11]MEHRA Kavita: *Indian capabilities in plant biotechnology: An Analysis-* Paper appearing in a book under publication, 1998.

products. No company has produced a high- tech biotechnology product. Studies reveal that in-house capabilities in basic research are generally at a low level and basic research in high risk areas is not yet being conducted.[3,8] To facilitate absorption of transferred technologies, some companies have established in-house R&D units. Recent changes in the industry's behaviour reveal that companies have started realising the need for innovative research and are allocating up to 5-6 per cent of their sales turnover for R&D. Another change is in the laboratory-industry relationship; private companies have started associating themselves with government laboratories.[3] This may prove of value in commercialisation of institutional knowhow.

The developmental capabilities of Indian industry include improvement of microbial strains by classical mutations, fermentation, process optimization and downstream product recovery, production and immobilization of enzymes. Capabilities also exist for fabrication of satellite reactor and process equipment for recovery of products. Basic strength in handling sterile operations has been developed.[3]

India is a big market for biotech products (Annexure VII). It is estimated that the market size was nearly Rs. 15,000 million during 1994 (Annexure VIII),[12] and it is expected to increase substantially. High risk and cost intensive features of the biotech industry pose limitations for the Indian industry. At present, conducting 200 million tests can generate sales up to Rs. 4000 million and it can go up many fold. The major component of the products in the market are based on conventional biotechnologies and additional requirements in the future could be met through new biotechnology. A part of these products are being produced indigenously.

In the field of agriculture, there are nearly 50 companies in India based on plant tissue culture technology. The size of the company varies in production from a few lakhs to more than 10 million plantlets per year. Some new companies have entered

[12] TECHNOLOGY INFORMATION, FORECASTING & ASSESMENT COUNCIL (TIFAC), Department of Science & Technology: *Life Sciences and Biotechnology* : Technology Vision—2020 New Delhi; DST, 1996.

in the business of plant tissue culture - based production, with the intention of exporting of products. Others are seed or horticultural plants producing companies; these are diversifying in plant tissue culture - based production for domestic and export market. The production was around 25 million plantlets during 1995, though the production capacity exists for more than 100 million plantlets per year. Plant tissue culture-raised plantlets are available in the domestic market and are also being exported. Cardamom, banana and a few flowering and green in-door plants are being consumed substantially in the domestic market.

Trends and Conclusions

Biotechnology research in India is primarily funded by government and the role of private sector in the promotion of biotechnology has been negligible. A separate department, DBT, was set up in the country to coordinate and promote biotechnology activities in the country. DBT has two main objectives : development of trained manpower and creation and strengthening of infrastructure. Capabilities in India exist mainly for medium level research, but there are a few exceptions which are contributing towards advanced biotechnology research. India still has to go a long way to take advantage of the potential of DNA technology.

In India, the existing R&D capabilities have also not been fully utilised at the industrial level. This could be due to three main reasons : (1) industry is not prepared to take risks and make high investments to make use of indigenous knowhow, (2) in-house R&D in the companies is not geared to take up innovative or adaptive R&D needed to absorb the technologies transferred from the research laboratories, and (3) there is a gap in the laboratory-industry relationship. Institutions are mostly unaware of industrial and market needs.

DBT has the overall responsibility to coordinate the national level biotechnology-related activities. It is playing the important role of developing the research efforts at product or process level irrespective of the agency where R&D is being undertaken in the country. However a better mechanism is required to eliminate the research component which is duplicate or irrevalent for industry. Though duplication in research is also essential, its value depends upon the purpose for which it is being conducted. In the Indian university laboratories, where capabilities exist say in plant tissue culture, a significant part of the research efforts gets repeated in several laboratories but with no significant benefit. The crops beings investigated also may not be important for basic understanding or industrial application.

Patenting of innovations is not very popular amongs the Indian scientists. In the changing world scenario, scientists can not, afford to lose their intellectual property rights. The scientific community needs to accept the concept of 'invent and patent' to exploit the genetic wealth and convert it into economic wealth in the 21st century.

Acknowledgement

Inputs through discussion with Dr. S. Visalakshi (Scientist, NISTADS) regarding commercialisation of healthcare biotechnology in India are gratefully acknowledged.

Annexure – 1

Thrust Areas in Life Sciences (SERC: 1980)

Basic Life Sciences

Molecular biophysics and theoretical biology.
Molecular and cellular biology.
Developmental biology of multi-cellular systems.
Neurobiology and mechanisms of behaviour.
Animal behaviour, ecology and evolution.
Biology of reproduction.

Medical Sciences

Immunological control of tropical and communicable diseases and modernisation of vaccine production technology.
Virology as related to hepatitis, Japanese Encephalitis, etc.
Human neurobiology in relation to mental health.
Fertility Control.
Applied Biological Sciences.
Genetic engineering.
Microbial productivity.
Biomass as a source of energy.
Physiology and biochemistry of plants.
Protection of endangered species and preservation of genetic diversity of living organisms.
Ecological balance for sustainable utilisation of biological resources, forests, grazing lands and fisheries.

Source: DEPARTMENT OF SCIENCE & TECHNOLOGYY (Ministry of Science and Technology, India), *Thrust area Programmes in Life Sciences*, 1980

Annexure II

Priorities Mentioned In Long - Term Plan of NBTB

Health

- (i) Prophylactic.
- (ii) Therapeutic.
- (iii) Diagnostic.
- (iv) Hygiene.
- (v) Population control.

Industry

- (i) Fermentation (Antibiotics, organic acids).
- (ii) Biofuels.
- (iii) Food and feed.
- (iv) Metallurgy & mining.
- (v) Oil recovery.

Agriculture

- (i) Soil fertility.
- (ii) Bio-fertilisers.
- (iii) New varieties.
- (iv) N_2 fixation.
- (v) Quick propagation through tissue culture.
- (vi) Improvements in animal health and productivity.

Energy

- (i) Biomass.
- (ii) Energy plantation.
- (iii) Biofuels and bioreactors.

Environment

- (i) Conservation of forests and afforestation.
- (ii) Pollution control.
- (iii) Waste recycling.

Cmmunication & Informatics

- (i) Computer - based information collection and dissemination.

Educatopm and Training

- (i) University level education.
- (ii) Specialised training programme.

Source: DEPARTMENT OF SCIENCE & TECHNOLOGYY (Ministry of Science and Technology, India), *Long Term Plan for Biotechnology*, 1983

Annexure III

Programmes of NBTB in the VII Plan (1985- 90)

— Manpower development.
— New projects/programmes in the multi-disciplinary area involving multi institution.
— Necessary linkages with industry.
— Involving techno-economic feasibility studies.
— Providing seed and risk capital in establishing large-scale applications using technologies developed within the country.
— Setting up manufacture (and associated R&D) of products based on modern biotechnology.
— The necessary infrastructure like germ plasm banks, pilot plant and other bio-engineering scale-up facilities, etc.
— Network of biotechnology information systems.
— Animal house facilities.

Source : *Seventh Five Year Plan* : Mehrotra & Tandon, 1989.

Annexure IV

Mandate of DBT

- — Support R&D Programmes and Biotechnology-related Manufacturing.
- — Identify and Set up Centres of Excellence for R&D.
- — Promote Large Scale use of Biotechnology.
- — Integrated Programme for Manpower Development
- — Establishment of Infrastructural Facilities to Support R&D and Production.
- — Evolve Safety Guidelines for Laboratory Resources, Production, Application.
- — Promote University-Industry Interaction.
- — To serve as Nodal Point for. Specific international Collaborations.
- — Manufacture and Application of Cell-based Vaccines.
- — Responsibility for the Autonomous Institutions, such as NII, NFATCC and ICGEB.

Source : DBT- Annual Reports 1986 - 87, 1988 - 89

Annexure V

Task Forces to Identify Thrust Areas in Biotechnology

— Aquaculture and Marine Biotechnology.
— Animal Biotechnology - Veterinary Sciences, Animal Husbandry and Leather Biotechnology.
— Basic Research, Emerging Areas and R&D Facilities.
— Biochemical Engineering, Downstream Processing and Instrumentation.
— Bioinformatics.
— Biological Pest Control.
— Environmental Biotechnology.
— Fuel, Fodder, Biomass, Horticulture and Plantation Crops, Sericulture.
— Industrial Biotechnology.
— Integrated Manpower Development.
— Medical Biotechnology.
— Microbial Biotechnology.
— Plant Molecular Biology and Agricultural Biotechnology

Source : DEPARTMENT OF BIOTECHNOLOGY (Ministry of Science and Technology, India):Based on Task Force recmomendations, *'Biotechnology-Research Development and Demonstration, Thrust areas'* 1991.

Annexure - VI

Technologies Transferred from Indian Institutions (Not exhaustive)

Name of the technology	Institution where developed	Place of transfer (Production Unit)
* Filariasis detection kit	Mahatma Gandhi Institute of Medical Science, Wardha	Cadila Laboratories, Ahmedabad
* Pregnancy slide test Latex agglutination	National Institute of Immunology, New Delhi	Ranbaxy Laboratories, New Delhi
* Pregnancy DOT-ELISA	National Institute of Immunology, New Delhi	Ranbaxy Laboratories, New Delhi
* Typhoid fever detection kit	National Institute of Immunology, New Delhi	Lupin Laboratories, Bombay
* Typhoid fever detection kit	All India Institute of Medical Science, New Delhi	Ranbaxy Laboratories, New Delhi
* Amoebic liver abscess	National Institute of Immunology, New Delhi	Cadila Laboratories, Ahmedabad
* Polypeptide P from bitter gourd	University of Rajasthan, Rajasthan	Lupin Laboratories, Bomaby
* Bamboo by tissue culture	University of Delhi, Delhi	Tata Energy Research Institute, New Delhi
* Animal birth control injection (TALSUR)	National Institute of Immunology, New Delhi	Karnataka Antibiotics & Pharmaceuticals Ltd., Bangalore
* Osmotolerant and high alcohol	Institute of Microbial Technology, Chandigarh and Vittal Malaya	United Breweries, Bangalore

tolerant yeast strain	Scientific Research Foundation, Bangalore	
*Blood grouping monoclonals	National Institute of Immunology, New Delhi	Cadila laboratories, Ahmedabad
* Microbial conversion of benzaldehyde into L-phenylacetyl-carbinol	Central Drug Research Institute, Lucknow	Atul Laboratories, Ambala
* F-MOC derivatives of 12 amino acids	Centre for Biochemical Technology, New Delhi	Atul Products, Bulsar
* Hepatitis B detection kit	National Institute of Immunology, New Delhi	Lupin Laboratories, Bombay
* Leprosy immunomodulator	National Institute of Immunology, New Delhi	Cadila Laboratories, Ahmedabad
* Leishmaniasis detection kit	Central Drug Research Institute, Lucknow	Span Diagnostics Ltd., Surat
* Monoclonals to. M 13 phage proteins III & VIII	University of Delhi, South Campus, New Delhi	Pharmacia Inc., USA.
* Cardamom	National Chemical Laboratory, Pune	A.V. Thomas & Co. Ltd., Cochin Cardamom Board Unicorn Biotek, Hyderabad.
* Eucalyptus globulus	National Chemical Laboratory, Pune	Tata Tea Ltd.
* E. tereticornis	National Chemical Laboratory, Pune	T.N. Forest Corporation
* E. camaldulensis	National Chemical Laboratory, Pune	Grasim Forest Research Institute, Dharwad
* Salvadora persica	National Chemical Laboratory, Pune	CSMCRI (CSIR), Bhavnagar

* Rubber	National Chemical Laboratory, Pune	Rubber Research Institute, Kottayam
* Pomegranate	National Chemical Laboratory, Pune	Bio-tissue Labs Pvt. Ltd., Hyderabad
* Turmeric	National Chemical Laboratory, Pune	Turmeric Research Station, Maharashtra
* Sugarcane	National Chemical Laboratory, Pune	Deccan Sugar Institute
* Dendrocalamus strictus	National Chemical Laboratory, Pune	Grasim Forest Research Institute, Dharwad
* Cymbopogon winterianus (a somaclonal variant - CIMAP/ BIO-1 3)	Central Institute of Medicinal and Aromatic Plants, Lucknow	Upto Jan. 1989 transferred to 9 growers (e.g. M/s Jallan Enterprises, Assam, M/s Rallies India Ltd. Bangalore, Mr. N.K.Joshi, CSIR Complex Palampur etc.)

Sources: P.K. Ghosh (1996) ; Kavita Mehra (1998)

Annexure VII

Consumption and Future Demand of Biotech Products in India

Rs. million

Product category	Consumption 1992	Estimated 1995	Demand 2000 AD
Human & Animal Health	13750	19590	35320
Agriculture*	680	1540	3850
Industrial products	4290	5700	15000
Other biotech products	20	300	1300
Total	18740	27130	55470
(in US $ million)	625	904	1849

*Excludes fisheries and processed food industry

Source : GHOSH P.K.:*'Biotech Industry Guide'*.New Delhi; Biotech Consortium India Limited (BCIL), 1995.

Annexure VIII

Estimated Indian Biotechnology Market (1994)

	Rs. million
Total	15000
Pharmaceuticals	5000**
Diagnostics (Reagent)	800
Diagnostics (instrumentation)	700
Vaccine (Private Sector) (Rabies, Mumps, Measles, Oral Typhoid)	1600
Plant Tissue Culture	200
Floriculture	100
Horticulture	1500
Others	5100

Notes : *Environment, Aqua Culture, Animal Husbandry, Agribiotech.

**Rs. 416 billion ($ 13 billion) World market.

Source - TIFAC, 1996.

PART VII : INFORMATION TECHNOLOGIES

18. China's Telecommunications Management: Features and Trends*

Ding Lu

Abstract

Provides some institutional facts about the managerial system of China's telecommunications industry. The study is divided into three parts. The first part deals with the decision-making structure in China's telecommunications industry. The second part describes the financial status of telecommunications companies and their investment strategies, particularly the financial relationship between the centre and the local telecommunications companies. The third part discusses the trends in the reform of telecommunications management and regulations. Points out that before the reforms, China's post and telecommunications sector was part of the government administration. After the initiation of reforms, efforts have been geared to make the post and telecommunications enterprises financially independent. Particularly, telecommunication companies have been separated from the postal service agents. Concludes that China's bid to join the World Trade Organization will surely have profound impact on the country's telecommunications industry. The WTO's acceptance of China hinges on whether China will agree on terms that mandate opening up its service sectors, including the

* This paper is based on "The Management of China's Telecommunications Industry: Some Institutional Facts", in Telecommunications Policy, 1994 18 (13). The author has revised the paper with updated information for publication in this book. The author appreciates the valuable assistance he received from Mr. Gao Yangzhi of P&T Economy, Ms. Wang Yu and Mr. Hu Zaichi of Shanghai P&T Administration, and Mr. Lu Wenda of Shanghai Urban Economics Society.

telecommunications market.

Introduction

As China opens its economy to the world, the country's telecommunications industry remains an exception. China's Ministry of Posts and Telecommunications (MPT), the regulator and dominant player of the country's telecommunications business, has virtually outlawed any form of foreign participation in China's domestic telecommunications management. Lack of foreign participation makes this stronghold of the state-owned and state-run industry a largely unexplored regime to the outside world.

This paper provides some institutional facts about the managerial system of China's telecommunications industry. The information, if not explicitly noted with publications, is derived from the records of the author's interviews with local telecommunications executives in China.

The first section presents the decision-making structure in China's telecommunications industry. The second section describes the financial status of telecommunications companies and their investment strategies, particularly the financial relationship between the centre and the local telecommunications companies. The third section discusses the trends in the reform of telecommunications management and regulation.

The Decision-making Structure

a. The Principles of Telecommunications Administration

In China, the Ministry of Posts and Telecommunications is the functional organ of the State Council responsible for nation-wide postal and telecommunications services. In the 1950s to 1970s, the management of China's state-owned telecommunications system was semi-military and highly centralized.[1,2] The telecommunications sector was seen purely as a tool for meeting administrative

---

[1]GAO Yangzhi: Shanghai's Telephone Network: status quo and development strategy. *P&T Economy* 1991, 16 (3), 22-28.

[2]YANG Peifang: On the Goals and Ways of a Harmonic Development of Telecommunication. *P&T Economy* 1991, 16 (3), 2-4.

needs. Subscribing a residential telephone line was a political privilege and a symbol of social status. Whenever the government found it necessary to reduce its budget, the telecommunications expenses were usually among the first to be classified as "non-productive" and cut.

When China embarked on its programme of economic modernization in the late 1970s, the nation's central planners were concerned with the backwardness of China's telecommunications infrastructure. Lack of development funds at the state level resulted in decentralization of decision-making and financial responsibilities to the local level. In 1979, a directive (No. 165) issued by the State Council initiated this process of decentralisation. It stipulated that in post and telecommunications industries, government administration and business management should be separated, postal business and telecommunications business should also be separately administrated, and enterprises should have their own accounts and should be financially independent.[+] Since then, pioneered by Shanghai, 19 provincial administrations have set up local post and telecommunications task forces, working with the MPT to enforce a "dual leadership" in the industry. In Shanghai, for instance, the municipal government's task force holds at least two meetings per year to hold discussions with MPT executives on policy issues.

In October 1984, the State Council stipulated a "six-point instruction" to the post and telecommunications industry.[3] The main ideas of this instruction are[*]:

* The overall planning of industrial development should be unified
* The ministerial administration should be coordinated with regional authorities.
* Responsibilities should be defined and shared among different administrative levels.

[+]In Chinese: *"zheng qi fen kai, you dian fen ying, qiye hesuan, zi ji ying kui"*
[*]In Chinese: *"tong chou guihua, tiao kuai jiehe, fen ceng fu ze, lianhe jianshe"*

[3]JIN Jiyuan. On Some Problems Concerning the Deepening of Reforms and the Transition of Business Administration. *P&T Economy* 1992, 20 (3), 4-6.

* The construction of infrastructure should mobilize resources from all concerned.

b. Features of Administrative Structure

In line with the above principles, the MPT system underwent a revamp in 1988 amidst a major state enterprise reform initiative taking place nation-wide.[4] The MPT granted its national manufacturing, construction, and purchasing departments, status of separate legal entities or greater independence in financial accounting and human resource management. Meanwhile, the MPT set up the Directorate General of Telecommunications (DGT) and the Directorate General of Posts (DGP) to incorporate business enterprise functions. The DGT is now known as China Telecom, which comprises 29 provincial post and telecommunications authorities (PTAs), all of which offer local and long-distance services. The PTAs in Beijing, Shanghai, and Guangdong also offer international services.

The MPT itself handles regulatory matters through its Department of Policy and Regulation. Its other governmental functions were distributed among its Telecommunications Administration Department, the Department of Science and Technology, and the Department of Finance.

The structure of China's telecommunications administration under the MPT is basically a vertically organized hierarchy. Mueller and Tan (1997, pp. 31-32) observed:

> "At the top is the national ministry, which is responsible of overall planning and management of the sector. At the next level, each of the 22 provinces, 5 autonomous regions, and 3 provincial-level municipalities has its own posts and telecommunications administration, or PTA. Below them are 364 municipal and prefectural bureaux of post and telecommunications (PTBs). Next, 2,182 county post and telecommunications enterprises (PTEs) operate local service networks in the county capitals and extend lines into the

[4]MUELLER M. and Tan Z: China in the Information Age: Telecommunications and the Dilemmas of Reform. Westport, Connecticut: Praeger Publishers, 1997.

surrounding rural areas. Below them are tens of thousands of branch offices operating exchanges at the village level. There are also many private branch exchanges (PBXs) owned and operated by work units and rural villages."

As observed by some Chinese experts, a major problem of the hierarchy of vertical responsibilities was the weak horizontal ties between different services. In cities like Beijing and Shanghai, the urban telephone network has been or used to be separated from the rural network and administered by different authorities. The connection between the two networks has been a bottleneck of local communications. Beijing's telecommunications administration was built on the model suggested by Soviet advisers in the early 1950s. Before the restructure suggested by Soviet advisers, Beijing telecommunications used to have three departments: business department, operation department, and service department. The Soviet advisers set up a number of departments to replace the old three: intra-city service department, long distance department, telegram department, wireless communication department, etc. In this system, every specialized department has to administer everything from infrastructure investment, network operation, to maintenance services and internal offairs. The vertical responsibilities sever the horizontal ties between different services. This structure of administration is extremely inappropriate for the development of modern telecommunications services that integrate different forms of informatics.[1,5,6] A recent development in the industry since 1995 has been the regionalization of networks. This development has broken down the barriers between the urban network and the rural network and greatly optimized the national network structure.[7]

[5]ZHANG Xiaotie: The Telecommunications Administration in Large Cities Urgently Needs Restructure. *P&T Economy* 1992, 18 (1), 13-16.

[6]ZHANG Xiaotie: An Inquiry into the Market-oriented Organization Structure of the Telecommunications Enterprises. *P&T Economy* 1993, 23 (2), 8-10.

[7]JIANG Shu: Strengthen Management of Local Telephone Network and Fully Utilize the Integrated Network Efficiency. *P&T Economy* 1996, 37 (4), 27-30.

The coordination problem in China's telecommunications industry can be observed not only in the bottleneck of local networks but also in other areas. For instance, in the early 1990s, there were reports of inconsistencies between the construction of trunk phone lines and that of users' phone lines. This was because housing construction was administered by housing development authorities, which followed the old codes of construction that did not define phone lines as a basic supply like gas, water and electricity. The build-up of switchboard capacity was often under-used due to the lagging construction of network facilities.[1] To rectify the problem, MPT and the Ministry of Construction jointly issued a directive (No. 488) in 1992 to redefine the codes of residential construction for telecommunications needs. The new construction code has, however, been ignored or poorly enforced in many cases.[8]

China's post and telecommunications enterprises, like the enterprises in other state owned sectors, are also overburdened by many administrative entities. The pre-reform structure of administration was very complicated and badly designed. For instance, a Chinese researcher noted that, in Beijing Telecommunications Administration of the early 1990s, there were 28 executive sections; 5 departments that handle Communist Party affairs; 6 departments in charge of labour union affairs; and more than 10 research institutions, trade societies, and associations.[3] In such a system, the decision-making power and responsibilities of different administrative entities are not well defined. Therefore, the top executive officers have to hold many meetings to coordinate conflicting decisions. It is also not surprising that decision-making tends to be slow in this system. Even worse, in China's state-owned industries, every department or enterprise is a mini welfare state that takes care of the employees' welfare ranging from child care services to housing construction. The so-called "logistic departments" in charge of these affairs are usually the largest and

[8]ZHOU Jingguang: Strengthen Scientific Management and Improve Intensive Operation of Posts and Telecommunications Business. *P&T Economy* 1996, 37 (4), 8-10.

most powerful in the administration. To tackle the overstaffing problem, MPT in 1995 decided to freeze the size of its employee population for a few years.[9]

c. The Status and Rights of Local Telecommunications Authorities

In China, it has long been stipulated that government administration and business management should be separated. In most areas, however, the local post and telecommunications bureau is a combined government branch and business company. Until recent years, at the city and county level, the local post and telecommunications bureaux in some provinces did not even have a business license, not to mention the status of a legal person. The definition of "enterprise" used to be very blurred in post and telecommunications industries. The ministerial departments, provincial/municipal bureaux, and city-county bureaux all claimed to be business enterprises.[10] Nowadays, most post and telecommunications enterprises are still in the process of "corporatization" to restructure themselves into independent accounting entities that conform with China's Corporation Law.[11]

Telephone companies above or at the county and township level are owned and run by the state. Companies at the county and township level are owned and run by the local governments. Exchanges at village level are collectively owned by peasants. Long-distance service revenue is redistributed nationally. In cities, telephone infrastructure development mainly comes from local telephone revenues. The fragmented administration and ownership have contributed to the problems of coordination in services and network development.[12]

---

[9]ZHU Fabin: Why did the Demand for P&T College Students Decline? *P&T Economy* 1995, 33 (4), 45-46.

[10]XIAO Jinxue: Views on the Transition of Operational Mechanism of Posts and Telecommunications Enterprises. *P&T Economy* 1992, 21 (4), 17–19.

[11]LI Guohua: Post and Telecommunications Business in New Transition. P&T Economy 1996, 37 (4), 10-15.

[12]LI ZHONGXI: Thoughts on the Reforms of Local Telephone Networks. *P&T Economy* 1992, 18 (1), 5-6.

The local companies' capital investment is restricted by planned quotas of credits and foreign exchanges.[3] The pricing of most services is controlled by the ministry. The policy before the 1980s was to under-price telephone services. Upward adjustment of rates was rare. Due to the under-priced rates, before 1980, the post and telecommunications sector was in the red for more than 10 years. During the period 1986 to 1990, the State Council authorized the MPT to adjust telecommunications service rates. Consequently, the rate of capital return in the industry increased from 9 per cent in 1986 to 17 per cent in 1990, somewhat higher than the average rate of capital return for all industries. According to the World Bank, the rate of return of China's telecommunications industry in 1989, adjusted for accounting differentials, was equivalent to 12 per cent by Western business standards. Under the existing price structure, the rate of return is lower for local (intra-city) city services and higher for long distance and international services.[13] An MPT source disclosed that in 1993, the profit margin on local service was only 2 - 3 per cent, while the margin on long distance calls was 25 per cent and on international calls, 75 per cent.[4]

Before 1980, telephone service price included monthly rent and communication charges. In 1980, local telephone companies were authorized to collect installation fees. In 1990, MPT adjusted its guidelines for telephone installation fees. The installation fee charges have then been based on line connection costs. Decision - making regarding installation charges has been decentralized to the local level. In the early 1990s, the installation charges varied from around 2000 yuan to more than 5000 yuan per line.[14] In comparison, the average annual wage in China was only 2140 yuan in 1990[15]. At

---

[13]MA Qiang: Ways to Estimate Telecommunications Rates and an International Comparison of rates. *P&T Economy* 1992, 18 (1),27-30.

[14]LIU Zaoshi: Reform-minded Views on "leased right" of Telephone. *P&T Economy* 1992, 18 (1), 11-12.

[15]The official exchange rate in 1990 was US$ 1.00 = 4.78 yuan. After the Chinese yuan was made partially convertible in 1994, the exchange rate was about US$ 1.00 = 8.6 yuan. Since then, the Chinese currency has gradually appreciated against US dollar. As of 30 April 1997, the exchange rate was about US$ 1.00 = 8.29 yuan.

the end of 1990, the intra-city telephone rates were widely adjusted. The ministry set a price cap according to local telephone companies' average costs with a mark-up for profit. Local telecommunications companies were authorized to set their own intra-city rates not exceeding the cap, subject to the approval of local government's price control authorities.[16]

The reform in 1988 shifted more decision-making authorities regarding procurement, operations, network development, and financing from the MPT headquarters to the municipal and county PTEs. As a result, the PTEs could choose among competing domestic and even foreign equipment suppliers. The purchasing and production of network equipment became highly competitive. MPT also largely deregulated the terminal equipment used by customers of the network.[4] These decentralization and deregulation schemes were accompanied by more financial accountability of the PTEs and PTAs at the local level (see the next section). This financial accountability provides stronger incentives for managers to improve efficiency of the operations.

d. Private Networks

In China, private networks have been developed mainly by ministries other than MPT. The major problem of private network development is lack of coordination with the public network. There are numerous examples of repetitive construction of network, signal interference and technological incompatibility.[17] It is said that China's private networks have grown so large that they dwarf the public network. The traffic generated by the private ones has overloaded the public network and the situation is just like "a small horse pulling a big wagon".[18] There is a severe imbalance between the utilization of private network and that of public network. Some private networks have plenty

---

[16]SUN Yaming: Issues Related to Pricing Policies Regarding P&T New Services. *P&T Economy* 1992, 18 (1), 36-37.

[17]NING Cong: Coordinate Project Planning and Develop Public and Private Networks Consistently. *P&T Economy* 1991, 17 (4), 25-27.

[18]LIU Zhaoquan: The Development of P&T and Reform of Asset Management. *P&T Economy* 1992, 18 (1), 2-4.

of unused capacity while public network is congested with traffic. The connection between the public network and private network is often blocked by technical or artificial factors.[1] The allocation of radio spectrum is under the administration of the MPT, which specifies the technical standards. At the provincial level, there is also a provincial radio wave commission in charge of spectrum allocation. The commission is composed of the local PTA officials. The commission in Shanghai is under the leadership of Shanghai's municipal government.

Lack of coordination of the development of private networks is part of the old problem of repetitive construction of many inefficient projects in China's inherited centrally planned economy. The system gave ministerial and local leaders an incentive to devote resources to "empire building" in order to expand or maintain their power base.[19] To solve this problem, the State Council in 1990 issued Directive No. 54 to enforce coordination between the public and private networks. Since then some private networks have been successfully connected to the public network.[17]

The existence of private networks gives rise to opportunities for competition. Starting in 1993, the State Council allowed the non-MPT enterprises to provide value-added services. In July 1994, the government awarded a basic telecommunications licence to China United Telecom (or Unicom Beijing). It was set up by the Ministry of Electronic Industries, the Ministry of Railway, the Ministry of Electric Power, and 13 other major companies. The carrier intends to compete with MPT in the long-distance and international services market. Another player is Ji Tong Communications Co. Ltd. (Beijing), a VAN (value-added network) operator set up by the State Economic and Trade Commission. Its charter is to build a nation-wide backbone linking the networks belonging to government ministries, universities, and research and state-owned organizations. It is also supposed to offer VAN services to government departments and the private sector. Both new players are shareholder based but are largely state-owned and controlled.

---

[19]KOHLER, Heinz: *Comparative Economic Systems*. Scott, Foresman and Company. Gleview and London. 1989.

Although the State Council has authorized the Unicom to enter the market of long distance service, MPT would like to restrict Unicom in domestic long distance service only. In domestic service market, the State Council allows the Unicom to offer services in regions "where there is a severe shortage in public telephone network capacity". MPT interprets this to mean rural areas only. Unicom, however, maintains that since "severe shortage" accurately describes the capacity situation in most of China, it should be free to operate just about anywhere.[20]

Financial Status of Telecommunications Companies

a. The Accounting System for Telephone Companies

In China's pre-reform centrally-planned state sector, the government used to manage state-owned assets in the manner of a bureaucracy. The post and telecommunications industries were no exception. The telecommunications enterprises were not treated as business firms but as government branches. Before 1985, there was virtually no independent accounting at the enterprise (local bureau or exchange) level.

From 1983 to 1985, MPT developed a business accounting system as part of its process to implement the management contract responsibility system. The accounting system, known as "Enterprise Own Revenue" (EOR), defined an elaborate methodology to determine each post and telecommunications enterprise's revenue.[21] The accounting system for long-distance service commenced in 1985. It was at first a simple record of the firm's long-distance revenue and operation costs. Later, it was found that local exchange was not properly motivated under this accounting system. As telecommunications service is a two-way exchange, the increase in real business revenue at one location may not necessarily reflect the growth of service quantity provided by the firm at that location. Therefore, the system was adjusted and improved in 1988 on a

[20]TO Chee Eng: *Chinese Telecom: Giant Steps, Giant Problems*. Data Communication on the Web, February 1997.

[21]XU Yan: Competition Without Privatization: The Chinese path. Paper presented at the Eleventh Biennial Conference of the International Telecommunications Society, Seville, Spain, 18 June 1996.

concept of network revenue. Currently, an enterprise' revenue is accounted as:

ERV = P(a) * Q * R

Where ERV is the enterprise revenue; Q is the firm's service output; P(a) is the unified national accounting rate; and R is the regional cost coefficient.

In the early 1990s, the accounting rate was calculated according to 1987's average cost and profit-tax rate in the industry. The accounting rates of more than 140 service products were listed in a catalogue issued by MPT. The regional cost coefficient was estimated by regressing 32 provincial telecommunications companies' real costs against seven factors, such as per capita income, weather conditions, telecommunications infrastructure investment, etc.[22]

Independent accounting has been used as a measure to contract responsibilities for business operations to enterprise managers and employees. In principle, 80 per cent of the after-tax profit calculated by the above method would be retained by the enterprise and 20 per cent would be handed over to the Ministry. In fact, there is usually a redistribution of revenue after the 20 per cent revenue is handed over. Because a portion of the ministry level revenue has actually been generated by local traffic, the ministry redistributes this portion to local companies that have been loaded with heavy traffic.

Each contract period between the ministry and the local company may last two or three years. By the mid-1990s, most local enterprises had completed their third contracting period. The more recent contracts included supplementary evaluative criteria, such as communication quality and capital investment. Beginning in 1991, some contracts have included terms of "asset responsibility", which define managers' responsibility of maintaining and improving the state assets under their charge.[18]

---

[22]ZHANG Xuan: A Discussion of Perfecting the Economic Mechanism of Telecommunications Enterprises. *P&T Economy* 1991, 16 (3), 5-7.

Settlement arrangement covering communication between any two domestic provincial or municipal telecommunications companies is still underdeveloped. It appears that the more wealthy coastal provinces have shown greater interest in a settlement procedure than the less wealthy hinterland provinces.

Determination of telecommunications employees' wages and salaries is a sophisticated process. At the national level, the state allowed MPT to increase 80 cents of its wage fund for every dollar of increased revenue. Then at the second level of the hierarchy, MPT allowed the local company to increase 70 cents of its wage fund for every dollar of the company's increased revenue index. Before 1993, the local company's increased revenue index was calculated as:

R = 0.5 * local revenue increase + 0.5* MPT revenue increase

In 1993, the quantum of local company's revenue increased to 60 per cent. The rationale of this arrangement is consideration of the fact that local company's service not only contributes to its own revenue but also to the revenue of the entire network.

In 1995, the state made more adjustments in the wage-salary formula. The proportion of the wage fund increase linked to revenue changes was reduced from 80 per cent to 45 per cent. Meanwhile, the proportion of the wage fund increase linked to pre-tax profit changes was increased from 20 per cent to 55 per cent.[9]

In 1980, the State Council adopted a policy of developing intra-city telephone service with intra-city telephone revenue. Since then, the intra-city telephone service revenue has been fully retained by the local company. In addition, local telephone companies have been authorized to charge installation fees. In 1990, they were allowed to determine the installation charge according to the full installation cost. The installation fee revenue became a major financial resource for intra-city telephone development.[1]

To facilitate local telephone development, in the 1980s, the State Council developed a policy of 'three-ninety-per cent".[1,23]

[23]WU Changgeng and Zhang Xuan: An analysis of the Seemingly High Profit in the Industry. *P&T Economy* 1992, 18 (1), 6-9.

* Ninety per cent of profit was to be retained by the local enterprise (in other words, the tax rate is 10 per cent, well below the 55 per cent tax rate for other industries);
* Ninety per cent of foreign exchange (hard currency) earnings were to be retained by the enterprise;
* Ninety per cent of central government investment is not considered as repayable loans.

In addition, telephone companies might also enjoy favourable interest rates when they get loans from state banks.[24]

Nevertheless, there were still complaints from the local telecommunications managers. One complaint was that these companies, like most other state-owned companies, had to pay a number of government levies and were sometimes ordered to purchase government bonds issued for some priority projects. It was asserted by the early 1990s that post and telecommunications enterprises still had to hand over nearly half of their profit to the state, if these non-tax levies were taken into account.[1]

In 1994, China introduced a major tax reform aiming to unify the tax rates into an across-board 33 per cent rate for all enterprises and to simplify the tax-levy structure. Meanwhile, with a reform to unify foreign exchange rates and make Chinese currency convertible, there is little benefit for enterprises to retain their foreign exchange earnings. The preferential "three-ninety-per cent" policy has since no longer been practised.[25]

Some experts have been worried that the current local business revenue has been too heavily dependent on the collection of installation fees. In the early 1990s, the new accounting system and the favourable official policy towards the telecommunications sector resulted in a tremendous increase in the industry's reported profits. This led some to conclude that some of the increased profits should be returned to consumers in the form of lowered

---

[24]LUO Yunchao: The Monopoly Nature of P&T Industry and its Effects on Operation. *P&T Economy* 1992, 21 (4),15-17.

[25]LUO Guilin: Correctly Deal with Several Issues in the Transformation of Telecommunications Growth Mode. *P&T Economy* 1996, 37 (4), 5 - 7.

installation fees.[23] From 1980 to 1991, one fourth of Shanghai's investment in network expansion came from installation fee fund.[26] Wu Jichuan, the Minister of MPT, estimated that the revenue from installation fee accounted for about half of the annual investment capital in 1995.[27] Relative to the country's per capita income, China's installation charge is the highest in the world. The extraordinary high installation fee could become an impediment to residential telephone development.[23] The slow increase in the number of subscribers may lead to under-utilization of network capacities and thus undermine network externalities.[28] Some suggested that compulsory bond purchase at installation should be an alternative way to finance the construction fund.[1]

b. The Policy on Local Telephony Investment

China's central planners are well aware of the importance of telecommunications investment to accelerate economic growth. Deng Xiaoping, China's paramount leader, was reported to have agreed that modernization of communications should be put before the so-called "four modernizations of China", namely modernization of industry, agriculture, national defence and science and techonogy.[29] The Centre for Economic and Technological Development of MPT carried a study based on econometric modelling which showed that the marginal contribution to national economy brought by adding one dollar of investment in information resources is 15.8 times that brought by investment in physical resources.[2] The centre used the data from 29 countries to regress telephone line density on factors

[26]WU Houzeng: Discussion of the Installation fee for Private Residential Telephones. *P&T Economy* 1992, 20 (3), 44 46.

[27]Wu Jichuan's interview, *Yazhou Zhoukan (The International Chinese News Weekly)*, 19 May 1996, p.58

[28]HU Shiliang: Some Thoughts on the Reform of Telecommunications Fees in China. *P&T Economy* 1996, 34 (1), 19-2 1.

[29]Gao Yangzhi: Insist on a combined mind of pragmatism and liberalization, develop P&T industries with persistence, speed, and harmony. *P&T Economy* 1992, 20 (3), 24.

such as per capita GNP, average education level and relative size of service industries. The regression result was compared with China's relevant data and it was concluded that only half of the social demand for telecommunications was met in the 1980s.

Development of a modern telecommunications network compatible with computer application was one of the major goals of China's Eighth Five-Year National Economic Plan (1991-95). The fixed capital investment in telecommunications sector during the Eighth Five-Year Plan period increased 12 times over that in the period of the Seventh Five-Year Plan (1986-90). From 1991 to 1995, China installed 73 million phone lines, more than those in the rest of the developing world combined. This has made China one of the largest telecommunications networks in the world. Still China aims to double the size of its network by the year 2000.[30] A footnote to this rapid growth is the equally rapid privatization of residential phone service subscription. Telephone service is no longer only an administrative tool of the state, it is rather a private good for business operation and household consumption. In 1991, of the total residential phone subscribers, private ones accounted for 75 per cent. In more than 10 provinces and municipals, residential phones accounted for more than one third of the total phones.[31] In 1995, more than 80 per cent of the increased telephone exchange capacity was for residential uses.[32]

Since the 1980s, MPT has followed the principle of "developing telecommunications infrastructure in harmony with economic growth, properly advancing the latter"[29] MPT interpretation of the idea of being "in harmony" emphasizes coordination between the ministry's public network and the other sectors' private

---

[30]Wu Jichuan's interview, *Yazhou Zhoukan (The Intenational Chinese News Weekly)*, 19 May 1996, p. 57-58; Diana Lady Dougan's "Forword" in Mueller and Tan (1997), pp. vii - xiii.

[31]HONG Xiuyi and Qian Yongwei: The Trend of Residential Telephone Development in China. *P&T Economy* 1992, 21 (4), 19-23.

[32]WU Jichuan's interview, *Yazhou Zhoukan [The International Chinese News Weekly]*, 19 May 1996, p. 57-58.

networks. A major concern is that a laissez faire competition between the public network and the private networks would result in repetitive construction and waste of resources. The private network may also "unfairly" take away profits, a concept close to the Western concern over "cream skimming".[29]

For local (intra-city) telephone investment, the State Council has clearly defined the division of responsibilities between the ministry and the local government. MPT should invest in equipment and machinery, while the local government should invest in cable and line construction.[1] One important policy is called "mobilizing four resources together" ["sige yiqi shang" in Chinese], namely, the resources of government fiscal expenditure, users' contribution, enterprises' finance, and domestic/foreign loans.[23] This policy is a follow-up of an instruction of the State Council's in 1984 (see Section 1) to "mobilize resources of all concerned in telecommunications infrastructure development". In 1990, less than 22 per cent of the total telecommunications investment came from PTEs themselves.[23] According to MPT, during the first half of the 1990s, the investment in telecommunications infrastructure was mainly financed through three sources. One was the installation fee collected from the users, which accounted for 40 per cent of the capital. The second came from domestic and foreign government loans, accounting for about 30 per cent of the total capital. The remaining 30 per cent came from PTE's profits and capital depreciation.[33]

Shanghai's telecommunications infrastructure build-up is a successful example of these policies. Of Shanghai's tetecommunications investment during 1986-90, 10.6 per cent is from government fiscal expenditure, 22.4 per cent is from telephone company's profit, 27.9 per cent is from user's installation fee, 3.5 per cent is collected from commercial users, 13.3 per cent is from domestic banks'loans, and 22.3 per cent is from foreign investment.[1] Entering the 1990s, the Shanghai PTA signed joint construction agreements with all district-level and county-level

[33]*Lianhe Zaobao* [United Mornings, Singapore], 8 April 1994, p. 23.

governments under the Shanghai Municipality to incorporate telecommunications infrastructure construction into their urban redevelopment plans. Joint investment between PTA and the local governments or big business entities played a pivotal role in rapid infrastructure development in recent years.[34] Users' installation fee and network connection fee (for local network) were one of the main sources of investment capital. Meanwhile, the municipal government returned the local taxes to the telephone company for investment, allocated hard currency quotas to facilitate telecommunications investment, and allowed local post and telecommunications rates to fluctuate with greater flexibility. In early 1992, to facilitate the development of a "teleport" in the Pudong area (a special economic zone in the east part of Shanghai), the Pudong P&T Bureau was set up. This bureau is in charge of designing and coordinating the development of public network and all private networks in the teleport area. Urban infrastructure development in Pudong zone is planned with the concept of a modern telecommunications hub.[35] Therefore, the Pudong PTA has been vested with an unusual amount of decision-making power regarding design, investment, and finance. In 1992, the Shanghai Guomai Share Corporation Limited was launched as China's first telecommunications company listed in the stock market.[36] To further facilitate investment finance, the Shanghai PTA set a Measure for Capital Finance in Telecommunications Projects in 1996. The document defines the principle of "joint investment, accountable capital uses, dividend distribution according to invested shares, and tax payment according to laws".[34]

During the period 1991-95, the telecommunications investment accounted for 1.4 per cent of Shanghai's GDP. The telecomm-

[34]HUA Ruiqiang: Joint Construction has Facilitated Shanghai's Telecommunications Development. *P&T Economy* 1996, 37 (4), 30-32.

[35]XU Ruoqing: The future of Shanghai's Pudong teleport. *P&T Economy* 1992, 20 (3), 37-40.

[36]FAN Boxing: Shanghai's telecommunications development must undergo two transitions. *P&T Economy* 1996, 36 (3), 14-16.

unications traffic grew at an annual rate of 42.2 per cent, almost three times higher than the annual GDP growth (13 per cent).[37] In the same period, Shanghai's teledensity (telephone lines per hundred persons) grew almost four times, reaching 17.2 by the end of 1995.[36]

China's policy towards foreign investment in the telecommunications sector is two-fold. On the one hand, the government is keen in using foreign capital to develop China's telecommunications industry. On the other hand, it bans foreign equity investment or direct operational control over their ventures in China's telecommunications sector. Most of the foreign capital comes from government loans and loans from international organizations, such as World Bank, Asian Development Bank, and the IMF. There are also cases of borrowing of foreign businesses' export credit. Meanwhile, the manufacturing of telecommunications equipment has been open for foreign direct investment. In this field, Bell Shanghai and Shenda Telecom (in Shenzhen) are the two earliest manufacturing joint ventures in China.[38] The 1994 reform in foreign exchange management made Chinese currency virtually convertible and greatly reduced the foreign exchange risk in using foreign capital.

Trends in Reform

From the above discussion, we observe that China's post and telecommunications sector used to be part of the government administration. In the late 1980s to the early 1990's efforts have been made to make the post and telecommunications enterprises financially independent. Reforms have separated telecommunications companies from the postal service agents. Meanwhile, measures have been taken to mobilize resources from non-MPT sectors to develop telecommunications infrastructure. Asset ownership of different investors is, however, still not well defined in many cases. Corporatization of PTEs is still in the initial stage.

---

[37]CHEN Liangan: A Study on the Evaluation System for the two Transitions in Telecommunications Sector. *P&T Economy* 1996, 36 (3), 16-19.

[38]ZHOU Zhiqun: The Usage of Foreign Investment in P&T Business. *P&T Economy* 1991, 16 (3), 36-37.

In the early 1990s, China's Association of Post and Telecommunications Business Administration prepared a "Guideline for Improving Post and Telecommunications Business Administration during the Period of the Eighth Five Year National Economic Plan (1991-95)". This guideline, although quite general, emphasizes the importance of market signals to the management. It requires that all business decisions should "come from the market and end up with the market". The guideline proposes that reforms should target "vitality of enterprise" as well as "vitality of the whole network". The post and telecommunications enterprises should develop "an investment mechanism with autonomous growth and autonomous constraints; a flexible mechanism of market-oriented competition; an incentive compatible reward mechanism; and a personnel mechanism with easy exit and entry".[39]

The future of China's telecommunications business lies in a new Telecommunications Law, which is being drafted by MPT. As pointed out by Liu Cai, director general of Policy and Regulation of the MPT, Chinese telecommunications development is hampered by outmoded legal and regulatory constraints.[40] According to him the legislation will address the following objectives:

* maintaining the current pace of business expansion, focusing on telecommunications' role as a multiplier of economic growth;
* decentralizing the present structure of telecommunications enterprises;
* assuring availability of advanced technology, both from foreign sources and expanded domestic R&D facilities;
* strengthening competition among foreign and domestic suppliers of network equipment, including the possible introduction of open bidding; and
* encouraging improvements in financing telecommunications expansion, including changes in user-pricing patterns.

---

[39]Guideline Task Force: Introduction to "The Guideline for Improving Post and Telecommunications Business Administration During the Period of the Eighth Five Year National Economic Plan (1991-95)". P&T Economy 1992, 21 (4), 6-11.

Regulatory reform at the central government level will endorse the process initiated in 1979 by the State Council's Directive No. 165 (see Section 1) by legally defining what Liu called "the three separations". They are (a) separating the MPT's postal and telecommunications units; (b) separating MPT's regulatory and operational functions, and (c) separating telecommunications manufacturing functions from service obligations. Therefore, MPT will be set up as solely a regulatory body and China Telecom as an operational body. The two organizations will be completely separate. Compatible with the new law, the Ministry of Electronics Industries is likely to be transformed into a trade association. More network development decisions will be devolved to provincial and municipal administrations. The legislation is expected to be submitted to China's State Council for approval in 1997. China's bid to join the World Trade Organization will surely have a profound impact on the country's telecommunications industry. WTO's acceptance of China will hinge on whether China will agree on terms that mandate opening up its service sectors, including the telecommunications market. If China prepares to commit itself to open its domestic service market and meet the standards set in WTO's hard-won Global Telecommunications Pact, the new telecommunications legislation must allow more room for reforms than that suggested by "the three separations".

---

[40]*International Communications Society Web Page*, China's MPT Director General Outlines Issues for New Telecom Law, 11 December 1996.

19. Political and Social Imperatives of Telecom Regulations in India

Parthasarathi Banerjee

Abstract

The integration of telecommunications with information technology and the media (ICT) has created a new situation, where the role of the regulator has to be redefined. Such a role must recognize the re-regulatory strategies of any deregulatory exercise. This redefining has to take into account a federated domestic political process, the economic rationale of the industry, the global strategic concerns of the state and a resurgence of the importance of social control. In fact, consequent to the integration in the field of ICT, the differences between the social and industrial spheres are diminishing. It is argued therefore, that redefining of regulations should be carried out at three levels, i.e. global, industrial and social levels. Regulation cannot be limited only to the quasi-juridical administration of business. The rule-making in regulation is necessarily case-based and natural law cannot remain as the guiding principle. The authorities of both politics and society must now reassert in any regulatory exercise. Indian case is cited as an example.

Introduction

Information-telecommunication is the key transforming agent for an economy at the cross-roads. Increasing returns on knowledge and information, growth through connectivity, increasing information-intensity in all kinds of activities, new international division of labour, resurgence of macro economic-instruments and freeing of finance are only a few facets of this new manifestations of the economy. The inside of economic organisation is now out onto the market-logistics-services;

manufacturing is being led by R&D-services-marketing, and the learning-entertainment-communication is but a big business. Communication and information, broadcasting and processing of information have merged together. These developments throw up serious challenges to the traditionally accepted notions on monopoly, industrial classifications, administrative control through tariffs and the principle of economies of scale. More important perhaps is the relevance of this new perspective of telecommunications for the society. Business is now out into the open, having broken out of the organizational walls of the "firm", the economics of business is now dependent on the increasing return derivable from investments in varieties of social institutions. New perspective of telecommunications has transformed business productivity into social productivity.

It is often argued that the driving force behind these changes in economies is the voluntary market, and that the state should, therefore, remain on the back seat. Such a market versus state situation, we argue, is wrong. The contemporary restructuring in economies is aimed at achieving the twin objectives of new distribution (hence allocation) of resources, and creating potentials for generating resources. Issues at stake are, therefore, intensely political and, we argue, that the reality of regulatory reform, which should indeed be called "re-regulation"[1] , is constrained by both intra- and inter-state political restructuring. The state is often indeed at the driver's seat[2]. The "extent" of market is being defined by the political process that operates through its instrumentality around social institutions. In order to comprehend the scope of regulation, it is necessary to consider deregulation, liberalisation and privatisation as well.

The regulator has then to be defined appropriately. Traditional arguments on regulator as a non-departmental business-rules based quasi-judicial social entity is an excuse for evasion that usurps power from the political authority and that also hides

---

[1]VOGEL S.K.: *Freer Markets, More Rules: Regulatory Reform in Advanced Industrial Countries*. Ithaca; Cornell University Press, 1996.

[2]PETRAZZINI B. A.: *The Political Economy of Telecommunications Reform in Developing Countries*. Westport, Conn.; Praeger, 1995.

the fact that the regulator itself is an administrator. In fact, the regulatory administration is being challenged by both, the technological convergence of telecommunication and information, and the political desire to achieve a new mode of control. A limit to regulation and consequently to regulate with the least "interfering" principles of natural law, such as the freedom of content and the rights to free speech[3], cannot provide a direction, at least to a developing country such as India, which we argue, needs to regulate through multiple institutions and using positive and pro-active contingencies of case-approach.

Such an approach, we argue, is contingent on the facts that: India is a vast, populous country with a complex social structure and a rich cultural heritage; Indian society has to be transformed into a knowledge society that would have to satisfy the basic needs of its population; knowledge-enterprises would have to come up even in villages and knowledge-intensity of all economic operations would have to be raised; demands would have to be made for capability, equity and justice; and telecommunications infrastructure should enable India to gain strategic international advantage. The case approach does not demand the existence of constitutional principle *a priori*; in contrast, it confers flexibility on the state authority and the social/business institutions. We argue for a case-based reasoning and a corresponding regulatory institution. Such an institution, we argue further, has to differentiate between the political and social-economic regulatory authorities. This separation strengthens the political authority through reregulation. It regulates further the social sphere through empowering itself with regulatory controls over economic decisions. We argue further that quasi-juridical control over economic decisions or regulations based on economic rationality are flawed, because it is either constitutionally-driven or driven by economic rationalities. The social regulation can look after the strategies for a knowledge-society, having satisfied in the process the social criteria of dignity, and basic necessities. The

---

[3]KLINGLER R.: *The New Information Industry: Regulatory Challenges and the First Amendment*. Washington D.C; Brooking Institution Press, 1996.

political regulation can look after the making of strategies for the international sphere and for such domestic concerns as enhancing inter-regional equity.

State and Extent Of Market

For a rather long time, the paradigmatic assumption that the extent of market went proportionately with the falling cost-line prevailed. This has often been described as Fordism[4]. The economy after Fordism, rich with information and a networked industrial organization, burdened least with inventory - supplying promptly the merchandise, such as through POS (points of sales), - is less dependent on mass-production and often caters to the needs for such goods that are chosen by the customers based not so much on the cost as on the criterion of quality. In addition, tele-density is an important factor. Indian tele-density is only 1.5, lower than that of both Pakistan and China. An increase in connectivity, even through PCOs (public call offices), enhances immediately the market exponentially. Additionally, if one considers the Integrated Services Digital Network (ISDN), the very large chunks of data that can be supplied through the National Information Infrastructure (NII) backbone, it is realised that the marginal cost of providing a connection is virtually zero. Given a basic backbone, the size and depth and the complexities of the market would depend on factors other than considerations of cost. What are these factors? Who can enrich a costless information and what are the structures and divisions in society that may enrich a market? Does the division of labour increase the extent of market when it is costless and can the state play a role in deepening and broadening the market?

Perhaps the ideology and normativity of competition has remained the guide to the actions of the state. Such a state that has also based itself on sanctions and economic-turf creations, has been drawing its legitimacy from positive economic decision-

[4]BOYER R. & DURAND J.: *After Fordism.* Houndmills; Macmillan Business, 1997.

making. The state, we would like to argue, needs to limit itself to setting up injunctions in putting up limitors to only certain proscribed areas. Competition has been a constitutive principle and it has never been an injunction. The principle of competition, even if it posses some economic merit, cannot be employed by the state, but if necessary has to be employed by perhaps a social authority. Moreover, competition presupposes a homogeneity, a scarcity, relevance of size, and considerations on entry. We know that scarcity does not prevail in the market of information, homogeneity kills the information, and the varieties of integrations now taking place amongst the diverse fields of Information-Communications Technologies (ICT) incorporating the media as well, have rendered the dominance of large companies a paper tiger and the umpteen number of barriers to entry a joke. Finally, a mode of action through injunction should call for different types of limitors, designed on non-constitutional positive case-based rules allowing for certain exceptions.

Currently, Indian telecom market is divided into about 20 circles or zones, roughly corresponding to the state boundaries, and 4 metro circles, covering the four corporations of Delhi, Mumbai, Calcutta and Chennai. Further, in each circle, for example, the services are divided into two broad categories, namely, the "basic" local, long distance and the international services; and secondly, the value-added services (VAS), implying primarily the cellular services. The former is essentially under government monopoly, while the latter is with the private sector. The operative principle is that of duopolistic separation, which often gets described as the competition. In the "basic" services for the circles, private provisioning is yet to emerge and hence parallel provisioning or quasi-competition would ensue. The Mahanagar Telephone Nigam Limited (MTNL), under the Department of Telecommunications (DoT), established in 1986, looks after the basic operations in Delhi and Mumbai. Two-thirds of MTNL equity is held by the government and the rest is vested with mutual funds and the private sector. In the circles, private sector companies have been allowed to build and operate the local and intra-circle long distance systems, including the wireless in local loop and the VSATs (very small aperture terminals). With the announcement of the National Telecommunications Policy (NTP) in 1994, granting of radio-paging licences to 27 operators and cellular telephony at four metropolitan cities were initiated.

The present cellular market in India is very small as compared to China or Japan. Some of the market consultants estimate that the number of subscribers will reach 3.5 million by 2002, 49.5% of the then Indian market; the rest will be shared by the circles. The present share of the latter in the Indian market is about 28.3%. Emergence of brand-name market is no longer limited to the OECD, and even in India there is such a niche. The cellular phone market is a prime example. In contrast to the assumption that the mobile phone brings in competition in the local-loop market, in India these two types are supplementary. However, this absence does not take us to the theory of natural monopoly on which, in particular, most part of the regulatory rationale has so far depended. Supplementarity and non-monopoly aspects are joined further by the absence of scale-economy. In fact, the small size of the partitioned telecom market in India was not considered as a deterrent by the prospective entrepreneurs. The total number of payphones in India was 0.1066 million in 1991-92, rising to 0.3002 million in 1994-95 and to an estimated 0.71 million in 1997-98. Out of this figure of payphones, the share of the villages as the Panchayat phones was 0.068 million in 1991-92, 0.1617 million in 1994-95 and an estimated 0.4662 million in 1997-98. The total number of telephone sets during the respective years are 6.7056, 10.7551 and an estimated 18.4753 millions.

India has also committed itself to the 'World Trade Organization (WTO) Basic Telecommunications Agreement 1997' on authorizing one competitive wireline operator in addition to the DoT in each service area for 10 years. Further commitment has been made regarding review of the national long distance competition policy by 1999. The international services are under the monopoly of the Videsh Sanchar Nigam Limited (VSNL), formed in 1986. Most of the services are thus separated. The value added services are primarily under the private sector, although about seven such services are with the DoT and the MTNL. The recent MTNL decision on providing value added services has been challenged by the private operators, who feel threatened on their local monopolies. This market is, therefore, partitioned into several turfs, divided on the lines of spatial divisions and type of services. In the case of duopoly too, what we may observe is market-partitioning. A partitioned market cannot offer a competition. The duopolies provide for supplementarity. Competition indeed is being put up by such other factors as the ICT integrations.

The supplementary nature of demand, especially for those goods that have little product-life, but preferably with brand-names, clearly defies assumptions on competition and monopoly. In fact, if we look at the entertainment market in India, we simply cannot overlook the extravaganza. This market in particular, with its varieties of multimedia features and umpteen number of distributions, is growing very fast and telecommunications has direct involvement there. The cable TV market is a case in point, which has a very high growth rate and the highest growth in towns of categories C and D (small towns). Urban dependence is now a story of the past. The number of channels varies from city to city and from provider to provider. Doordarshan, the national authority, provides for three primary and fourteen other subsidiary channels. At least 25 other private channels also compete in this segment. The estimated viewers are about 145 million; however, only a fraction of them own sets. Regarding cable television, only about 10 million households have satellite TV service through cables. The Direct to Home (DTH) is yet to come and has been operationalised only as experiments by the Department of Space (DOS). A few indigenous broadcasters are likely to come up soon, following the government initiatives and inducements in this regard. The Broadcasting Bill, having been tabled and the Prasar Bharati Board that has come up, are but the advancing footsteps towards deregulation.

This, however, is not the full story. There are the expanding segments of software, multimedia, use of CD-ROMS, and varieties of convergences, as also differentiations that have begun dotting the ICT sector as such. From a narrow perspective, it appears to be determined by the technological supplies. Such supplies, however, are but part of the strategic operations and are not the outcome of the static competitive scenario. Moreover, a large number of innovations are but user-induced or are user-led; and in neither case can one attribute the technological stand-alone supply as the cause. This expanding "extent" of market, it can be argued, is not determined *per se* by the technological supplies, but is governed by the Smithian dictum on division of labour, the latter invariably influenced by the governing institutions and their rules. State being nothing other than an equilibrium between such governing institutions, the extent of market in India today is being determined by the balance between such institutional forces. Privatization and deregulation are but the names of the media of exchanges between these institutions.

Social Productivity

This extended market is no different from the society. The immediate corollary of it is that the standard rate of return from productive investment in a manufacturing establishment gets substituted by not only the rate of return from social investment but also by the endogenous social increasing rate of return. The latter constitutes the strategic social thrust. In fact, the returns from investment in firm-level ICT are otherwise not well established. In the early years, governmental initiatives in telecommunications were legitimized by the universal service obligation and in the enforcement of regulations on an otherwise claimed-to-be "natural monopolies". Universal service obligation can no longer enjoy the legitimizing aspect, as argued above. Moreover, the role of a proactive state too cannot be explained unless we recognize the new socialized market. Policies and programmes, therefore, acquire new meaning. Privatization would connote transformations into social capital being controlled by social groups/interests; or, deregulation would connote a social re-regulation; and liberalization would imply loosening the firm-centered operationality. These three slogans in the case of socialized ICT would imply a set of integrated policies and actions to be undertaken by the executives looking after all wings of ICT, the industry in general, the education and the wings of S&T and R&D. Of immediate relevance are the policies and actions relating to software, multimedia, media, manpower, national information infrastructure (NII), and surely the hardware.

The software, for example, should be considered more as a basic input to all kinds of knowledge services than as an item of export, fetching foreign exchange. Software industry in India has been growing steadily, from 243 US $ million in 1990-91 to 388 US $ million in 1992-93 and to 835 US $ million in 1994-95. Export of software has all through maintained higher figures. The software export growth rate of India in 1990 was 34% compared to 43% for China, 40% for Korea, and 39% for Israel. The Indian share of US software market was 0.82% in 1990-91, rising to 1.06% in 1992-93 and to 1.70% in 1994-95. However, the domestic penetration of ICT appears to be lagging. Measured in terms of domestic software market and the domestic hardware and maintenance, we observe that the total of these three in 1990-91 was 644 US $ million, 623 US $ million in 1992-93 and 1082 US $ million in 1994-95. This is not an impressive growth, indicating a rather subdued ICT penetration into the Indian society as such, although in this figure we did not take into account the contribution from telecom. The telecom data for India (Table 1) should, however, be read along with the data on literacy rate, on number of PCs per capita or number of TVs per capita, etc. Another interesting observation from WTO data on telecom revenue and on the share of total world revenue, is that the US is first, having 29.70 per cent share followed by the European Union with 28.27 per cent, Japan with 15.59 per cent, and India securing the 12th position with only 0.63 per cent of the world revenue. Indian penetration rate defined in terms of the above parameters, namely, literacy rate, per capita telephone - TV - PC, etc., software and hardware consumption, media channels, multimedia uses, and the like, is surely much below that for even most of the developing countries, such as Brazil, Malaysia or China.

Table 1: Data on Growth of Telecom Services in India

Sectors	1991-92	1994-95	1997-98
Main lines in service(000)	5809.9	9795.6	16795.8
New main lines added(000)	735.2	1770	2860.6
Digital switching capacity, local(000)	1880	7235.7	16688.2
Telephones sets(000)	6705.6	10755.1	18475.3
Total equipment market, million US $	1025.3	2001	3512.9

The socialization of ICT and in particular the telecommunications, albeit being important, has received little attention. Moreover, this socialized ICT and an information literate population alone can generate a social space and consequently the demand for a new order of regulation, liberalization and privatization. The making of strategy towards such a society, we argue further, can be undertaken only non-constitutionally. What is needed is the contingent case-based positive reasoning/ approach that can be decided by multiple authorities, ranging from the political to the executive or quasi-executive to the social authorities.

Political Balance and Regulator

National level regulator and the national policies regarding telecom are in a predicament. They have to face global supplies of telecom services, products and rules regarding such services and products, although their instrumentality does not enable them to even effectively control the dynamics of the events. Policy instruments and regulation are mostly industry-specific, for example, while the events in contrast are mostly inter-industries. A telecom regulator is supposed to look into the tariffs of long-distance calls, but is hardly empowered to examine cases of such calls through the internet, for example. Again, universal service as a political obligation, which has often been subsidized by the long-distance revenue, is now under severe peril. The logic of universal service was derived from the economics of telephony and now since such a logic is defunct, the obligation to serve everyone has to necessarily combine a number of other value-added services with the basic purpose of communication. Hence, the regulator works towards the enrichment of VAS, despite the fact that it is beyond its jurisdiction.

The Telecom Regulatory Authority of India (TRAI) was set up in 1997 under the TRAI of India Act. The mandate of TRAI covers among others the following: regulating arrangements among service providers for sharing revenues, protecting the consumer's interest, settling disputes among the service providers, recommending the need and timing of introduction of new service providers, including revocation of licenses, ensuring technical compatibility and effective interconnection, ensuring compliance to terms and conditions of the license as well as to universal service obligation, facilitating competition and promoting efficiency in the operation of telecommunication services so as to ensure growth in such services, etc. The TRAI, and later the Prasar Bharati Board do not however challenge, overrule and modify the existing legal provisions, which were framed taking into consideration the ideals of separated terrains of ICT and the ideal of competition-led growth, equity and service-obligations. A far more important lacuna is that the regulator in a country such as India has indeed a short-arm and a blinded-vision. Arm of the regulator cannot reach all the events even inside the national territory. Policy instruments of the other wings of the state, for example, on finance or education or on culture, are often not related to the regulator's instruments. Perhaps the most important aspect of ICT-led knowledge society is education; and surely, excepting some occasional reflections on contents by the Prasar Bharati Board (the TRAI in fact is not empowered to look into), the regulators in this country cannot look into ICT-led education.

Such a blindness or ineffectual arm is, however, not the case for a number of OECD countries[5] ; a few of the OECD states

---

[5]LEE K.: *Global Telecommunications Regulation: A Political Economy Perspective*. London; Pinter, 1996.

are indeed extremely powerful in effecting their desired modes of delivery of services, terms relating to such transactions, in setting standards[6] and in ensuring a market for their products, even through such negotiating fora as the ITU or the GATT. The knowledge society perspective requires the ICT policy to redefine sectoral policies, boundaries, institutions and regulations[7]; and these sectors are the domains of technology policy, industrial policy, media policy and the telecommunication policy. It has been argued that the telecommunication policy should create and shape the transmission infrastructure; the media policy should define the framework for provisioning of electronic media content; the technology policy should stimulate the development of new technologies; and the industrial policy should shape the industrial structures, stimulate structural changes and support competition. In terms of spheres of participation and coordinated action, it has been argued that the ICT ought to act as a process of participatory interactive learning and planning between the spheres of industry, science and technology community, civil society and the state executive and the legislature. The contemporary structure of regulations in India, as well as the other executive and legislative structures that exist, have not evolved out of a recognition of these interpenetrating co-spheres. Such an evolution cannot ensue from an executive fiat, nor can it emerge from mere legislative actions, especially given the absence of demands from a college, interest-groups or the society. The social sphere is the mother and till such time as it is not infused with telecommunication, especially the ICT, the politics of ICT cannot emerge.

Another aspect of hiding the political bankruptcy in the weak developing nations has hardly been discussed. The rationale

---

[6]BESEN M. & SALONER G.: "Compatibility standards and the market for telecommunications services" in T. J. Allen & M.S. Scott Morton (Eds.) *Information technology and the corporation of the 1990's: Research studies* New York; Oxford University Press, 1994.

[7]MANSELL R. & WEHN U. (Eds.) : *Knowledge societies: Information technologies for sustainable development.* Published for & on behalf of The United Nations; Oxford University Press, 1998.

behind much of the international negotiating fora as also behind the re - regulation in the nations is that the business should be allowed to prosper voluntarily, following its own logic. However, international negotiations are directly or indirectly controlled politically, whereas the national level political authority has very little, if any, control even over the process of national regulation. The latter is a pure business administration. This weakness stems from the fact that, as in India, the state and the government have been made identical while implementing an electoral democracy over a long period of time. A non-departmental regulator with the least control by the legislators is a rather new phenomenon. There is hardly any political institution through which the legislature can keep a control over the process of re-regulation. It implies the presence of a hegemonic international power, through proxy - the regulator re-regulates, as per international decree.

There is another sinister dimension in India. It relates to the federated and local political powers. The Indian state departments on telecommunications and broadcasting (PTT - Public Telephone and Telegraph) remained under central control by dint of the same "Fordistic" logic, viz., scale-economy, natural monopoly, common-carrier regulation and universal access. However, indirectly, and albeit being weak, the provincial legislatures could negotiate the quantitative parameters of allocation and access. Above all, in the entertainment and cultural areas and in the formation of socialized professional groups, etc., the provincial and sometimes even the local powers could negotiate, arbitrate and control the rules and events. With the separation of national-level regulator from the central ministry and centralized institutions of political negotiations, the provincial legislatures lose power, whatever little they had so far with them. Moreover, unlike the earlier regime of telecommunication, the contemporary communication events encompass virtually all walks of life - education, entertainment, social and business transactions, manufacturing and negotiations. A re-regulation of this sphere is indicative of a thorough rupture and delinking of most of the political institutions. The thrust of the current changes is towards pauperizing and de-legitimizing of the elected legislatures. The vision of an internationalized society without a base of local society and localized power is likely to be nightmarish.

The Three Instruments

Privatization, liberalization and deregulation are the three slogans of a new ideology. As it happens in any ideology, the purports of these three are often confounded and mixed-up with each other. The placement of public ownership into private hands is the purport of the first instrument. However, the contemporary "private" does not represent private ownership; it represents privately-controlled (or, decided) social capital. In the absence of political institutions, especially of the local electoral college, such a shift signifies a transfer of local wealth as it were in a tribute, to a hegemon, whose locus is extra-local or extra-national. Therefore, any privatization ought to imply re-strengthening of and creation of a set of political institutions. Moreover, in the context of the Indian PTT, a privatization on the model of the Baby Bells cannot bring in competition, especially at the local loop. The logic of efficient service at the local level hence cannot arise. The intra-regional-entity competition is indeed a competition among the regional growth-forces and therefore concerns of allocation are necessarily determined historically, more according to the needs of the hegemonic international capital; and owing to the weakening of the local electoral institutions, the growth-oriented increasing-return aspects of investment decisions are relegated to a secondary status.

Resource-generating aspects of investments require a long-term perspective and that can be taken care of only by either the political authority or a social institution controlled by politics. The instrument called liberalization does not refer to any laissez-faire but it refers to loosening of the departmental administrative control and its substitution by another administrative or quasi-judicial control. A proper liberalization, we may argue, should bring in the twin elements of, first a business-domain regulator normatively controlled by a set of institutions and, second social regulation. In either case, electoral voice and electoral principles are needed to set the norms and the rules, or at least guide a contingent event.

The regulator, being a non-departmental administrator, has limits on its actions. These limits are defined by the office of the administrator. However, as described above, the post-Fordistic industrialization based on privately auctioned social-capital and new classes of products and services is required to be guided by a social network of mutually-dependent decisions. This cannot be provided by the regulator. We then require mutually interpenetrated social organizations, in short, a new institution. In order that the decisions are convergent and stable, presence of the political authority supervenient on the decisions is a necessity. This we describe as re-regulation. The slogan of de-regulation tacitly implies the substituting of the old regulatory structure by a set of new interdependent structures, being controlled ultimately by the political authority.

Re-regulation needs to be precedent to liberalization, which in turn should be precedent to privatization. Non-observance of this, as has happened in India, leads to failure in attaining the goal. This is a relational structure, and we may observe that this also contains the principle of structural reform.

A Positive Principle

This principle should logically provide the rationale of regulatory decisions. Demands on regulatory decisions are being made by the convergence of technology exhibited in the form of overlapping of industrial classifications, and a disproportionately large externality, (i.e., information literacy significantly influences the market-outcomes); competition and price-quantity calculations are now being substituted often by considerations on quality, inter-operability and availability; etc.

These are new demands and administrative precedences as guides do not exist. A liberal programme, therefore, asks for limits to administrative oversight and a firm "no" to content regulation. The principle, this programme argues, should be based on freedom of speech and of association. A principle of natural law, it argues, should in time receive new definitions of industry and a new code of practice. We have doubts over the emergence of such voluntary state of affairs. Indian courts too recently argued for similar freedom.

The principal objection is that providing this freedom does not constitute the primary legitimation of state and whatever be the state of affairs and consequently its outcome, the political authority can never be loosened during the period of transformations and of restructuring. If that be the case, the state of affairs is a contingent outcome-dependent on the interpenetrating organizations and on the expectations of the other party. Perhaps this is not yet a norm. However, it provides for a positive case, and we argue, there is no need to bring in a principle of natural guidance. Contemporary dynamics in information-telecommunication can be regulated rationally based on the positive facts of synergy-competition in the social arena and the marketplace.

Multiple Regulators

Facts build upon facts, and this evolving dynamics, not being dependent upon a natural principle or a routine/norm, structures itself around the structures of power. Moreover, since there are several such power structures, the dynamics too shapes itself severally and differentially. Three broad types can be identified and we can imagine the necessity of three types of regulators.[8] The first regulator at the national level needs to be concerned with strategic and international issues on the information-telecommunication-entertainment events. Regulator at the next level needs to look after issues relating to industrial dynamics. The third and the last level of regulation needs to be concerned with education, information literacy, formation of social groups and their effectiveness, especially in science and technology discourses and communication, etc. The first type of regulation is primarily political and strategic. The second type is limited to administrative and positive case-based regulation. The third regulator is active as a social institution. Physical locations of these three regulators should encompass authorities from the cabinet to the telecom-regulator to the level of provincial/local

[8]BANERJEE P. & BHATTACHARYA S.: *Advanced Telecommunication in India: Regulation and Strategy.* New Delhi; Har-Anand Publications, 1996.

social and political bodies. The issue of content- regulation should, therefore, remain well within the scope of positive case-based approaches of at least the third-level regulator.

Conclusions

In adumbration, the fast-changing events in telecommunication and related fields signify a change in the economy and the setting forth of new divisions of labour. Simple administrative regulation cannot guide this dynamics. There is an urgency about inducting the political authority in the guidance of these evolving events. Privatization, liberalization and re-regulation are the three instruments in the hands of political authority to be exercised either through the regulator or through social institutions. The affairs of regulation need not be based on the principle of freedom of speech, but can be considered on case-basis contingencies. As a result, political and social authorities can appear as regulators at multiple levels of regulations. It is important not to forget the abiding role of political and social institutions and authorities, especially if the country is populous-diverse and poor, as is the case of India, in matters pertaining to regulating and guiding the information-telecommunication-entertainment events.

PART VIII : REGIONAL COOPERATION

20. Pathways to Asian Cooperation in Science and Technology

Sunil Sondhi

Abstract

This paper explores the need and conditions for successful cooperation in science and technology across national boundaries in the Asian region. It focuses on (1) the different types of strengths that cooperation may capitalize on; (2) the issues to be considered in establishing institutional linkages; and (3) the supportive platforms that may need to be built up for the most effective cooperation. The paper stresses the need for developing prior S&T information resources for the region.

Introduction

The Asian region consists of a distinctly mixed bag of countries. Centered in (and emulating) the economic success of Japan and South Korea, caught in the economic reverse cycles of Bangladesh and Burma or the political insecurities of the Philippines or Sri Lanka, and varying in national population from Papua-New Guinea to China, the region is more economically, culturally and politically diverse than any other region in the world. As a group of countries, by the end of the present decade and according to its present trajectory, the Asian region is likely to grow in economic power to exceed the economic wealth of the present European Economic Community (EEC) - generating a collective GNP of US$6.7 trillion versus US$6.0 trillion for the EEC nations.

The Asian continent now accounts for more than 25 per cent of global exports, 22 per cent of global imports and 33 per cent of global international reserves. And it is by far the fastest growing region in the world. In the 1980s, East Asia grew by 7.8 per cent and South Asia by 5.2 per cent-compared with 3.1 per cent for the OECD countries, 1.6 per cent for Latin America and 2.1 per cent for sub - Saharan Africa. Meanwhile, some of the world's largest ocean, agricultural and mineral resources, as well as information resources, lie within the region. The region is, therefore, potentially as powerful as it is diverse.

Post - Cold War Change

The new world situation ushered in by the end of the Cold War distinguishes itself by three major characteristics. First, the bipolarity that dominated the world system for more than four decades came to an end, yielding to a new multipolarity. Multipolarity augurs less well for systemic stability for a number of reasons. With increase in polarity, the number of dyads and alliance permutations also increases.

The second major characteristic of the post-Cold War era is the decline in the salience of strategic nuclear weapons. The world is in a transition from nuclear to conventional deterrence at the central (global) level. In the Cold War era, the strategic pillar of mutual assured destruction (MAD) made conquest difficult and expansion futile for either camp. The return of conventional deterrence spells instability for international relations, mainly because of an irrepressible power imbalance problem. As there is no similar overkill restraint with conventional weapons, endless arms races will create endless power imbalances.

The third, and final, characteristic of the post-Cold War period is that economic security will rival military security in importance. Strategic analysts have forecast that by the year 2010 China will fill in as the world's number three economic power, after the United States and Japan. As economic security intrudes into the decision -makers consciousness, geo-economics will weigh more heavily in global strategic thinking during the twenty-first century, as geopolitics used to be in the twentieth.

These changes have significant implications for the Asian continent. Already, Asia is witnessing a shopping spree for arms. While West Asia has long been a heavy arms importing region, lately both East and South Asia seem to be following in its footsteps. Need for regional security structures was never more urgent to prevent such shopping spree from becoming a shooting spree. Increased importance of geo-economics implies that henceforth Asia will have to increasingly look within the region for engines of economic growth, rather than depend on Europe or America for external stimulus.

The end of the cold war has thus created conditions in which the Asian countries, particularly the Asian giants, India and China, need to look away from the ideological rigidities of the cold war years and adopt more pragmatic policies for cooperation. Rather than remaining as economic and political powers in the games of superpowers, the countries of Asia can now cooperate for mutual benefit and build up their strength vis-a-vis the developed countries of the North.

Need for Regional Cooperation

During the past decade, Europe and North America have been moving towards unity of identity, and therefore of interest. The EEC is already on a trajectory to dissolve national identities, as well as political and economic boundaries; unified Germany is bracing up to compete globally with renewed vigour; the East European countries are teetering on the edge of free market openness to European trade. The Free Trade Agreement signed between the US and Canada in 1987 has now been extended to include Mexico as well, leading to the emergence of a unified North American economic bloc.[1,2]

The Asian region, on the other hand, has responded lowly, not recognizing, perhaps, the competitive imperative that a collective identity and common purpose must produce within this region in order to survive within the context of a refashioned Europe and unified North American market. There are a variety of collective regional organizational communities that have been generated to encourage interaction among countries of the region-ASEAN, APEC, PECC, SAPTA, and the Pacific Forum, for example. But while these organizations enhance generalized cooperation and awareness, none is seriously breaking down national boundaries and forging bonds of common identity yet.

---

[1] *BUSINESS AMERICA*. Special issue on NAFTA. October 18, 1993.

[2] THUROW Lester: *Head to Head-The Coming Economic Battle between Japan, Europe and America*. New York; William Morrow, 1992.

Interestingly enough, however, one of the key leading edges of cooperation throughout Asia has been in science, with the ASEAN's Council for Science and Technology (ASEAN-COST) being one of the most successful enterprises within the ASEAN. Scientific cooperation could well provide the means of capitalizing on the divided science capabilities throughout the region, nurturing collective economic strength against the new power of European and North American economic dominance.[3]

Beyond the politics of regional economic integration, the fragility of the world ecosystem as a whole is now impinging on the shape of Asian development. The Asian continent confronts ecological constraints as a region, not as individual countries that can continue to exploit their surrounding world without direct national cost.

Within this adversity lies opportunity for considerable scientific cooperation. Global (or regional) problems require global (or ragional) solutions. Research on ecological and sustainable resource issues is in the interest of all nations. Monitoring of ecological impact and identification of means of redressing common ecological problems are of common concern. Collaborative environmental research, therefore provides a very positive starting point for regional research cooperation. Even smaller countries like Maldives and Brunei need to have access to scientific information in order to monitor and exploit sustainable marine and agricultural resources, and to circumvent impending pollution problems. Scientific cooperation is, therefore, particularly important for the many smaller countries of Asia. Specific areas of cooperation among countries of the region which need particular attention are environmental, ecological and sustainable resource issues.

Barriers to Cooperation

Given the change in the essential character of international science from pursuit of knowledge to pursuit of material benefits over the last half of the twentieth century, we can forget, as a

[3]SOESATRO Hadi & PANGESTU Mari: *Technological Change in the Asia-Pacific Economy*. London; Allen Unwin, 1990.

justification for scientific cooperation, the warm friendly glow of scientists sharing secrets of the universe or battling back the frontiers of knowledge together. Science's leading edge today is ruled by more mundane and privatized vested interests; its currency is increasingly intellectual property rights rather than recognition for contributions to knowledge.

Therefore, while there may well be political reasons for generalized scientific cooperation (such as using science to generate more general political alignments), we need to be extremely cautious with the assumption that scientific cooperation across international boundaries, like going to the opera, is intrinsically a good thing or the natural path of contemporary science.

As a first caution, we need to be conscious of the context of power. Science application is in the hands of capital interests that are effective according to their organizational capability to surround scientific opportunity and exploit it. Unless the smaller partner in collaboration has a watertight agreement about intellectual property rights benefits, this party is very likely to be exploited.

As a second caution, we need to be conscious of the context of technical environment. Often overlooked in the choice of topics for a national research effort, is a careful assessment of the technical "receptivity" of the industrial environment into which innovations must "bed-in" for application to be successful.[(4,5)] There are many examples of a national research effort producing eminently useful innovations that cannot be produced nationally and to national advantage and that are, therefore, immediately susceptible to international capital interests.[(6,7)] The impact of this

---

[4]BELL R. M. HILL Stephen C.: "Research on Technology Transfer and Innovation: Some Limitations and New Directions." In *Transfer Processes in Technical Change*, edited by F.R. Bradbury, R. Johnston, and A.W. Pearson. London; Noardhoff, 1978. p 225–74.

[5]HILL Stephen C.: "Basic Design Principles for National Research in Developing Countries." *Technology in Society* 1987, 9, p.63-73.

[6]HILL Stephen C.: "Contrary Meanings of Science. In *Perspectives in the Sociology of Science*, edited by Stuart S. Blume. Chichester, Great Britain; John Wiley and Sons, 1977. p 195–230.

[7]RAHMAN A. & HILL Stephen: Science, Technology and Development in Asia and the Pacific. Paris; UNESCO, 1982.

factor is likely to be even more important in international cooperative scientific programmes, where the connection between research choice and local commercial application environment is likely to be even more remote. Therefore, unless the partners in collaboration make cautious assessments of the technical receptivity of their own commercial enterprise and link their research choices into these assessments, they may both be seriously at risk to external interests that capitalize on the 'disconnection', between research and commercialization.

As a third caution, we need to be conscious of the context of prestige. In many of the developing countries of the region, participation in international science still remains a key to local social status and escape from deprivation. An internal brain drain can easily follow, where the goals of national scientists are oriented towards the scientific interests of the advanced nations, rather than of their own national situation, and where prestige is conferred on researchers who work on exotic and advanced science topics, the choice of which is often fashion or novelty and has little to do with the more pedestrian scientific and technical problems faced by a developing nation's economy.[8] Consequently, international scientific collaboration can work against the interests of the smaller developing national scientific establishments through distracting effort towards larger nation interest.

As a fourth caution, we need to be conscious of culture. Although it is often assumed that scientific research and practice is universalistic - a citizen of one internationalized culture, this simply is not true. The institutional and organizational practices, the goals and rewards, and the meaning of scientific research vary across different national and research cultures. An Indian researcher might find difficulty with the more patrimonial organization structures of Thai research while a Thai researcher might find difficulty adjusting to the planning structures of Indian research. An Australian scientist may find some difficulty within

[8]SAGASTI Fransisco: "Underdevelopment, Science and Technology: The Point of View of the Underdeveloped Countries.' *Science Studies*, 1973, 3, p.47-59.

the more formal organization expectations of Japan; similarly a Japanese scientist may find difficulty in the more participative structures of Australian research. South Korean researchers may find difficulty jumping from the practical, commercial "digestion,' of much current South Korean work into the more anarchic turbulence of pure basic research, as can be found more readily within Indian institutions, and so on. This cultural dimension of science and technology is rarely taken into account, yet it could provide both the source of potential major advantage or, alternatively, the source of potential major failure-through mutually reinforcing solutions to separate national weaknesses in research culture.

Pathways to Cooperation

Indeed, it may well be that to capitalize on what appears to be a potential weakness - different research cultures - could afford major advantages to nations participating in international scientific relations. It is worth noting that a major reason for the productive shape and direction of post-world War II electronics research was the "interbreeding" of national cultural research styles between the European and Anglo-American research species.[9] In solid state physics, the German style was pedantic, basic and rigorous, and it approached a problem from the point of view of an all - embracing pragmatism; the French school was elegant and pure, where the realism of assumptions was less important than the clarity of logic, and the aesthetics of theory overshadowed the possible lack of agreement with facts; the English school was like its common law: theory less concerned with elegance and rigorous derivation from general principles, but made to work.[10] Similarly, within the Asian region, capitalizing on variation in cultural research style could well produce a productivity advantage. For example, India has a

---

[9]HOCH Paul K. "John Clarke Slater and the Science for the Transistor." *European Journal of Physics,* 1990, 11, p.283-91.

[10]SMOLUCHOWSKI Roman: "Random Comments on the Early Days of Solid State Physics." In *The Beginnings of Solid State Physics,* edited by N.F. Mott. London; Carlton House, 1980, p.100-101.

research culture that emphasizes excellence in basic research, but is weakly integrated into application; South Korea has a research culture that emphasizes digestion and application, but is right now having to cross into unfamiliar longer-term, more innovative research. There could be major advantages to both countries in selective collaboration.

Also, while national research remains isolated behind national boundaries-of language, research priorities, or whatever-the quality of research remains somewhat weakened. Opening up the national research to international scrutiny, through international collaboration, allows the expectations of researcher productivity to be measured against international standards. Direct laboratory-to-laboratory cooperation is particularly important in this internationalization of research standards of countries that largely publish in languages other than English; alternatively, such international collaboration allows assessment of unpublished practical research according to international standards of quality.

Furthermore, there may well be quite pragmatic economic reasons for regional research cooperation, such as the sharing of costs for expensive basic research equipment, access to super-computer time through satellite linkages, or the provision of expensive gene banks of plant species (as is the case for Japan's RIVKEN laboratory collecting and providing gene banks for Asian flora).

In this context, one area in which regional cooperation can benefit Asian countries a great deal is space research. India and China have already signed a memorandum of understanding (MOU) "for bilateral cooperation in the area of space research, space technology and space applications. Both countries have developed sufficient expertise in satellite fabrication. Joint fabrication of communication and earth observation satellites is a potential area of collaboration. The area of satellite subsystems is where Indian strength lies, whereas Chinese expertise in launch vehicle technology is by now well established. Apart from space research, there are other areas like nuclear science, biotechnology and lasers, where cooperation between India and China could benefit not only the two countries but could be of great help to the entire Asian region.

However, we should be quite realistic about the assumption that scientific cooperation is likely to lead to a mutually beneficial commercial advantage. The cold reality is that, given the power structure that surrounds today's application of research, this will happen only when linkages into agreed commercial application are prepared from the start.

What this suggests is that regional research cooperation is most likely to work if the shared activity is placed further back from the immediate practical end-towards strategic basic research, that is, research where *a priori* decision suggests the field's relevance but not specific applications. Entry into collaboration is relatively easy, as relationships are established prior to the stage where national interests and competition come to dominate international relations and possibilities. Furthermore, as with the earlier observations about the advantage of bridging research cultural styles, there may be real advantages that can follow-in escaping limitations to vision that immediate pragmatic assumptions might foster. In retrospect, it is curious, for example, that the transistor was not developed during the time of World War II, especially as the electron theory of metals and semiconductors at places like MIT was, even in the 1930's, a fashionable subject, ripe for application. What prevented inventive vision was the necessarily short time scale for useful wartime scientific work and the diversion of electronic talent to the more pressing wartime uses of microwave techniques.[11]

It should also be recognized that there are spin-off benefits to directly national commercial activity that can follow from strategic basic research collaboration:

* Since technology is a global commodity, the transfer of technology and access to new technologies is enhanced through movement of people to learn the skills involved in new technological developments.

* Direct laboratory and workshop experience is invaluable, particularly in front-line areas, such as biotechnology, information

---

[11]BULLARD E.: "The Effect of World War II on the Development of Knowledge in the Physical Sciences." *Proceedings of Rural Sociology* 1975, A342, p.519-36.

technology, advanced manufacturing techniques, and so on, where the value of intellectual property exercises control over the free spread of new techniques.

* Collaboration schemes help introduce research scientists and engineers to colleagues and business leaders overseas, thus building up a nations S&T visibility as well as the nation's credibility for overseas investment in national industry.

* International S&T linkages provide extra opportunity for national industry to gain early access to research discoveries and technology development overseas, and through that to strategic association with overseas companies.[12]

However, while the case can be made for a focus of attention on strategic basic research in international collaboration, one essential factor yet to be included in the establishment of agreements is an assessment of the level of potential contribution of each of the research partners to subsequent possible commercial outcomes; or alternatively an assessment of the level of innovativeness that has contributed when carving up the commercial payoffs. The point is that in any collaborative research relationship one partner is likely to be stronger in providing the creative or "breakthrough" research (as distinct from the more programmable and routine investigation of alternatives). Yet, when the project is nearing completion and is closer to commercial application, the level of this contribution may well not be rewarded if the feet of the other partner are more solidly planted in the commercial world.

It is, therefore, necessary that international scientific collaboration should be conducted with overviews, foresight, and information rather than left either to individual scientists taking responsibility for initiative or to passive government acceptance (or disregard) of initiatives taken by organizations or commercial interests from elsewhere. To follow this point further, nations participating in regional collaboration bring a range of different strengths to internationally collaborative scientific enterprise, and

[12]PITMAN Michael: "Support for International Science and Technology.' Ascent", 1989, 18 (August).

a range of different weaknesses that international S&T collaboration may be able to redress.

At present it is difficult to identify precisely where these strengths and weaknesses lie in specific areas of research and in the industrial innovation process and commercial exploitation capabilities of specific industries and nations, because there is no consistent "S&T management information system" throughout the region. Considerable work still needs to be done to develop such an information system. Without such information, however, all collaborations chosen for support will, to some extent, be sporadic.[13]

Indeed, there is an even more compelling reason for the development of this kind of information base. That is, presently many of the countries of the region are directly competing in very similar areas of research. This is particularly the case in the more "fashionable" areas of high-technology-related science: biotechnology, electronics and so on. Preferable would be the ability for each of the nations of the region to identify precisely which research niches are likely to be most productive for them, given the present (and potential) distribution of strengths and opportunities within their own nation and internationally. An appropriate S&T management information system could also enhance the relative power of the weakest of the region's scientific capabilities as they engage in international collaboration- by allowing them to focus very specifically on where they can contribute to (perhaps) quite small interests of the leading scientific and technological powers. There is, therefore, considerable need to build an effective science and technology management information system across the region which can be used by the concerned nations to make informed decisions about entering into S&T collaborations.

Conclusions

Attention to enhanced scientific research cooperation throughout the Asia-Pacific region is timely. It may well serve a highly constructive purpose in fostering the regions political and

[13]NATIONAL SCIENCE FOUNDATION: *Human Resources for Science and Technology: The Asian Region*. Washington D.C.; NSF, 1993.

economic identity against a powerfully refashioned Europe and an integrated North America in strengthening the separately weak S&T capabilities to nurture collective economic power within the new world context. Furthermore, there are regional, environmental, sustainable development, and social equity issues that could only benefit from sustained collaborative research action on an international basis across the region.

In addition, it is in the interest of all nations of the region to foster the scientific and technical capabilities of the weaker members of the regional community-for such S&T development assistance fosters regional identity and commitment and at the same time, it enhances the indigenous capability of recipient nations to choose and provide resources for their development paths more autonomously.

However, in developing research relationships, we should be informed by the changed nature of today's scientific production process, where power to use leading edge science now lies largely in the hands of organizationally powerful private interests. If mutual advantage is to follow where collaboration is enjoined close to the edge of commercial exploitation, particular attention must be placed on prior specific agreement about what will be done with intellectual property rights particularly by the junior partner in the exchange; and close attention must be paid from the outset to the specific technical capabilities to absorb research results within the production system targeted. Lack of precision is most likely to be in the interest of the most powerful partner or of even more powerful interest that have finely tuned information systems watching where the collaborative research is heading.

Collaboration close to the edge of profit is risky though: secretiveness and vested interest can all too easily intrude on a complex cross-cultural relationship and destroy it. Collaboration directed towards building platforms to prepare the way for national commercial research is, therefore, likely to be more fruitful. The platform-building position implies entering international collaborative research relationships at the level of strategic basic research, that is, at the stage in research prior to the point where national interest and competition comes to dominate the process. Benefits that can accrue to the collaborating nations are all ones that strengthen their respective research production systems: widening of research horizons; use of international quality standards; cost sharing; mobility of research personnel; and strategic relationship between research and international business.

To build these platforms with strength and direction, however, requires the development of prior S&T management information resources for the region, as it also requires the re-evaluation at national levels of the specific purposes of present international research collaboration policy.

Selected References

INSTITUTE OF ELECTRICAL AND ELECTRONICS ENGINEERS, Inc (IEEE) *Asia Power. IEEE Spectrum(Special Report),* 1991 (June).

MELLOR Bill: "A Year for Crowing." *Time* 1993 (February) p.23.

MINDEN Karen(Ed): *Pacific Cooperation in Science and Technology.* Honolulu, Hawaii; East West Centre, 1991.

SONDHI Sunil: "Triumph of the Asian Way. *The Pioneer* 1995, 16 July.

Index